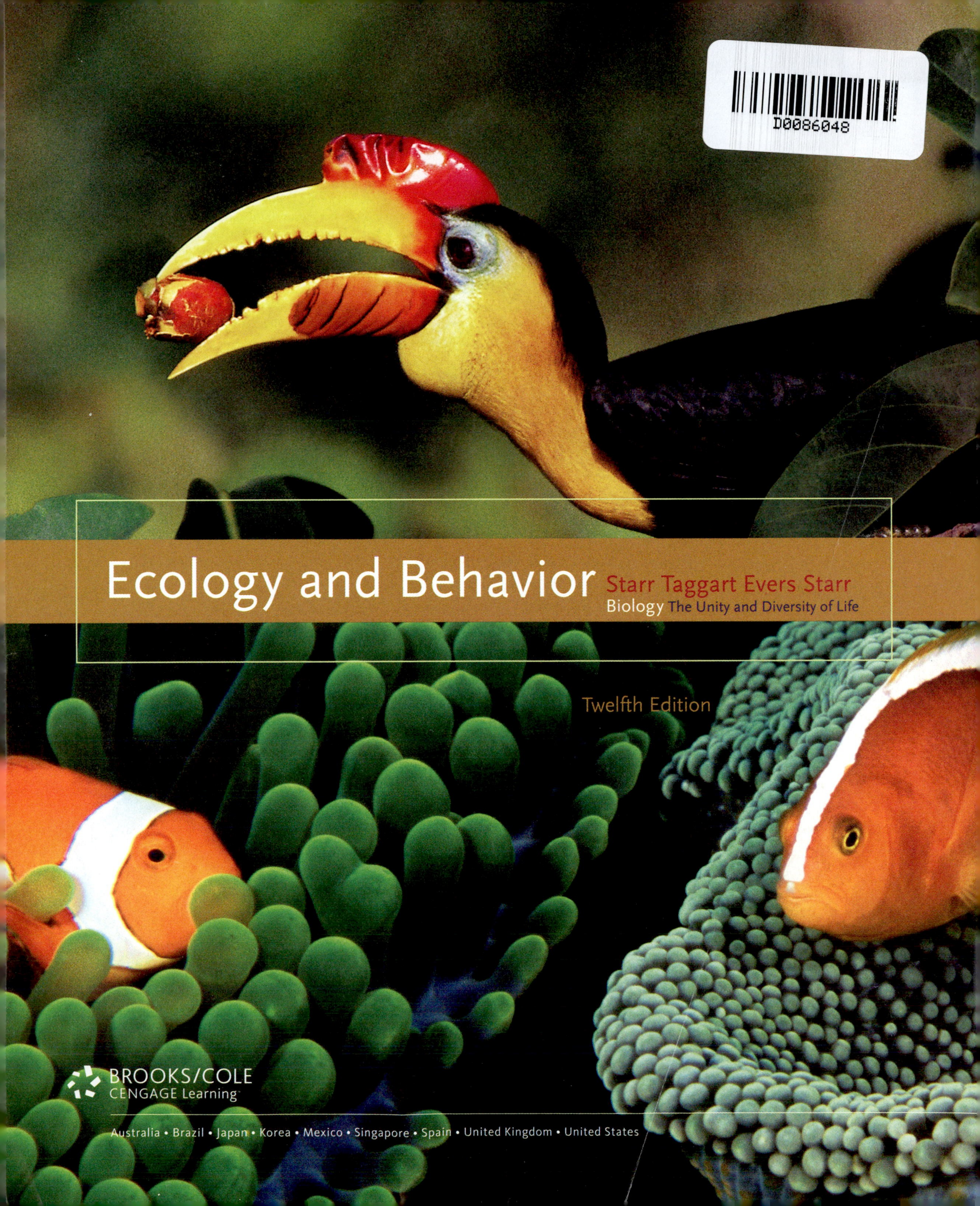

D0086048
Ecology and Behavior
Starr Taggart Evers Starr
Biology The Unity and Diversity of Life
Twelfth Edition
BROOKS/COLE
CENGAGE Learning
Australia • Brazil • Japan • Korea • Mexico • Singapore • Spain • United Kingdom • United States

Ecology and Behavior
Biology: The Unity and Diversity of Life, Twelfth Edition
Cecie Starr, Ralph Taggart, Christine Evers, Lisa Starr

Publisher: Yolanda Cossio

Managing Development Editor: Peggy Williams

Assistant Editor: Elizabeth Momb

Editorial Assistant: Samantha Arvin

Technology Project Manager: Kristina Razmara

Marketing Manager: Amanda Jellerichs

Marketing Assistant: Katherine Malatesta

Marketing Communications Manager: Linda Yip

Project Manager, Editorial Production: Andy Marinkovich

Creative Director: Rob Hugel

Art Director: John Walker

Print Buyer: Karen Hunt

Permissions Editor: Bob Kauser

Production Service: Grace Davidson & Associates

Text and Cover Design: John Walker

Photo Researcher: Myrna Engler Photo Research Inc.

Copy Editor: Anita Wagner

Illustrators: Gary Head, ScEYEnce Studios, Lisa Starr

Compositor: Lachina Publishing Services

Cover Image: Biologist/photographer Tim Laman took these photos of mutualisms in Indonesia. *Top:* A wrinkled hornbill (*Aceros corrugatus*) eats fruits of a strangler fig (*Ficus stupenda*). The plant provides food for the bird, and the bird disperses its seeds. *Below:* Two species of sea anemone, each with its own species of anemone fish. Anemones provide a safe haven for anemonefish, who chase away other fish that would graze on the anemone's tentacles. www.timlaman.com

Library of Congress Control Number: 2008930421

ISBN-13: 978-0-495-55803-3

ISBN-10: 0-495-55803-6

Brooks/Cole
10 Davis Drive
Belmont, CA 94002
USA

Printed in the United States of America
1 2 3 4 5 6 7 12 11 10 09 08

CONTENTS IN BRIEF

Highlighted chapters are not included in Ecology and Behavior

INTRODUCTION

1 Invitation to Biology

UNIT I PRINCIPLES OF CELLULAR LIFE

2 Life's Chemical Basis
3 Molecules of Life
4 Cell Structure and Function
5 A Closer Look at Cell Membranes
6 Ground Rules of Metabolism
7 Where It Starts—Photosynthesis
8 How Cells Release Chemical Energy

UNIT II PRINCIPLES OF INHERITANCE

9 How Cells Reproduce
10 Meiosis and Sexual Reproduction
11 Observing Patterns in Inherited Traits
12 Chromosomes and Human Inheritance
13 DNA Structure and Function
14 From DNA to Protein
15 Controls Over Genes
16 Studying and Manipulating Genomes

UNIT III PRINCIPLES OF EVOLUTION

17 Evidence of Evolution
18 Processes of Evolution
19 Organizing Information About Species
20 Life's Origin and Early Evolution

UNIT IV EVOLUTION AND BIODIVERSITY

21 Viruses and Prokaryotes
22 Protists—The Simplest Eukaryotes
23 The Land Plants
24 Fungi
25 Animal Evolution—The Invertebrates
26 Animal Evolution—The Chordates
27 Plants and Animals—Common Challenges

UNIT V HOW PLANTS WORK

28 Plant Tissues
29 Plant Nutrition and Transport
30 Plant Reproduction
31 Plant Development

UNIT VI HOW ANIMALS WORK

32 Animal Tissues and Organ Systems
33 Neural Control
34 Sensory Perception
35 Endocrine Control
36 Structural Support and Movement
37 Circulation
38 Immunity
39 Respiration
40 Digestion and Human Nutrition
41 Maintaining the Internal Environment
42 Animal Reproductive Systems
43 Animal Development

UNIT VII PRINCIPLES OF ECOLOGY

44 Animal Behavior
45 Population Ecology
46 Community Structure and Biodiversity
47 Ecosystems
48 The Biosphere
49 Human Impacts on the Biosphere

DETAILED CONTENTS

UNIT VII PRINCIPLES OF ECOLOGY

44 Animal Behavior

IMPACTS, ISSUES My Pheromones Made Me Do It *780*

44.1 Behavioral Genetics *782*
- How Genes Affect Behavior *782*
- Studying Variation Within a Species *782*
- Comparisons Among Species *783*
- Knockouts and Other Mutations *783*

44.2 Instinct and Learning *784*
- Instinctive Behavior *784*
- Time-Sensitive Learning *784*
- Conditioned Responses *785*
- Other Types of Learned Behavior *785*

44.3 Adaptive Behavior *786*

44.4 Communication Signals *786*

44.5 Mates, Offspring, and Reproductive Success *788*
- Sexual Selection and Mating Behavior *788*
- Parental Care *789*

44.6 Living in Groups *790*
- Defense Against Predators *790*
- Improved Feeding Opportunities *790*
- Dominance Hierarchies *791*
- Regarding the Costs of Group Living *791*

44.7 Why Sacrifice Yourself? *792*
- Social Insects *792*
- Social Mole-Rats *792*
- Evolution of Altruism *792*

44.8 FOCUS ON SCIENCE Human Behavior *793*
- Hormones and Pheromones *793*
- Morality and Behavior *793*

45 Population Ecology

IMPACTS, ISSUES The Numbers Game *796*

45.1 Population Demographics *798*

45.2 FOCUS ON SCIENCE Elusive Heads to Count *799*

45.3 Population Size and Exponential Growth *800*
- Gains and Losses in Population Size *800*
- From Zero to Exponential Growth *800*
- What Is the Biotic Potential? *801*

45.4 Limits on Population Growth *802*
- Environmental Limits on Growth *802*
- Carrying Capacity and Logistic Growth *802*
- Two Categories of Limiting Factors *803*

45.5 Life History Patterns *804*
- Life Tables *804*
- Survivorship Curves *804*
- Reproductive Strategies *805*

45.6 FOCUS ON SCIENCE Natural Selection and Life Histories *806*
- Predation on Guppies in Trinidad *806*
- Overfishing and the Atlantic Cod *807*

45.7 Human Population Growth *808*
- The Human Population Today *808*
- Extraordinary Foundations for Growth *808*
  - Geographic Expansion *808*
  - Increased Carrying Capacity *808*
  - Sidestepped Limiting Factors *808*

45.8 Fertility Rates and Age Structure *810*
- Some Projections *810*
- Shifting Fertility Rates *810*

45.9 Population Growth and Economic Effects *812*
- Demographic Transitions *812*
- Resource Consumption *812*

45.10 Rise of the Seniors *813*

46 Community Structure and Biodiversity

IMPACTS, ISSUES Fire Ants in the Pants *816*

46.1 Which Factors Shape Community Structure? *818*
- The Niche *818*
- Categories of Species Interactions *818*

46.2 Mutualism *819*

46.3 Competitive Interactions *820*
- Effects of Competition *820*
- Resource Partitioning *821*

46.4 Predator-Prey Interactions *822*
- Models for Predator-Prey Interactions *822*
- The Canadian Lynx and Snowshoe Hare *822*
- Coevolution of Predators and Prey *823*

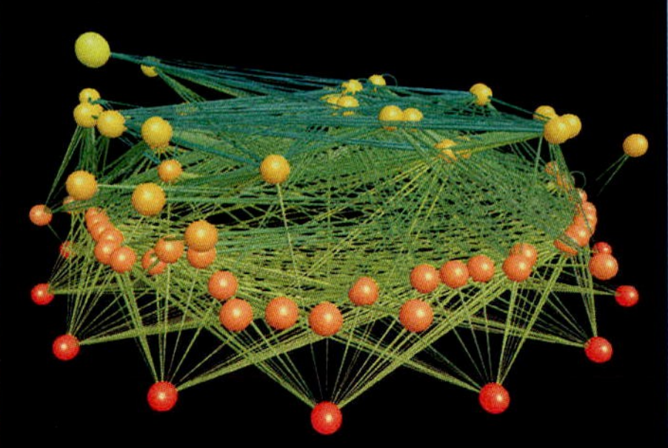

46.5 FOCUS ON EVOLUTION An Evolutionary Arms Race *824*

Prey Defenses *824*

Adaptive Responses of Predators *825*

46.6 Parasite-Host Interactions *826*

Parasites and Parasitoids *826*

Biological Control Agents *827*

46.7 FOCUS ON EVOLUTION Strangers in the Nest *827*

46.8 Ecological Succession *828*

Successional Change *828*

Factors Affecting Succession *828*

46.9 Species Interactions and Community Instability *830*

The Role of Keystone Species *830*

Species Introductions Can Tip the Balance *831*

46.10 FOCUS ON THE ENVIRONMENT Exotic Invaders *832*

Battling Algae *832*

The Plants That Overran Georgia *832*

The Rabbits That Ate Australia *833*

Gray Squirrels Versus Red Squirrels *833*

46.11 Biogeographic Patterns in Community Structure *834*

Mainland and Marine Patterns *834*

Island Patterns *834*

47 Ecosystems

IMPACTS, ISSUES Bye-Bye, Blue Bayou *838*

47.1 The Nature of Ecosystems *840*

Overview of the Participants *840*

Trophic Structure of Ecosystems *840*

47.2 The Nature of Food Webs *842*

Interconnecting Food Chains *842*

How Many Transfers? *843*

47.3 Energy Flow Through Ecosystems *844*

Capturing and Storing Energy *844*

Ecological Pyramids *844*

Ecological Efficiency *845*

47.4 FOCUS ON THE ENVIRONMENT Biological Magnification *846*

DDT and Silent Spring *846*

The Mercury Menace *846*

47.5 Biogeochemical Cycles *847*

47.6 The Water Cycle *848*

How and Where Water Moves *848*

A Global Water Crisis *848*

47.7 Carbon Cycle *850*

47.8 FOCUS ON THE ENVIRONMENT Greenhouse Gases and Climate Change *852*

47.9 Nitrogen Cycle *854*

Inputs Into Ecosystems *854*

Natural Losses From Ecosystems *855*

Disruptions by Human Activities *855*

47.10 The Phosphorus Cycle *856*

48 The Biosphere

IMPACTS, ISSUES Surfers, Seals, and the Sea *860*

48.1 Global Air Circulation Patterns *862*

Air Circulation and Regional Climates *862*

Harnessing the Sun and Wind *863*

48.2 FOCUS ON THE ENVIRONMENT Something in the Air *864*

Swirling Polar Winds and Ozone Thinning *864*

No Wind, Lots of Pollutants, and Smog *864*

Winds and Acid Rain *865*

Windborne Particles and Health *865*

48.3 The Ocean, Landforms, and Climates *866*

Ocean Currents and Their Effects *866*

Rain Shadows and Monsoons *866*

48.4 Biogeographic Realms and Biomes *868*

48.5 Soils of Major Biomes *870*

48.6 Deserts *871*

48.7 Grasslands, Shrublands, and Woodlands *872*

Grasslands *872*

Dry Shrublands and Woodlands *873*

48.8 More Rain, Broadleaf Forests *874*

Semi-Evergreen and Deciduous Broadleaf Forests *874*

Tropical Rain Forests *874*

48.9 **FOCUS ON BIOETHICS** You and the Tropical Forests *875*

48.10 Coniferous Forests *876*

48.11 Tundra *877*

48.12 Freshwater Ecosystems *878*

Lakes *878*

Nutrient Content and Succession *878*

Seasonal Changes *878*

Streams and Rivers *879*

48.13 **FOCUS ON HEALTH** "Fresh" Water? *880*

48.14 Coastal Zones *880*

Wetlands and the Intertidal Zone *880*

Rocky and Sandy Coastlines *881*

48.15 **FOCUS ON THE ENVIRONMENT** The Once and Future Reefs *882*

48.16 The Open Ocean *884*

Oceanic Zones and Habitats *884*

Upwelling—A Nutrient Delivery System *885*

48.17 Climate, Copepods, and Cholera *886*

49 Human Impacts on the Biosphere

IMPACTS, ISSUES A Long Reach *890*

49.1 The Extinction Crisis *892*

Mass Extinctions and Slow Recoveries *892*

The Sixth Great Mass Extinction *893*

49.2 Current Threats to Species *894*

Habitat Loss, Fragmentation, and Degradation *894*

Overharvesting and Poaching *894*

Species Introductions *895*

Interacting Effects *895*

49.3 **FOCUS ON RESEARCH** The Unknown Losses *896*

49.4 Assessing Biodiversity *896*

Conservation Biology *896*

Monitoring Indicator Species *896*

Identifying Regions at Risk *896*

49.5 Effects of Development and Consumption *898*

Effects of Urban and Suburban Development *898*

Effects of Resource Consumption *898*

49.6 The Threat of Desertification *900*

49.7 The Trouble With Trash *901*

49.8 Maintaining Biodiversity and Human Populations *902*

Bioeconomic Considerations *902*

Sustainable Use of Biological Wealth *902*

Using Genetic Diversity *902*

Discovering Useful Chemicals *902*

Ecotourism *902*

Sustainable Logging *903*

Responsible Ranching *903*

Appendix I Classification System

Appendix II Annotations to A Journal Article

Appendix III Answers to Self-Quizzes and Genetics Problems

Appendix IX Units of Measure

Preface

In preparation for this revision, we invited instructors who teach introductory biology for non-majors students to meet with with us and discuss the goals of their courses. The main goal of almost every instructor was something like this: "To provide students with the tools to make informed choices as consumers and as voters by familiarizing them with the way science works." Most students who use this book will not become biologists, and many will never take another science course. Yet for the rest of their lives they will have to make decisions that require a basic understanding of biology and the process of science.

Our book provides these future decision makers with an accessible introduction to biology. Current research, along with photos and videos of the scientists who do it, underscore the concept that science is an ongoing endeavor carried out by a diverse community of people. The research topics include not only what the researchers discovered, but also how the discoveries were made, how our understanding has changed over time, and what remains undiscovered. The role of evolution is a unifying theme, as it is in all aspects of biology.

As authors, we feel that understanding stems mainly from making connections, so we are constantly trying to achieve the perfect balance between accessibility and level of detail. A narrative with too much detail is inaccessible to the introductory student; one with too little detail comes across as a series of facts that beg to be memorized. Thus, we revised every page to make the text in this edition as clear and straightforward as possible, keeping in mind that English is a second language for many students. We also simplified many figures and added tables that summarize key points.

CHANGES IN THIS EDITION

Impacts, Issues To make the *Impacts, Issues* essays more appealing, we shortened and updated them, and improved their integration throughout the chapters. Many new essays were added to this edition.

Key Concepts Introductory summaries of the *Key Concepts* covered in the chapter are now enlivened with eye-catching graphics taken from relevant sections. The links to earlier concepts now include descriptions of the linked concepts in addition to the section numbers.

Take Home Message Each section now concludes with a *Take Home Message* box. Here we pose a question that reflects the critical content of the section, and we also provide answers to the question in bulleted list format.

Figure It Out *Figure It Out Questions* with answers allow students to check their understanding of a figure as they read through the chapter.

Data Analysis Exercise To further strengthen a student's analytical skills and provide insight into contemporary research, each chapter includes a *Data Analysis Exercise*. The exercise includes a short text passage—usually about a published scientific experiment—and a table, chart, or other graphic that presents experimental data. The student must use information in the text and graphic to answer a series of questions.

Chapter-Specific Changes Every chapter was extensively revised for clarity; this edition has more than 250 new photos and over 300 new or updated figures. A page-by-page guide to content and figures is available upon request, but we summarize the highlights here.

- *Chapter 1, Invitation to Biology* New essay about the discovery of new species. Greatly expanded coverage of critical thinking and the process of science; new section on sampling error.
- *Chapter 2, Life's Chemical Basis* Sections on subatomic particles, bonding, and pH simplified; new pH art.
- *Chapter 3, Molecules of Life* New essay about *trans* fats. Structural representations simplified and standardized.
- *Chapter 4, Cell Structure and Function* New essay about foodborne *E. coli*; microscopy section updated; new section on cell theory and history of microscopy; two new focus essays on biofilms and lysosome malfunction.
- *Chapter 5, A Closer Look at Cell Membranes* Membrane art reorganized; new figure illustrating cotransport.
- *Chapter 6, Ground Rules of Metabolism* Energy and metabolism sections reorganized and rewritten; much new art, including molecular model of active site.
- *Chapter 7, Where It Starts—Photosynthesis* New essay about biofuels. Sections on light-dependent reactions and carbon fixing adaptations simplified; new focus essay on atmospheric CO_2 and global warming.
- *Chapter 8, How Cells Release Chemical Energy* All art showing metabolic pathways revised and simplified.
- *Chapter 9, How Cells Reproduce* Updated micrographs of mitosis in plant and animal cells.
- *Chapter 10, Meiosis and Sexual Reproduction* Crossing over, segregation, and life cycle art revised.
- *Chapter 11, Observing Patterns in Inherited Traits* New essay about inheritance of skin color; mono- and dihybrid cross figures revised; new Punnett square for coat color in dogs; environmental effects on *Daphnia* phenotype added.
- *Chapter 12, Chromosomes and Human Inheritance* Chapter reorganized; expanded discussion and new figure on the evolution of chromosome structure.
- *Chapter 13, DNA Structure and Function* New opener essay on pet cloning; adult cloning section updated.
- *Chapter 14, From DNA to Protein* New art comparing DNA and RNA, other art simplified throughout; new micrographs of transcription Christmas tree, polysomes.
- *Chapter 15, Controls Over Genes* Chapter reorganized; eukaryotic gene control section rewritten; updated X chromosome inactivation photos; new lac operon art.
- *Chapter 16, Studying and Manipulating Genomes* Text extensively rewritten and updated; new photos of *bt* corn, DNA fingerprinting; sequencing art revised.
- *Chapter 17, Evidence of Evolution* Extensively revised, reorganized. Revised essay on evidence/inference; new

focus essay on whale evolution; updated geologic time scale correlated with grand canyon strata.

- *Chapter 18, Processes of Evolution* Extensively revised, reorganized. New photos showing sexual selection in stalk-eyed flies, mechanical isolation in sage.
- *Chapter 19, Organizing Information About Species* Extensively revised, reorganized. New comparative embryology photo series; updated tree of life.
- *Chapter 20, Life's Origin and Early Evolution* Information about origin of agents of metabolism updated. New discussion of ribozymes as evidence for RNA world.
- *Chapter 21, Viruses and Prokaryotes* Opening essay about HIV moved here, along with discussion of HIV replication. New art of viral structure. New section describes the discovery of viroids and prions.
- *Chapter 22, Protists—The Simplest Eukaryotes* New opening essay about malaria. New figures show protist traits, how protists relate to other groups.
- *Chapter 23, The Land Plants* Evolutionary trends revised. More coverage of liverworts and hornworts.
- *Chapter 24, Fungi* New opening essay about airborne spores. More information on fungal uses and pathogens.
- *Chapter 25, Animal Evolution—The Invertebrates* New summary table for animal traits. Coverage of relationships among invertebrates updated.
- *Chapter 26, Animal Evolution—The Chordates* New section on lampreys. Human evolution updated.
- Material previously covered in the *Biodiversity in Prespective* chapter now integrated into other chapters.
- *Chapter 27, Plants and Animals—Common Challenges* New section about heat-related illness.
- *Chapter 28, Plant Tissues* Secondary structure section simplified; new essay on dendroclimatology.
- *Chapter 29, Plant Nutrition and Transport* Root function section rewritten and expanded; new translocation art.
- *Chapter 30, Plant Reproduction* Extensively revised. New essay on colony collapse disorder; new table showing flower specializations for specific pollinators; new section on flower sex; many new photos added.
- *Chapter 31, Plant Development* Sections on plant development and hormone mechanisms rewritten.
- *Chapter 32, Animal Tissues and Organ Systems* Essay on stem cells updated. New section on lab-grown skin.
- *Chapter 33, Neural Control* Reflexes integrated with coverage of spinal cord. Section on brain heavily revised.
- *Chapter 34, Sensory Perception* New art of vestibular apparatus, image formation in eyes, and accommodation. Improved coverage of eye disorders and disease.
- *Chapter 35, Endocrine Control* New section about pituitary disorders. Tables summarizing hormone sources now in appropriate sections, rather than at end.
- *Chapter 36, Structural Support and Movement* Improved coverage of joints and joint disorders.
- *Chapter 37, Circulation* Updated opening essay. New section about hemostasis. Blood cell diagram simplified. Blood typing section revised for clarity.
- *Chapter 38, Immunity* New essay on HPV vaccine; new focus essays on periodontal-cardiovascular disease and allergies; vaccines and AIDS sections updated.
- *Chapter 39, Respiration* Better coverage of invertebrate respiration and of Heimlich maneuver.
- *Chapter 40, Digestion and Human Nutrition* Nutritional information and obesity research sections updated.
- *Chapter 41, Maintaining the Internal Environment* New figure of fluid distribution in the human body. Improved coverage of kidney disorders and dialysis.
- *Chapter 42, Animal Reproductive Systems* New essay on intersex conditions. Coverage of reproductive anatomy, gamete production, intercourse, and fertilization.
- *Chapter 43, Animal Development* Information about principles of animal development streamlined.
- *Chapter 44, Animal Behavior* More on types of learning.
- *Chapter 45, Population Ecology* Exponential and logistic growth clarified. Human population material updated.
- *Chapter 46, Community Structure and Biodiversity* New table of species interactions. Competition section heavily revised.
- *Chapter 47, Ecosystems* New figures for food chain and food webs. Updated greenhouse gas coverage.
- *Chapter 48, The Biosphere* Improved coverage of lake turnover, ocean life, coral reefs, and threats to them.
- *Chapter 49, Human Impacts on the Biosphere* Covers extinction crisis, conservation biology, ecosystem degradation, and sustainable use of biological wealth.

Appendix V, Molecular Models New art and text explain why we use different types of molecular models.

Appendix VI, Closer Look at Some Major Metabolic Pathways New art shows details of electron transport chains in thylakoid membranes.

ACKNOWLEDGMENTS

No list can convey our thanks to the team of dedicated people who made this book happen. The professionals who are listed on the following page helped shape our thinking. Marty Zahn and Wenda Ribeiro deserve special recognition for their incisive comments on every chapter, as does Michael Plotkin for voluminous and excellent feedback. Grace Davidson calmly and tirelessly organized our efforts, filled in our gaps, and put all of the pieces of this book together. Paul Forkner's tenacious photo research helped us achieve our creative vision. At Cengage Learning, Yolanda Cossio and Peggy Williams unwaveringly supported us and our ideals. Andy Marinkovich made sure we had what we needed, Amanda Jellerichs arranged for us to meet with hundreds of professors, Kristina Razmara continues to refine our amazing technology package, Samantha Arvin helped us stay organized, and Elizabeth Momb managed all of the print ancillaries.

CECIE STARR, CHRISTINE EVERS, AND LISA STARR
June 2008

CONTRIBUTORS TO THIS EDITION: INFLUENTIAL CLASS TESTS AND REVIEWS

MARC C. ALBRECHT
University of Nebraska at Kearney

ELLEN BAKER
Santa Monica College

SARAH FOLLIS BARLOW
Middle Tennessee State University

MICHAEL C. BELL
Richland College

LOIS BREWER BOREK
Georgia State University

ROBERT S. BOYD
Auburn University

URIEL ANGEL BUITRAGO-SUAREZ
Harper College

MATTHEW REX BURNHAM
Jones County Junior College

P.V. CHERIAN
Saginaw Valley State University

WARREN COFFEEN
Linn Benton

LUIGIA COLLO
Universita' Degli Studi Di Brescia

DAVID T. COREY
Midlands Technical College

DAVID F. COX
Lincoln Land Community College

KATHRYN STEPHENSON CRAVEN
Armstrong Atlantic State University

SONDRA DUBOWSKY
Allen County Community College

PETER EKECHUKWU
Horry-Georgetown Technical College

DANIEL J. FAIRBANKS
Brigham Young University

MITCHELL A. FREYMILLER
University of Wisconsin - Eau Claire

RAUL GALVAN
South Texas College

NABARUN GHOSH
West Texas A&M University

JULIAN GRANIRER
URS Corporation

STEPHANIE G. HARVEY
Georgia Southwestern State University

JAMES A. HEWLETT
Finger lakes community College

JAMES HOLDEN
Tidewater Community College - Portsmouth

HELEN JAMES
Smithsonian Institution

DAVID LEONARD
Hawaii Department of Land and Natural Resources

STEVE MACKIE
Pima West Campus

CINDY MALONE
California State University - Northridge

KATHLEEN A. MARRS
Indiana University - Purdue University Indianapolis

EMILIO MERLO-PICH
GlaxoSmithKline

MICHAEL PLOTKIN
Mt. San Jacinto College

MICHAEL D. QUILLEN
Maysville Community and Technical College

WENDA RIBEIRO
Thomas Nelson Community College

MARGARET G. RICHEY
Centre College

JENNIFER CURRAN ROBERTS
Lewis University

FRANK A. ROMANO, III
Jacksonville State University

CAMERON RUSSELL
Tidewater Community College - Portsmouth

ROBIN V. SEARLES-ADENEGAN
Morgan State University

BRUCE SHMAEFSKY
Kingwood College

BRUCE STALLSMITH
University of Alabama - Huntsville

LINDA SMITH STATON
Pollissippi State Technical Community College

PETER SVENSSON
West Valley College

LISA WEASEL
Portland State University

DIANA C. WHEAT
Linn-Benton Community College

CLAUDIA M. WILLIAMS
Campbell University

MARTIN ZAHN
Thomas Nelson Community College

VII PRINCIPLES OF ECOLOGY

Lioness and her cub at sunset on the African savanna. What are the consequences of their interactions with each other, with other kinds of organisms, and with their environment? By the end of this last unit, you might find worlds within worlds in such photographs.

44 Animal Behavior

IMPACTS, ISSUES My Pheromones Made Me Do It

One spring day as Toha Bergerub was walking down a street near her Las Vegas home, she felt a sharp pain above her right eye—then another, and another. Within a few seconds, hundreds of stinging bees covered the upper half of her body. Firefighters in protective gear rescued her, but not before she was stung more than 500 times. Bergerub, who was seventy-seven years old at the time, spent a week in the hospital, but recovered fully.

Bergerub's attackers were Africanized honeybees, a hybrid between gentle European honeybees and a more aggressive subspecies native to Africa (Figure 44.1). Bee breeders had imported African bees to Brazil in the 1950s. The breeders thought cross-breeding might yield a mild-tempered but more active pollinator for commercial orchards. However, some African bees escaped and mated with European honeybees that had become established in Brazil before them.

Then, in a grand example of geographic dispersal, some descendants of the hybrids buzzed all the way from Brazil to Mexico and on into the United States. So far, they have settled in Texas, New Mexico, Nevada, Utah, California, Oklahoma, Louisiana, Alabama, and Florida.

Africanized honeybees became known as "killer bees," although they rarely kill humans. They have been in the United States since 1990, yet no more than fifteen people have died after being attacked.

All honeybees defend their hives by stinging. Each can sting only once, and all make the same kind of venom. Even so, compared with European honeybees, Africanized ones get riled up more easily, attack in greater numbers, and stay agitated longer. Some are known to have chased people for more than a quarter of a mile.

What makes Africanized bees so testy? Part of the answer is that they have a heightened response to alarm pheromone. A pheromone is a social cue, a type of chemical signal that is emitted by one individual and influences another individual of the same species. For instance, when a honeybee worker guarding the entrance to a hive senses an intruder, it releases alarm pheromone. Pheromone molecules diffuse through the air and excite other bees, which fly out and sting the intruder.

In one study, researchers tested hundreds of colonies of Africanized honeybees and European honeybees to quantify their responses to alarm pheromone. The researchers positioned a seemingly threatening object, such as a scrap of black cloth, near the entrance of each hive. Then they released a small quantity of an artificial alarm pheromone. The Africanized bees flew out of their hive and zeroed in on the perceived threat much faster. Those bees plunged six to eight times as many stingers into the target.

The two strains of honeybees also show other behavioral differences. Africanized bees are less picky about where they establish a colony. They are more likely to abandon their hive after a disturbance. Of greater concern to beekeepers, the Africanized bees are less interested in storing large amounts of honey.

Such differences among honeybees lead us into the world of animal behavior—to coordinated responses that animal species make to stimuli. We invite you to reflect first on behavior's genetic basis, which is the foundation for its instinctive and learned mechanisms. Along the way, you will also come across examples of the adaptive value of behavior.

See the video! Figure 44.1 Two Africanized honeybees stand guard at their hive entrance. If a threat appears, they will release an alarm pheromone that stimulates hivemates to join an attack.

Key Concepts

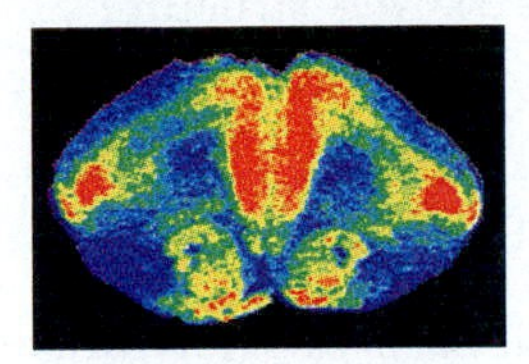

Foundations for behavior

Behavioral variations within or among species often have a genetic basis. Behavior can also be modified by learning. When behavioral traits have a heritable basis, they can evolve by way of natural selection. **Sections 44.1–44.3**

Animal communication

Interactions between members of a species depend on evolved modes of communication. Communication signals hold clear meaning for both the sender and the receiver of signals. **Section 44.4**

Mating and parental care

Behavioral traits that affect the ability to attract and hold a mate are shaped by sexual selection. Males and females are subject to different selective pressure. Parental care can increase reproductive success, but it has energetic costs. **Section 44.5**

Costs and benefits of social behavior

Life in social groups has reproductive benefits and costs. Self-sacrificing behavior has evolved among a few kinds of animals that live in large family groups. Human behavior is influenced by evolutionary factors, but humans alone make moral choices. **Sections 44.6–44.8**

Links to Earlier Concepts

- This chapter builds on your knowledge of sensory and endocrine systems (Sections 34.1, 35.3). We will discuss the role of hormones in lactation (43.12) and other behaviors. We will also look in more detail at pheromones 35.1.
- You may wish to review the concepts of adaptation (17.3) and sexual selection (18.6). You will see another example of the use of knockout experiments (15.3).
- You will be reminded again of the limits of science (1.5), and the rise of cultural traits (26.13).

How would you vote? Africanized bees are expanding their range in North America. Learning more about them may help us devise ways to protect ourselves. Should research into the genetic basis of their behavior be a high priority? See CengageNOW for details, then vote online.

44.1 Behavioral Genetics

- Variations in behavior within or among species often have their basis in genetic differences.
- Links to Knockout experiments 15.3, Sensory systems 34.1, Pituitary hormones 35.3, Lactation 43.12

How Genes Affect Behavior

Animal behavior requires a capacity to detect stimuli. A **stimulus**, recall, is some type of information about the environment that a sensory receptor has detected (Section 34.1). Which types of stimuli an animal is able to detect and the types of responses it can make start with the structure of its nervous system. Differences in genes that affect the structure and activity of the nervous system cause many differences in behavior.

Keep in mind, however, that mutations that affect metabolism or structural traits also influence behavior. For example, suppose you notice that some birds routinely eat large seeds and others focus on small seeds. Those that eat large seeds might do so because they cannot detect the smaller seeds. Or, they might see but ignore small seeds because the structure of their beaks allows them to easily open larger ones.

Figure 44.2 (**a**) Banana slug, the food of choice for adult garter snakes of coastal California. (**b**) A newborn garter snake from a coastal population, tongue-flicking at a cotton swab that had been drenched with fluids from a banana slug.

Characteristics	Rover	Sitter
Foraging behavior	Switches feeding area frequently	Tends to feed in one area
Genotype	*FF* or *Ff*	*ff*
PKG (enzyme) level	Higher	Lower
Speed of learning olfactory cues	Faster	Slower
Long-term memory for olfactory cues	Shorter	Longer

Figure 44.3 Characteristics of rovers and sitters, two behavioral phenotypes that occur in wild fruit fly populations. The two types differ in foraging behavior, learning, and memory, but not in general activity level. When food is not present, rovers and sitters are equally likely to move about.

Studying Variation Within a Species

One way to investigate the genetic basis of behavior is to examine behavioral differences among members of a single species. For example, Stevan Arnold studied feeding behavior in two populations of garter snakes. Some garter snakes live in coastal forests of the Pacific Northwest and their preferred food is banana slugs, which are common on the forest floor (Figure 44.2*a*). Farther inland, there are no banana slugs and the garter snakes prefer to eat fishes and tadpoles. Were these prey preferences inborn? To find out, Arnold offered newborn garter snakes of both populations a banana slug as their first meal. Most offspring of coastal snakes ate it. Offspring of inland snakes usually ignored it.

Newborn coastal snakes also flicked their tongue more often at a cotton swab soaked in slug juices, as in Figure 44.2*b*. (Tongue-flicking pulls molecules into the mouth.) Arnold hypothesized that inland snakes lack the genetically determined ability to associate the scent of slugs with "FOOD!" He predicted that if coastal garter snakes were crossed with inland snakes, the resulting offspring would make an intermediate response to slug odors. Results from his experimental crosses confirmed this prediction. Hybrid baby snakes tongue-flick at cotton swabs with slug juices more than newborn inland snakes do, but not as often as newborn coastal snakes do. Exactly which gene or genes underlie this difference has not been determined.

We do know about one gene that influences feeding behavior in fruit flies (*Drosophila melanogaster*). Marla Sokolowski showed that in wild fruit fly populations about 70 percent of the flies are "rovers"; they tend to move from place to place when food is present. About 30 percent of flies are "sitters"; they tend to feed in one place. Genotype at the *foraging* (*for*) locus determines whether a fly is rover or a sitter. Flies that have the dominant allele (*F*) are rovers. Those homozygous for the recessive allele (*f*) are sitters.

Sokolowski went on to uncover the molecular basis for the observed differences in behavior. She showed that the *for* gene encodes a cGMP-dependent protein kinase (PKG). This enzyme activates other molecules by donating a phosphate group to them, and it plays a role in many intercellular signaling pathways. Rovers make a bit more PKG than sitters. Having more PKG in the brain allows rovers to learn about new odors faster than sitters, but it also makes rovers forget what they learned faster. Figure 44.3 summarizes genotypes and behaviors of the rover and sitter phenotypes.

Examples such as this one, in which researchers can point to a single gene as the predominant cause of natural variations in behavior, are extremely rare. More typically, differences in many genes and exposure to different environmental factors cause members of a species to differ in their behavior.

Comparisons Among Species

Comparing behavior of related species can sometimes help clarify the genetic basis of a behavior. For instance, all mammals secrete the pituitary hormone oxytocin (OT), which acts in labor and lactation (Section 35.3). In many mammals, OT also influences pair bonding, aggression, territoriality, and other forms of behavior.

Among small rodents called prairie voles (*Microtus ochrogaster*), OT is the hormonal key that unlocks the female's heart. The female bonds with a male after a night of repeated matings, and she mates for life. In one experimental test of OT's influence, researchers injected pair-bonded female prairie voles with a drug that blocks OT action. Females that got the injection immediately dumped their partners.

Genetic differences in the number and distribution of OT receptors may help explain differences in mating systems among vole species. For example, prairie voles, which are monogamous and mate for life, have more OT receptors than mountain voles (*M. montanus*), which are highly promiscuous (Figure 44.4).

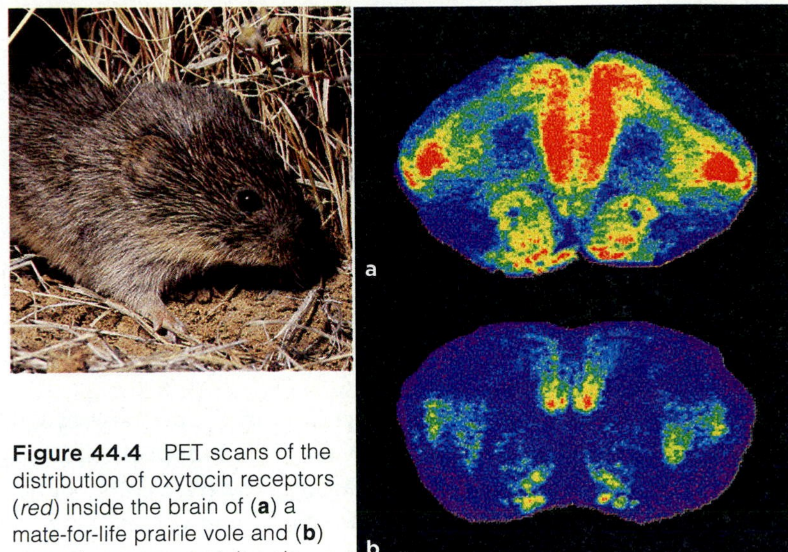

Figure 44.4 PET scans of the distribution of oxytocin receptors (*red*) inside the brain of (**a**) a mate-for-life prairie vole and (**b**) a promiscuous mountain vole.

Compared to males of promiscuous vole species, males of monogamous species also have more antidiuretic hormone (ADH) receptors in their forebrain. To test the effect of this difference, scientists isolated the gene for the ADH receptor in prairie voles. They then used a virus to add copies of this gene into the forebrain of some naturally promiscuous male meadow voles (*M. pennsylvanicus*). Results confirmed the role of ADH receptors in monogamy. Experimentally treated males preferred a female with whom they had mated over a new one. Control males that received the gene in a different brain region or virus with a different gene showed no preference for a familiar partner.

Knockouts and Other Mutations

Study of mutations can also help researchers understand behavior. As an example, fruit fly males with a mutation in the *fruitless* (*fru*) gene do not perform normal courtship movements and they court males in addition to females. When researchers compared the brains of male *fru* mutants to brains of normal males, they found the mutants—like normal females—lacked a certain set of neurons. Apparently development of that set of neurons has an integral role in governing typical male mate preference and courtship behavior.

As another example, knockout experiments (Section 15.3) confirmed the importance of oxytocin in mouse maternal behavior. Researchers produced female mice in which the gene for the OT receptor was knocked out. Lacking a functional receptor for OT, these mice could not respond to this hormone. As expected, these females did not lactate; oxytocin is required for contraction of milk ducts (Section 43.12). Knockout females also were less likely than normal mice to retrieve pups that researchers moved out of the nest. Based on these results, researchers concluded that oxytocin is required for normal maternal behavior in mice.

Take-Home Message

How do researchers study the effect of genes on animal behavior?

- Studying variations in behavior within a species or among related species allows researchers to determine whether the variation has a genetic basis. Such differences are rarely caused by variation in a single gene; many genes affect behavior.
- Researchers sometimes can determine the effect of a gene on a specific behavior by studying individuals in which the gene is nonfunctional.

44.2 Instinct and Learning

- Some behaviors are inborn and can be performed without any practice.
- Most behaviors are modified as a result of experience.

Instinctive Behavior

All animals are born with the capacity for **instinctive behavior**—an innate response to a specific and usually simple stimulus. A newborn coastal garter snake behaves instinctively when it attacks a banana slug. A male fruit fly instinctively waves its wings during courtship of a female.

The life cycle of the cuckoo bird provides several examples of instinct at work. This European bird is a social parasite. Females lay eggs in nests of other birds. A newly hatched cuckoo is blind, but contact with an egg laid by its foster parent stimulates an instinctive response. That hatchling maneuvers the egg onto its back, then shoves it out of the nest (Figure 44.5*a*). This behavior removes any potential competition for the foster parent's attention.

A cuckoo's egg-dumping response is a **fixed action pattern**: a series of instinctive movements, triggered by a specific stimulus, that—once started—continues to completion without the need for further cues. Such fixed behavior has survival advantages when it permits a fast response to an important stimulus. However, a fixed response to simple stimuli has limitations. For example, the cuckoo's foster parents are not equipped to note color and size of offspring. A simple stimulus —a chick's gaping mouth—induces the fixed action pattern of parental feeding behavior (Figure 44.5*b*).

Figure 44.5 Instinctive behavior. (**a**) A young cuckoo shoves its foster parent's eggs out of the nest. (**b**) The foster parent feeds the cuckoo chick in response to one simple cue: a gaping mouth.

Figure 44.6 Nobel laureate Konrad Lorenz with geese that imprinted on him. The smaller photograph shows results of a more typical imprinting episode.

Time-Sensitive Learning

Learned behavior is behavior that is altered by experience. Some instinctive behavior can be modified with learning. A garter snake's initial strikes at prey are instinctive, but the snake learns to avoid dangerous or unpalatable prey. Learning may occur throughout an animal's life, or be restricted to a critical period.

Imprinting is a form of learning that occurs during a genetically determined time period. For example, baby geese learn to follow the large object that bends over them in response to their first peep (Figure 44.6). With rare exceptions, this object is their mother. When mature, the geese will seek out a sexual partner that is similar to the imprinted object.

A genetic capacity to learn, combined with actual experiences in the environment, shapes most forms of behavior. For example, a male songbird has an inborn capacity to recognize his species' song when he hears older males singing it. The young male uses these overheard songs as a guide to fill in details of his own song. Males reared alone sing a simplified version of their species' song. So do males exposed only to the songs of other species.

Many birds must learn their species-specific song during a limited period early in life. For example, a male white-crowned sparrow will not sing normally if he does not hear a male "tutor" of his own species during his first 50 or so days. Hearing a same-species tutor later in life will not influence his singing.

Most birds must also practice their song to perfect it. In one experiment, researchers temporarily paralyzed throat muscles of zebra finches who were beginning to sing. After being temporarily unable to practice, these birds never mastered their song. In contrast, temporary paralysis of throat muscles in very young birds or adults did not impair later song production. Thus, in this species, there is a critical period for song practice, as well as for song learning.

Conditioned Responses

Nearly all animals are lifelong learners. Most learn to associate certain stimuli with rewards and others with negative consequences.

With **classical conditioning**, an animal's involuntary response to a stimulus becomes associated with another stimulus that is presented at the same time. In the most famous example, Ivan Pavlov rang a bell whenever he fed a dog. Eventually, the dog's reflexive response to food—increased salivation—was elicited by the sound of the bell alone.

With **operant conditioning**, an animal modifies its voluntary behavior in response to consequences of that behavior. This type of learning was first described for conditions in the lab. For example, a rat that presses a lever in a laboratory cage and is rewarded with a food pellet becomes more likely to press the lever again. A rat that receives a shock when it enters a particular area of a cage will quickly learn to avoid that area.

Other Types of Learned Behavior

With **habituation**, an animal learns by experience not to respond to a stimulus that has neither positive nor negative effects. For example, pigeons in cities learn not to flee from the large numbers of people who walk past them.

Many animals learn about the landmarks in their environment and form a sort of mental map. This map may be put to use when the animal needs to return home. For example, a fiddler crab foraging up to 10 meters (30 feet) away from its burrow is able to scurry straight home when it perceives a threat.

Many animals also learn the details of their social landscape; they learn to recognize mates, offspring, or competitors by appearance, calls, odor, or some combination of cues. For example, when two male lobsters meet up for the first time they will fight (Figure 44.7). Later, they will recognize one another by scent and behave accordingly, with the loser actively avoiding the winner. A lobster also recognizes its mate's scent.

With **observational learning**, an animal imitates the behavior of another individual. For example, Ludwig Huber and Bernhard Voelkel allowed marmoset monkeys to watch another marmoset demonstrate how to open a plastic container and retrieve the treat inside it. Marmosets who had seen the demonstrator open the container with its hands imitated this behavior, using their hands in the same way. In contrast, those who had watched a demonstrator open the box with its teeth attempted to do the same (Figure 44.8).

Figure 44.7 Getting to know one another. Two male lobsters battle at their first meeting. Later, the loser will remember the odor of the winner and avoid him. Without another meeting, memory of the defeat lasts up to two weeks.

Figure 44.8 Observational learning. A marmoset opens a container using its teeth. After watching one individual successfully perform this maneuver, other marmosets used the same technique. Analysis of videos of their movements showed that the observers closely imitated the behavior they had seen earlier.

Take-Home Message

How do instinct and learning shape behavior?

- Instinctive behavior can initially be performed without any prior experience, as when a simple cue triggers a fixed action pattern. Even instinctive behavior may be modified by experience.
- Certain types of learning can only occur at particular times in the life cycle.
- Learning affects both voluntary and involuntary behaviors.

44.3 Adaptive Behavior

- If a behavior varies and some of that variation has a genetic basis, then it will be subject to natural selection.
- Link to Adaptive traits 17.3

Behavior that increases an individual's reproductive success is adaptive. For example, Larry Clark and Russell Mason studied the nest decorating behavior of starlings. These birds tuck sprigs of aromatic plants such as wild carrot into their nests. Clark and Mason suspected that the plant bits control parasitic mites that feed on nestlings. To test their hypothesis, the researchers replaced natural starling nests with man-made ones that either had wild carrot sprigs or were sprig-free. They predicted that the decorated nests would have fewer mites than undecorated ones.

After the starling chicks left the nests, Clark and Mason recorded the number of mites left behind. The number was greater in sprig-free nests (Figure 44.9). Why? As it turns out, one organic compound in the leaves of wild carrot prevents mites from maturing.

Mason and Clark concluded that decorating a nest with sprigs deters bloodsucking mites. They inferred that this nest-decorating behavior is adaptive because it promotes nestling survival, increasing reproductive success for the nest-decorating birds.

As you will learn in Section 44.7, some behavior that increases the reproductive success of relatives at the expense of the individual can also be adaptive.

Take-Home Message

What makes a behavior adaptive?

- Most behavior is adaptive because it increases the reproductive success of the individual performing it. Some is adaptive because it benefits relatives.

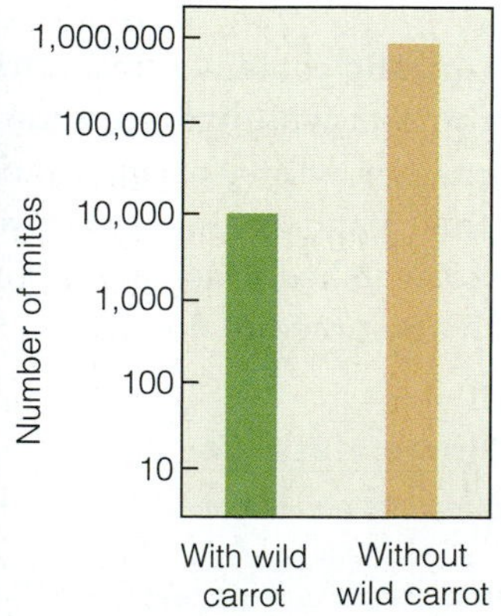

Figure 44.9 Results of an experiment to test the effect of wild carrot sprigs on the number of mites in starling nests. Nests with wild carrot pieces had significantly fewer mites than those with no greenery. There may be a selective advantage to using wild carrot and other aromatic plants as nest materials.

44.4 Communication Signals

- Cooperating to mate or in other ways requires individuals to share information about themselves and their environment.
- Link to Pheromones 35.1

Communication signals are cues for social behavior between members of a species. Chemical, acoustical, visual, and tactile signals transmit information from signalers to signal receivers.

Pheromones are chemical cues. Signal pheromones make a receiver alter its behavior fast. The honeybee alarm pheromone is an example. So are sex attractants that help males and females find each other. Priming pheromones cause longer-term responses, as when a chemical dissolved in the urine of certain male mice triggers ovulation in females of the same species.

Many acoustical signals, such as bird song, attract mates or define a territory. Others are alarm signals, such as a prairie dog's bark that warns of a predator.

One visual signal is a male baboon threat display, which communicates readiness to fight a rival (Figure

Figure 44.10 Visual signals. (**a**) A male baboon shows his teeth in a threat display. (**b**) Penguins engaged in a courtship display. (**c**) A wolf's play bow tells another wolf that behavior that follows is play, not aggression.

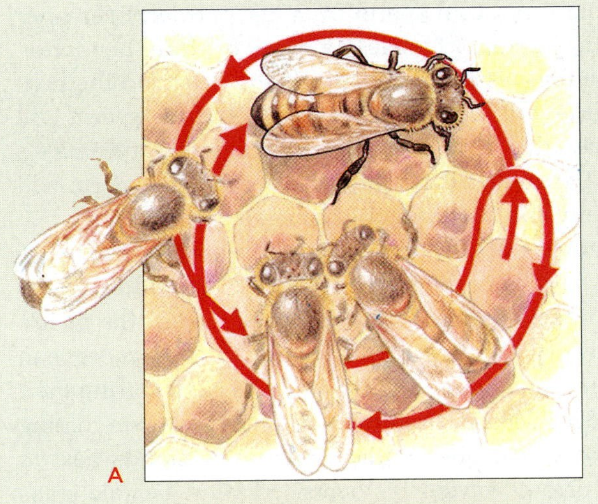

A

B

Figure 44.11 **Animated** Honeybee dances, an example of a tactile display. (**a**) Bees that have visited a source of food close to their hive return and perform a *round* dance on the hive's vertically oriented honeycomb. The bees that maintain contact with the dancer later fly out and search for food near the hive.

(**b**) A bee that visits a feeding source more than 100 meters (110 yards) from her hive performs a *waggle* dance. Orientation of an abdomen-waggling dancer in the straight run of her dance informs other bees about the direction of the food.

(**c**) If the food is in line with the sun, the dancer's waggling run proceeds straight up the honeycomb. (**d**) If food is in the opposite direction from the sun, the dancer's waggle run is straight down. (**e**) If food is 90 degrees to the right of the direction of the sun, the waggle run is offset by 90 degrees to the right of vertical.

The speed of the dance and the number of waggles in the straight run provide information about distance to the food. A dance inspired by food that is 200 meters away is much faster and has more waggles per straight run, than a dance inspired by a food source that is 500 meters away.

Figure It Out: Do the dances shown in parts c–e indicate different distances from the hive? *Answer: No. The number of waggles in the straight run does not vary.*

When bee moves straight up comb, recruits fly straight toward the sun.

C

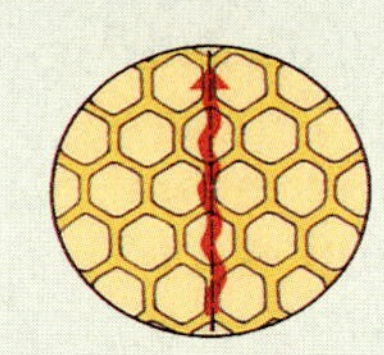

When bee moves straight down comb, recruits fly to source directly away from the sun.

D

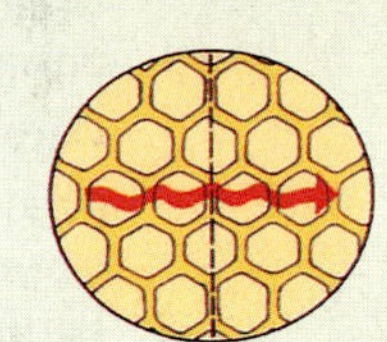

When bee moves to right of vertical, recruits fly at 90° angle to right of the sun.

E

44.10*a*). Visual signals are part of courtship displays that often precede mating in birds (Figure 44.10*b*). Unambiguous signals work best, so movements often get exaggerated and body form evolves in ways that draw attention to the movements.

With tactile displays, information is transmitted by touch. For example, after discovering food, a foraging honeybee worker returns to the hive and performs a complex dance. The bee moves in a defined pattern, jostling a crowd of other bees that surround her. The signals give other bees information about the distance and the direction of the food source (Figure 44.11).

The same signal sometimes functions in more than one context. For example, dogs and wolves solicit play behavior with a play bow (Figure 44.10*c*). A play bow informs an animal's prospective playmate that signals that follow, which would ordinarily be construed as aggressive or sexual, are friendly play behavior.

A communication signal evolves and persists only if it benefits both sender and receiver. If the signal has disadvantages, then natural selection will tend to favor individuals that do not send or respond to it. Other factors can also select against signalers. For example, male tungara frogs attract females with complex calls, which also make it easier for frog-eating bats to zero in on the caller. When bats are near, male frogs call less, and usually with less flair. The subdued signal is a trade-off between locating a partner for mating and the need for immediate survival.

There are illegitimate signalers, too. For example, fireflies attract mates by producing flashes of light in a characteristic pattern. Some female fireflies prey on males of other species. When a predatory female sees the flash from a male of the prey species, she flashes back as if she were a female of his own species. If she lures him close enough, she captures and eats him.

Take-Home Message

What are the benefits and costs of communication signals?

- A communication signal transfers information from one individual to another individual of the same species. Such signals benefit both the signaler and the receiver.
- Signals have a potential cost. Some individuals of a different species benefit by intercepting signals or by mimicking them.

44.5 Mates, Offspring, and Reproductive Success

- In studying behavior, we expect that each sex will evolve in ways that maximize its benefits, and minimize its costs, which can lead to conflicts.
- Link to Sexual selection 18.6

Sexual Selection and Mating Behavior

Males or females of a species often compete for access to mates, and many are choosy about their partners. Both situations lead to **sexual selection**. As explained in Section 18.6, this microevolutionary process favors characteristics that provide a competitive advantage in attracting and often holding on to mates.

But whose reproductive success is it—the male's or the female's? Male animals, remember, produce many small sperm, and females produce far larger but fewer eggs. For the male, success generally depends on how many eggs he can fertilize. For the female, it depends more on how many eggs she produces or how many offspring she can raise. Usually, the most important factor in a female's sexual preference is the quality of the mate, not the quantity of partners.

Female hangingflies (*Harpobittacus*) will mate only with males that supply food. A male hunts and kills a moth or some other insect. Then he releases a sex pheromone, which attracts females to him and his "nuptial gift" (Figure 44.12*a*). The female begins to eat the male's offering and copulation begins. Only after the female has been eating for five minutes or so does she start to accept sperm from her partner. Even after mating begins, a female can break off from her suitor, if she finishes eating his gift. If she does end the mating, she will seek out a new male and his sperm will replace the first male's. Thus, the larger the male's gift, the greater the chance that mostly his sperm will actually end up fertilizing the eggs of his mate.

Females of certain species shop around for males who have appealing traits. Consider the fiddler crabs that live along many sandy shores. One of the male's two claws is enlarged; it often accounts for more than half his total body weight (Figure 44.12*b*). During their breeding season, hundreds of males excavate mating burrows near one another. Each male stands next to his burrow, waving his oversized claw. Female crabs stroll along, checking out males. If a female likes what she sees, she inspects her suitor's burrow. Only when a burrow has the right location and dimensions does she mate with its owner and lay eggs in his burrow.

Some female birds are similarly choosy. Male sage grouse (*Centrocercus urophasianus*) converge at a lek, a type of communal display ground, where each stakes out a few square meters. With tail feathers erect, the males emit booming calls by puffing and deflating big neck pouches (Figure 44.12*d*). As they do, they stamp about on their patch of prairie. Females tend to select and mate with one male sage grouse. Afterward, they

Figure 44.12 (**a**) Male hangingfly dangling a moth as a nuptial gift for a potential mate. Females of some hangingfly species choose sexual partners that offer the largest gift to them. By waving his enlarged claw, a male fiddler crab (**b**) may attract the eye of a female fiddler crab (**c**). A male sage grouse (**d**) showing off as he competes for female attention at a communal display ground.

go off to nest and raise any offspring by themselves. Often, many females favor the same few males, and most males never have an opportunity to mate.

In another behavioral pattern, the sexually receptive females of some species cluster in defendable groups. Where you come across such a group, you are likely to observe males competing for access to the clusters. Competition for ready-made harems has resulted in combative male lions, sheep, elk, elephant seals, and bison, to name a few examples (Figure 44.13).

Figure 44.13 Male bison locked in combat during the breeding season.

Parental Care

When females fight for males, we can predict that the males provide more than sperm delivery. Some, such as the male midwife toad, help with parenting. The male holds strings of fertilized eggs around his legs until the eggs hatch (Figure 44.14*a*). Once her eggs are being cared for, a female can mate with other males, if she can find some that are not already caring for eggs. Late in the breeding season, males without strings of eggs are rare, and females fight for access to them. The females even attempt to pry mating pairs apart.

Parental behavior uses up time and energy, which parents otherwise might spend on living long enough to reproduce again. However, for some animals, the benefit of increased survival of the young outweighs the cost of parenting.

Few reptiles provide care for young. Crocodilians, the reptiles most closely related to birds, are a notable exception. Crocodile parents bury their eggs in a nest. When young are ready to hatch, they call and parents dig them out and care for them for some time.

Most birds are monogamous, and both parents often care for the young (Figure 44.14*b*). In mammals, males typically leave after mating. Females raise the young alone, and males attempt to mate again or conserve energy for the next breeding season (Figure 44.14*c*). Mammalian species in which males help care for the young tend to be monogamous, at least over the course of a breeding season. Only about 5 percent of mammals are monogamous.

Figure 44.14 (**a**) Male midwife toad with developing eggs wrapped around his legs. (**b**) A pair of Caspian terns cooperate in the care of their chick. (**c**) A female grizzly will care for her cub for as long as two years. The male takes no part in the cub's upbringing.

Take-Home Message

How does natural selection affect mating systems?

- Males and females each behave in ways that will maximize their own reproductive success.
- Most males compete for females and mate with more than one. Monogamy and male parental care are not common.

44.6 Living in Groups

- Survey the animal kingdom and you find evolutionary costs and benefits across a range of social groups.
- Link to Culture 26.13

Defense Against Predators

In some groups, cooperative responses to predators reduce the net risk to all. Vulnerable individuals can be on the alert for predators, join a counterattack, or engage in more effective defenses (Figure 44.15).

Birds, monkeys, meerkats, prairie dogs, and many other animals make alarm calls, as in Figure 44.15*a*. A prairie dog makes a particular bark when it sights an eagle and a different signal when it sights a coyote. Others dive into burrows to escape an eagle's attack or stand erect and observe the coyote's movements.

Sawfly caterpillars feed in clumps on branches and benefit by coordinated repulsion of predatory birds. When a potential predator approaches, the caterpillars rear up and vomit partly digested eucalyptus leaves (Figure 44.15*b*). Birgitta Sillén-Tullberg demonstrated that predatory birds prefer individual caterpillars to a wiggling group. When offered caterpillars one at a time, the birds ate an average of 5.6. Birds offered a cluster of twenty caterpillars ate an average of 4.1.

Whenever animals cluster, some individuals shield others from predators. Preference for the center of a group can create a **selfish herd**, in which individuals hide behind one another. Selfish-herd behavior occurs in bluegill sunfishes. A male sunfish builds a nest by scooping out a depression in mud on the bottom of a lake. Females lay eggs in these nests, and snails and fishes prey on eggs. Competition for the safest sites is greatest near the center of a group, with large males taking the innermost locations. Smaller males cluster around them and bear the brunt of the egg predation. Even so, the nests of small males are safer at the edge of the group than they would be alone in the open.

Improved Feeding Opportunities

Many mammals, including wolves, lions, wild dogs, and chimpanzees, live in social groups and cooperate in hunts (Figure 44.16). Are cooperative hunters more efficient than solitary ones? Often, no. In one study, researchers observed a solitary lion that caught prey about 15 percent of the time. Two lions cooperatively hunting caught prey twice as often but had to share it, so the amount of food per lion balanced out. When more lions joined a hunt, the success rate per lion fell. Wolves show a similar pattern. Among carnivores that hunt cooperatively, hunting success does not seem to be the major advantage of group living. Individuals hunt together, but they also may fend off scavengers, care for one another's young, and protect territory.

Group living also allows transmission of cultural traits, or behaviors learned by imitation. For example, chimpanzees make and use simple tools by stripping leaves from branches. They use thick sticks to make holes in a termite mound, then insert long, flexible "fishing sticks" into the holes (Figure 44.17). The long stick agitates the termites, which attack and cling to it.

Figure 44.15 Group defenses. (**a**) Black-tailed prairie dogs bark an alarm call that warns others of predators. Does this call put the caller at risk? Not much. Prairie dogs usually act as sentries only after they finish feeding and happen to be standing next to their burrows. (**b**) Australian sawfly caterpillars form clumps and regurgitate a fluid (the *yellow* blobs) that predators find unappealing. (**c**) Musk oxen adults (*Ovibos moschatus*) form a ring of horns, often around their young.

Figure 44.16 Members of a wolf pack (*Canis lupus*). Wolves cooperate in hunting, caring for young, and defending territory. Benefits are not distributed equally. Only the highest ranking individuals, the alpha male and alpha female, breed.

Chimps withdraw the stick and lick off termites, as a high-protein snack. Different groups of chimpanzees use slightly different tool-shaping and termite-fishing methods. Youngsters of each group learn by imitating the adults.

Dominance Hierarchies

In many social groups, subordinate individuals do not get an equal share of resources. Most wolf packs, for instance, have one dominant male that breeds with just one dominant female. The others are nonbreeding brothers and sisters, aunts and uncles. All hunt and carry food back to individuals that guard the young in their den.

Why would a subordinate give up resources and often breeding privileges? It might get injured or die if it challenges a strong individual. It might not be able to survive on its own. A subordinate might even get a chance to reproduce if it lives long enough or if its dominant peers are taken out by a predator or old age. As one example, some subordinate wolves move up the social ladder when the opportunity arises.

Figure 44.17 Chimpanzees (*Pan troglodytes*) using sticks as tools for extracting tasty termites from a nest. This behavior is learned by imitation.

Regarding the Costs of Group Living

If social behavior is advantageous, then why are there so few social species? In most habitats, costs outweigh benefits. For instance, when individuals are crowded together they compete more for resources. Cormorants and other seabirds form dense breeding colonies, as in Figure 44.18. All compete for space and food.

Large social groups also attract more predators. If individuals are crowded together, they are vulnerable to parasites and contagious diseases that jump from host to host. Individuals may also be at risk of being killed or exploited by others. Given the opportunity, a pair of breeding herring gulls will cannibalize the eggs and even the chicks of their neighbors.

Figure 44.18 Nearly uniform spacing in a crowded cormorant colony.

Take-Home Message

What are the benefits and costs of social groups?

- Living in a social group can provide benefits, as through cooperative defenses or shielding against predators.
- Group living has costs: increased competition, increased vulnerability to infections, and exploitation by others.

44.7 Why Sacrifice Yourself?

- Extreme cases of sterility and self-sacrifice have evolved in only a few groups of insects and one group of mammals. How are genes of the nonreproducers passed on?

Social Insects

Animals that are eusocial live together for generations in a group that has a reproductive division of labor. Eusocial insects include the honeybees, termites, and ants. In all of these groups, sterile workers care cooperatively for the offspring produced by just a few breeding individuals. Such workers often are highly specialized in their form and function (Figure 44.19).

A queen honeybee is the only fertile female in her hive. She is larger than other females, partly because of her enlarged ovaries (Figure 44.20*a*). She secretes a pheromone that makes all other female bees sterile.

All of the 30,000 to 50,000 worker bees are females that develop from fertilized eggs laid by the queen. They feed the larvae, maintain the hive, and construct honeycomb from wax they secrete. Workers also gather nectar and pollen that feeds the colony. They guard the hive and will sacrifice themselves to repel intruders.

In spring and summer, the queen lays unfertilized eggs that develop into drones. These male bees are stingless and subsist on food gathered by their worker sisters. Each day, drones fly in search of a mate. If one is lucky, he will meet a virgin queen on her one flight away from a colony. He dies after mating. A young queen mates with many males, and stores their sperm for use over her lifetime of several years.

Like honeybees, termites live in enormous family groups with a queen specialized for producing eggs (Figure 44.20*b*). Unlike the honeybee hive, a termite mound holds sterile individuals of both sexes. A king supplies the female with sperm. Winged reproductive termites of both sexes develop seasonally.

Social Mole-Rats

Sterility and extreme self-sacrifice are uncommon in vertebrates. The only eusocial mammals are African mole-rats. The best studied is *Heterocephalus glaber*, the naked mole-rat. Clans of this nearly hairless rodent build and occupy burrows in dry parts of East Africa.

A mole rat clan consists of a reproductive "queen" (Figure 44.20*c*), the one to three "kings," with whom she mates, and their nonbreeding worker offspring. Workers care for the queen, the king(s), and the young. Some workers serve as diggers that excavate tunnels and chambers. When a digger finds an edible root, it hauls a bit back to the main chamber and chirps. Its chirps recruit other workers to help carry food back to the chamber. Still other workers function as guards. When a predator appears, they chase and attack it at great risk to themselves.

Evolution of Altruism

A sterile worker in a social insect colony or a naked mole-rat clan shows **altruistic behavior**: behavior that enhances another individual's reproductive success at the altruist's expense. How did this behavior evolve? According to William Hamilton's **theory of inclusive fitness**, genes associated with altruism are selected if they lead to behavior that promotes the reproductive success of an altruist's closest relatives.

A sexually reproducing, diploid parent caring for offspring is not helping exact genetic copies of itself.

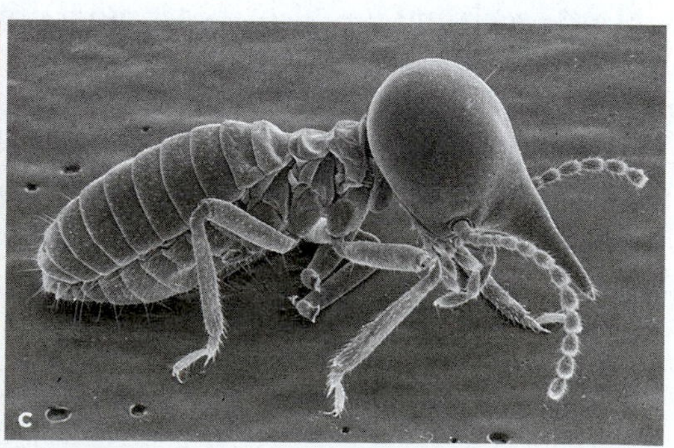

Figure 44.19 Specialized ways of serving and defending the colony. (**a**) An Australian honeypot ant worker. This sterile female is a living container for her colony's food reserves. (**b**) Army ant soldier (*Eciton burchelli*) with formidable mandibles. (**c**) Eyeless soldier termite (*Nasutitermes*). It bombards intruders with a stream of sticky goo from its nozzle-shaped head.

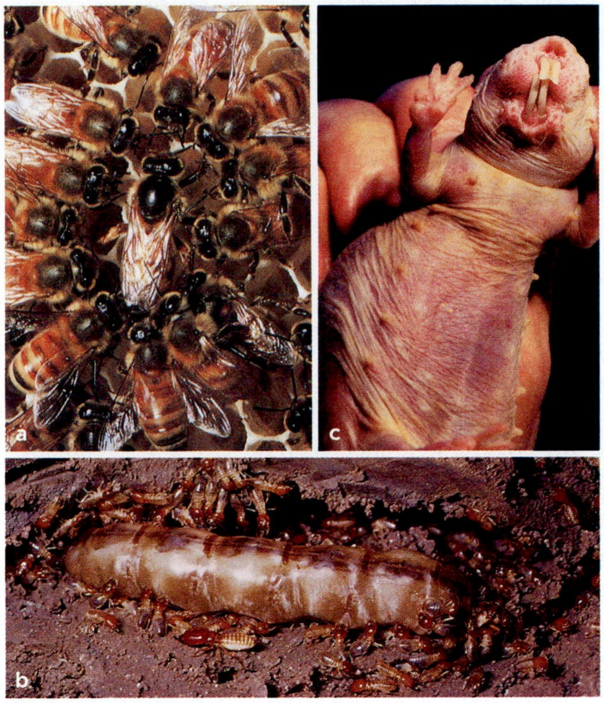

Figure 44.20 Three queens. (**a**) Queen honeybee with her sterile daughters. (**b**) A termite queen (*Macrotermes*) dwarfs her offspring and mate. Ovaries fill her enormous abdomen. (**c**) A naked mole-rat queen.

Each of its gametes, and each of its offspring, inherits one-half of its genes. Other individuals of the social group that have the same ancestors also share genes. Siblings (brothers or sisters) are as genetically similar as a parent and offspring. Nephews and nieces share about one-fourth of their uncle's genes.

Sterile workers promote genes for self-sacrifice by helping close relatives survive and reproduce. In honeybee, termite, and ant colonies, sterile workers assist fertile relatives with whom they share genes. A guard bee will die after she stings, but her sacrifice preserves many copies of her genes in her hivemates.

Inbreeding increases the genetic similarity among relatives and may play a role in mole-rat sociality. A clan is highly inbred as a result of many generations of sibling, mother–son, and father–daughter matings. Dry habitats and patchy food sources also may favor cooperation in digging, locating food, and fending off competitors and predators.

Take-Home Message

How can altruistic behavior be selectively advantageous?

- Altruistic behavior may be favored when individuals pass on genes indirectly, by helping relatives survive and reproduce.

44.8 Human Behavior

- Evolutionary forces shaped human behavior—but humans alone can make moral choices about their actions.
- Link to Limits of science 1.5

Hormones and Pheromones Are humans, too, influenced by hormones that contribute to bonding behavior in other mammals? Perhaps. Consider that autism, a developmental disorder in which people have trouble making social contacts, is often associated with low oxytocin levels. Oxytocin is known to affect bonding behavior in other mammals.

Pheromones in sweat may also affect human behavior. Women who live together often have synchronized menstrual cycles and experiments have shown that a woman's menstrual cycle will lengthen or shorten after she has been exposed to sweat from a woman who was in a different phase of the cycle. Other experiments have shown that exposure to male sweat can alter a woman's cortisol level.

Morality and Behavior If we are comfortable with studying the evolutionary basis of behavior of termites, naked mole-rats, and other animals, why do some people resist the idea of analyzing the evolutionary basis of human behavior? A common fear is that an objectionable behavior will be defined as "natural." To evolutionary biologists, however, "adaptive" does not mean "morally right." It simply means a behavior increases reproductive success. Scientific studies do not address moral issues (Section 1.5).

For example, infanticide is morally repugnant. Is it unnatural? No. It happens in many animal groups and all human cultures. Male lions often kill the offspring of other males when they take over a pride. Thus deprived of parenting tasks, the lionesses can now breed with the infanticidal male and increase that male's reproductive success.

Biologists would predict that unrelated human males are a threat to infants. Evidence supports the prediction. The absence of a biological father and the presence of an unrelated male increases risk of death for an American child under age two by more than sixty times.

What about parents who kill their own offspring? In her book on maternal behavior, primatologist Sarah Blaffer Hrdy cites a study of one village in Papua New Guinea in which parents killed about 40 percent of the newborns. As Hrdy argues, when resources or social support are hard to come by, a mother's fitness might increase if a newborn who is unlikely to survive is killed. The mother can allocate child-rearing energy to her other offspring or save it for children she may have in the future.

Do most of us find such behavior appalling? Yes. Can considering the possible evolutionary advantages of the behavior help us prevent it? Perhaps. An analysis of the conditions under which infanticide occurs tells us this: When mothers lack the resources they need to care for their children, they are more likely to harm them. We as a society can act upon such information.

IMPACTS, ISSUES REVISITED My Pheromones Made Me Do It

When a European queen bee mates with an Africanized drone, her worker offspring are just as aggressive as workers in a pure Africanized colony. In contrast, a cross between an Africanized queen and a European drone yields workers with an intermediate level of aggression. Unfortunately, European queen–Africanized male pairings occur far more frequently than the reciprocal cross. Africanized males outcompete European males for matings.

How would you vote?

Africanized honeybees continue to increase their range. Should study of their genetics be a high priority? See CengageNOW for details, then vote online.

Summary

Section 44.1 Behavior refers to coordinated responses that an animal makes to a **stimulus**. Genes that affect the nervous system often affect behavior, but other genes may also influence it. Studies of natural behavioral variations within and among species provide information about the genetic basis for behaviors, as does the study of induced or natural mutations.

Section 44.2 **Instinctive behavior** can occur without having been learned by experience. A **fixed action pattern** is an instinctive series of responses to a simple cue.

Learned behavior is altered by experience. **Imprinting** is one form of learning that happens only during a sensitive period early in life. With **classical conditioning**, an animal learns to associate an involuntary response to one stimulus with another stimulus. With **operant conditioning**, an animal modifies a voluntary behavior in response to the behavior's consequences. With **habituation**, an animal stops responding to an ongoing stimulus. With **observational learning**, it imitates another's actions.

Section 44.3 A behavior that has a genetic basis is subject to evolution by natural selection. Adaptive forms of behavior evolved as a result of individual differences in reproductive success in past generations.

Section 44.4 **Communication signals** allow animals of the same species to share information. Such signals evolve and persist only if they benefit both senders and receivers of the signal.

Chemical signals such as **pheromones** have roles in social communication, as do acoustical signals, visual signals that are part of courtship and threat displays, and tactile signals.

- *Use the animation on CengageNOW to explore the honeybee dance language.*

Section 44.5 **Sexual selection** favors traits that give an individual a competitive edge in attracting and often holding on to mates. The females of many species select males that have traits or engage in behaviors they find attractive. When large numbers of females cluster in a defensible area, males may compete with one another to control the areas.

Parental care has reproductive costs in terms of future reproduction and survival. It is adaptive when benefits to a present set of offspring offset the costs.

Section 44.6 Animals that live in social groups may benefit by cooperating in predator detection, defense, and rearing the young. A **selfish herd** forms when animals hide behind one another. Benefits of group living are often distributed unequally. Species that live in large groups incur costs, including increased disease and parasitism, and increased competition for resources.

Section 44.7 Ants, termites, and some other insects as well as two species of mole-rats are eusocial. They live in colonies with overlapping generations and have a reproductive division of labor. Most colony members do not reproduce; they assist their relatives instead. According to the **theory of inclusive fitness**, such **altruistic behavior** is perpetuated because altruistic individuals share genes with their reproducing relatives. Altruistic individuals help perpetuate the genes that led to their altruism by promoting the reproductive success of close relatives that also carry copies of these genes.

Section 44.8 Hormones and possibly pheromones influence human behavior. A behavior that is adaptive in the evolutionary sense may still be judged by society to be morally wrong. Science does not address morality.

Self-Quiz

Answers in Appendix III

1. Genes affect the behavior of individuals by ______ .
   a. influencing the development of nervous systems
   b. affecting the kinds of hormones in individuals
   c. determining which stimuli can be detected
   d. all of the above

2. Stevan Arnold offered slug meat to newborn garter snakes from different populations to test his hypothesis that the snakes' response to slugs ______ .
   a. was shaped by indirect selection
   b. is an instinctive behavior
   c. is based on pheromones
   d. is adaptive

3. A behavior is defined as adaptive if it ______ .
   a. varies among individuals of a population
   b. occurs without prior learning
   c. increases an individual's reproductive success
   d. is widespread across a species

4. The honeybee dance language transmits information about distance to food by way of ______ signals.
   a. tactile c. acoustical
   b. chemical d. visual

Data Analysis Exercise

Honeybees disperse by forming new colonies. An old queen leaves the hive along with a group of workers. These bees fly off, find a new nest site, and set up a new hive. Meanwhile, at the old hive, a new queen emerges, mates, and takes over. A new hive can be several kilometers from the old one.

Africanized honeybees form new colonies more often than European ones, a trait that contributes to their spread. Africanized bees also spread by taking over existing hives of European bees. In addition, in areas where European and Africanized hives coexist, European queens are more likely to mate with Africanized males, thus introducing Africanized traits into the colony. Figure 44.21 shows the counties in the United States in which Africanized honeybees became established from 1990 through 2006.

1. Where in the United States did Africanized bees first become established?
2. In what states did Africanized bees first appear in 2005?
3. Why is it likely that human transport of bees contributed to the spread of Africanized honeybees to Florida?
4. Based on this map, would you expect Africanized honeybees to colonize additional states in the next five years?

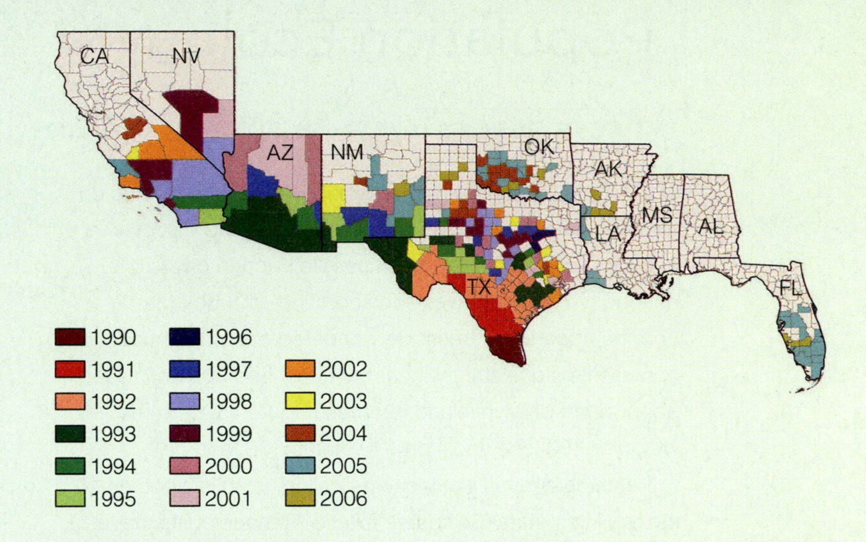

Figure 44.21 The spread of Africanized honeybees in the United States, from 1990 through 2006. The USDA adds a county to this map only when the state officially declares bees in that county Africanized. Bees can be identified as Africanized on the basis of morphological traits or analysis of their DNA.

5. A ______ is a chemical that conveys information between individuals of the same species.
   a. pheromone
   b. neurotransmitter
   c. hormone
   d. all of the above
6. In ______ , males and females typically cooperate in care of the young.
   a. mammals
   b. birds
   c. amphibians
   d. all of the above
7. Generally, living in a social group costs the individual in terms of ______ .
   a. competition for food, other resources
   b. vulnerability to contagious diseases
   c. competition for mates
   d. all of the above
8. Social behavior evolves because ______ .
   a. social animals are more advanced than solitary ones
   b. under some conditions, the costs of social life to individuals are offset by benefits to the species
   c. under some conditions, the benefits of social life to an individual offset the costs to that individual
   d. under most conditions, social life has no costs to an individual
9. Eusocial insects ______ .
   a. live in extended family groups
   b. include termites, honeybees, and ants
   c. show a reproductive division of labor
   d. a and c
   e. all of the above
10. Helping other individuals at a reproductive cost to oneself might be adaptive if those helped are ______ .
   a. members of another species
   b. competitors for mates
   c. close relatives
   d. illegitimate signalers
11. True or false? Some mammals live in colonies and act as sterile workers that serve close relatives.
12. Match the terms with their most suitable description.

___ fixed action pattern	a. time-dependent form of learning requiring exposure to key stimulus
___ altruism	b. genes plus actual experience
___ basis of instinctive and learned behavior	c. series of responses that runs to completion independently of feedback from environment
___ imprinting	d. assisting another individual at one's own expense
___ pheromone	e. one communication signal

■ *Visit CengageNOW for additional questions.*

Critical Thinking

1. For billions of years, the only bright objects in the night sky were stars or the moon. Night-flying moths used them to navigate in a straight line. Today, the instinct to fly toward bright objects causes moths to exhaust themselves fluttering around streetlights and banging against brightly lit windowpanes. This behavior is not adaptive, so why does it persist?

2. Damaraland mole-rats are relatives of naked mole-rats (Figure 44.19). In their clans, too, nonbreeding individuals of both sexes cooperatively assist one breeding pair. Even so, breeding individuals in wild Damaraland mole-rat colonies usually are unrelated, and few subordinates move up in the hierarchy to breeding status. Researchers suspect that ecological factors, not genetic ones, were the more important selective force in Damaraland mole-rat altruism. Explain why.

45 Population Ecology

IMPACTS, ISSUES The Numbers Game

In 1722, on Easter morning, a European explorer landed on a small volcanic island in the South Pacific and discovered a few hundred hungry, skittish people living in caves. He noticed withered grasses and scorched, shrubby plants—and the absence of trees. He wondered about the hundreds of massive stone statues near the coast and 500 unfinished, abandoned ones in inland quarries (Figure 45.1). Some weighed 100 tons and stood 10 meters (33 feet) high.

Easter Island, as it came to be called, is no larger than 165 square kilometers (64 square miles). Archaeologists have determined that voyagers from the Marquesas discovered this eastern outpost of Polynesia more than 1,650 years ago. The place was a paradise. Its volcanic soil supported dense forests and lush grassland. The colonists used long, straight palms to build canoes that were strengthened with rope made of fibers from hauhau trees. They used wood as fuel to cook fishes and dolphins. They cleared forests to plant crops. They had many children.

By 1440, as many as 15,000 people were living on the island. Crop yields declined; ongoing harvests and erosion had depleted the soil of nutrients. Fish vanished from the waters close to the island, so fishermen had to sail farther and farther out on the open ocean.

Those in power built statues to appeal to the gods. They directed others to carve images of unprecedented size and move the new statues to the coast. Wars broke out and by 1550, no one ventured offshore to fish. They could not build any more canoes because there were no more trees.

As central authority crumbled, the dwindling numbers of islanders retreated to caves and launched raids against one another. Winners ate the losers and tipped over statues. Even if the survivors had wanted to, they had no way to get off the island. The once-flourishing population collapsed.

Any natural population has the capacity to increase in number, given the right conditions. In North America, white-tailed deer are behaving like early settlers on Easter Island. With plenty of food and few predators, deer numbers are soaring. Deer overpopulation harms forests, damages crops, and increases the incidence of highway accidents.

With this chapter, we begin a survey of principles that govern the growth and sustainability of all populations. The principles are the bedrock of ecology—the systematic study of how organisms interact with one another and with their environment. Those interactions start within and between populations and extend to communities, ecosystems, and the biosphere.

See the video! Figure 45.1 Row of massive statues on Easter Island. Islanders set them up long ago, apparently as a plea for help after their once-large population wreaked havoc on their tropical paradise. Their plea had no effect whatsoever on reversing the loss in biodiversity on the island and in the surrounding sea. The human population did not recover, either.

Key Concepts

The vital statistics

Ecologists explain population growth in terms of population size, density, distribution, and number of individuals in different age categories. Field studies allow ecologists to estimate population size and density. **Sections 45.1, 45.2**

Exponential rates of growth

A population's size and reproductive base influence its rate of growth. When the population is increasing at a rate proportional to its size, it is undergoing exponential growth. **Section 45.3**

Limits on increases in number

Over time, an exponentially growing population typically overshoots the carrying capacity—the maximum number of individuals of a species that environmental resources can sustain. Some populations stabilize after a big decline. Others never recover. **Section 45.4**

Patterns of survival and reproduction

Resource availability, disease, and predation are major factors that can restrict population growth. These limiting factors differ among species and shape their life history patterns. **Sections 45.5, 45.6**

The human population

Human populations sidestepped limits to growth by way of global expansion into new habitats, cultural interventions, and innovative technology. Even so, no population can continue to expand indefinitely. **Sections 45.7–45.10**

Links to Earlier Concepts

- Earlier chapters defined and explored the evolutionary history and genetic nature of populations, including those of humans (Sections 18.1 and 26.15). Now you will consider factors that limit population growth, including contraception (42.9).
- You will be reminded of the effects of infectious disease (Chapter 21 introduction, 21.8), and the stunning reproductive capacity of prokaryotes (21.5).
- Gene flow (18.8) and directional selection (18.4) are discussed in the context of evolving populations. We also consider how sampling error (1.8) affects population studies.

How would you vote? Soaring numbers of white-tailed deer threaten forest plants and the animals that depend on them. Is encouraging deer hunting in regions where their overabundance is a threat to other species the best solution? See CengageNOW for details, then vote online.

45.1 Population Demographics

- A population's size, density, distribution, and age structure are shaped by ecological factors, and may shift over time.
- Link to Population genetics 18.1

Ecologists typically use the term "population" to refer to all members of a species within an area defined by the researcher. Studies of population ecology start with **demographics**: statistics that describe population size, age structure, density, distribution, and other factors.

Population size is the number of individuals in the population. **Age structure** is the number of individuals in each of several age categories. Individuals are often grouped as pre-reproductive, reproductive, or post-reproductive. Those in the pre-reproductive category have the capacity to produce offspring when mature. Together with individuals in the reproductive group, they make up the population's **reproductive base**.

Population density is the number of individuals in a specified portion of a habitat. A habitat, remember, is the type of place where a species lives. We characterize a habitat by its physical and chemical features, and its array of species.

Density refers to how many individuals are in an area but not how they are dispersed through it. Even a habitat that looks uniform, such as a sandy shore, has variations in light, moisture, and many other variables. A population may live in only a small part of the habitat, and it may do so all of the time or only some of the time.

The pattern in which individuals are dispersed in their habitat is the **population distribution**. It may be clumped, nearly uniform, or random (Figure 45.2).

A clumped distribution is most common, for several reasons. First, the conditions and resources that most species require tend to be patchy. Animals cluster at a water hole, seeds sprout only in moist soil, and so on. Second, most seeds and some animal offspring cannot disperse far from their parents. Third, some animals spend their lives in social groups that offer protection and other advantages.

With a nearly uniform distribution, individuals are more evenly spaced than we would expect on the basis of chance alone. Such distribution is relatively rare. It happens when competition for resources or territory is fierce, as in a nesting colony of seabirds.

We observe random distribution only when habitat conditions are nearly uniform, resource availability is fairly steady, and individuals of a population or pairs of them neither attract nor avoid one another. Each wolf spider does not hunt far from its burrow, which can be almost anywhere in forest soil (Figure 45.2*b*).

The scale of the study area and timing of a study can influence the observed pattern of distribution. For example, seabirds often are spaced almost uniformly at a nesting site, but nesting sites are clustered along a shoreline. Also, these birds crowd together during the breeding season, but disperse when breeding is over.

Figure 45.2 Three patterns of population distribution: (**a**) clumped, as in squirrelfish schools; (**b**) random, as when wolf spiders dig their burrows almost anywhere in forest soil; and (**c**) more or less uniform, as in a royal penguin nesting colony.

Take-Home Message

How do we describe a natural population?

- Each population has characteristic demographics, such as size, density, distribution pattern, and age structure.
- Environmental conditions and species interactions shape these characteristics, which may change over time.

45.2 Elusive Heads to Count

- Ecologists carry out field studies to test hypotheses about populations and to monitor the status of populations that are threatened or endangered.
- Link to Sampling error 1.8

Many white-tailed deer (*Odocoileus virginianus*) live in the forests, fields, and suburbs of North America. How could you find out how many deer live in a particular region?

A full count would be a careful measure of absolute population density. In the United States, census takers attempt such a count of human populations every ten years, although not everyone answers the door. Ecologists sometimes make counts of large species in small areas, such as fur seals at their breeding grounds, and sea stars in a tidepool.

More often, a full count would be impractical, so they sample part of a population and estimate its total density. For instance, you could divide a map of your county into small plots, or quadrats. **Quadrats** are sampling areas of the same size and shape, such as rectangles, squares, and hexagons. You could count individual deer in several plots and, from that, extrapolate the average number for the county as a whole. Ecologists often make such estimates for plants and other species that stay put (Figure 45.3). Such estimates run the risk of sampling error (Section 1.8), if the number of sampled plots is not large.

Ecologists use **capture-recapture methods** to estimate the population sizes of deer and other animals that do not stay put. First, they trap and mark some individuals. Deer get collars, squirrels get tattoos, salmon get tags, birds get leg rings, butterflies get wing markers, and so forth (Figure 45.4). Marked animals are released at time 1. At time 2, traps are reset. The proportion of marked animals in the second sample is then taken to be representative of the proportion marked in the whole population:

$$\frac{\text{marked individuals in sampling at time 2}}{\text{total captured in sampling 2}} = \frac{\text{marked individuals in sampling at time 1}}{\text{total population size}}$$

Ideally, both marked and unmarked individuals of the population are captured at random, no marked animal is overlooked, and marking does not affect whether animals die or otherwise depart during the study interval.

In the real world, recaptured individuals might not be a random sample; they might over- or underrepresent their population. Squirrels marked after being attracted to bait in boxes might now be trap-happy or trap-shy. Instead of mailing tags of marked fish to ecologists, a fisherman may keep them as souvenirs. Birds lose leg rings.

Estimates of population size may also vary depending on the time of year they are made. The distribution of a population may change seasonally. Many types of animals move between different parts of their range in response to seasonal changes in resource abundance.

As with other population data, the accuracy of size estimates can be increased by repeated samplings. The more data that can be accumulated, the lower the risk of sampling error.

Figure 45.3 Easy-to-count creosote bushes near the eastern base of the Sierra Nevada. They are an example of a relatively uniform distribution pattern. Individual plants compete for scarce water in this desert, which has extremely hot, dry summers and mild winters.

Figure 45.4 Two individuals marked for population studies. (**a**) Florida Key deer and (**b**) Costa Rican owl butterfly (*Caligo*).

45.3 Population Size and Exponential Growth

- Populations are dynamic units. They are continually adding and losing individuals. All populations have a capacity to increase in number.
- Link to Bacterial reproduction 21.5

Gains and Losses in Population Size

Populations continually change size. They increase in size because of births and **immigration**, the arrival of new residents from other populations. They decrease in size because of deaths and **emigration**, departure of individuals that then take up permanent residence elsewhere. For example, a freshwater turtle population changes size in the spring when young turtles emigrate from their home pond. The young emigrants typically become immigrants at another pond some distance away.

What about the individuals of species that migrate daily or seasonally? A **migration** is a recurring round-trip between regions, usually in response to expected shifts or gradients in environmental resources. Some or all members of a population leave an area, spend time in another area, then return. For our purposes, we may ignore these recurring gains and losses, because we can assume that they balance out over time.

From Zero to Exponential Growth

Zero population growth is an interval during which the number of births is balanced by an equal number of deaths. Population size remains stable, with no net increase or decrease in the number of individuals.

We can measure births and deaths in terms of rates per individual, or per capita. *Capita* means head, as in a head count. Subtract a population's per capita death rate (d) from its per capita birth rate (b) and you have the **per capita growth rate**, or r:

$$\underset{\text{(per capita growth rate)}}{r} = \underset{\text{(per capita birth rate)}}{b} - \underset{\text{(per capita death rate)}}{d}$$

As long as r remains constant and greater than zero, **exponential growth** will continue: Population size will increase by the same proportion in every successive time interval.

Imagine a population of 2,000 mice living in a field. If 1,000 mice are born each month, the birth rate is 0.5 per mouse per month (1,000 births/2,000 mice). If 200 mice die each month, the death rate is 200/2,000 = 0.1 per mouse per month. Given these birth and death rates, r is 0.5 – 0.1 = 0.4 per mouse per month. In other words, the mouse population grows by 4 percent each

	Starting Population Size		Net Monthly Increase	New Population Size
G = r ×	2,000	=	800	2,800
r ×	2,800	=	1,120	3,920
r ×	3,920	=	1,568	5,488
r ×	5,488	=	2,195	7,683
r ×	7,683	=	3,073	10,756
r ×	10,756	=	4,302	15,058
r ×	15,058	=	6,023	21,081
r ×	21,081	=	8,432	29,513
r ×	29,513	=	11,805	41,318
r ×	41,318	=	16,527	57,845
r ×	57,845	=	23,138	80,983
r ×	80,983	=	32,393	113,376
r ×	113,376	=	45,350	158,726
r ×	158,726	=	63,490	222,216
r ×	222,216	=	88,887	311,103
r ×	311,103	=	124,441	435,544
r ×	435,544	=	174,218	609,762
r ×	609,762	=	243,905	853,667
r ×	853,667	=	341,467	1,195,134

A

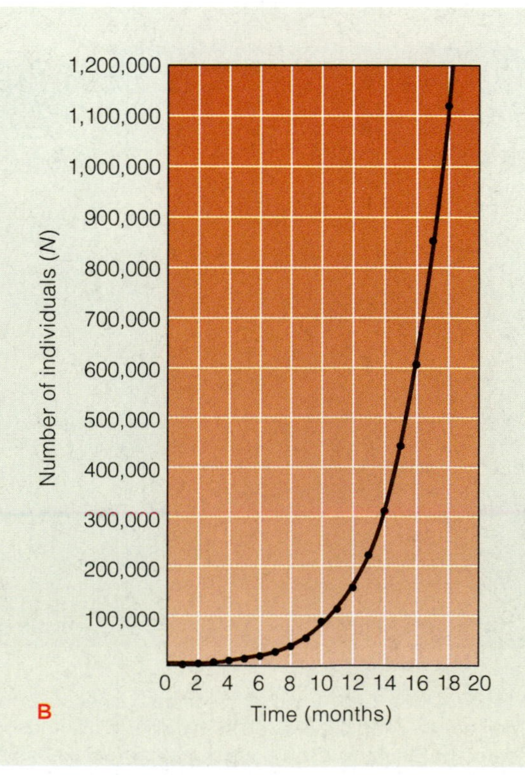

Figure 45.5 Animated (**a**) Net monthly increases in a hypothetical population of mice when the per capita rate of growth (r) is 0.4 per mouse per month and the starting population size is 2,000.

(**b**) Graph these numerical data and you end up with a J-shaped growth curve.

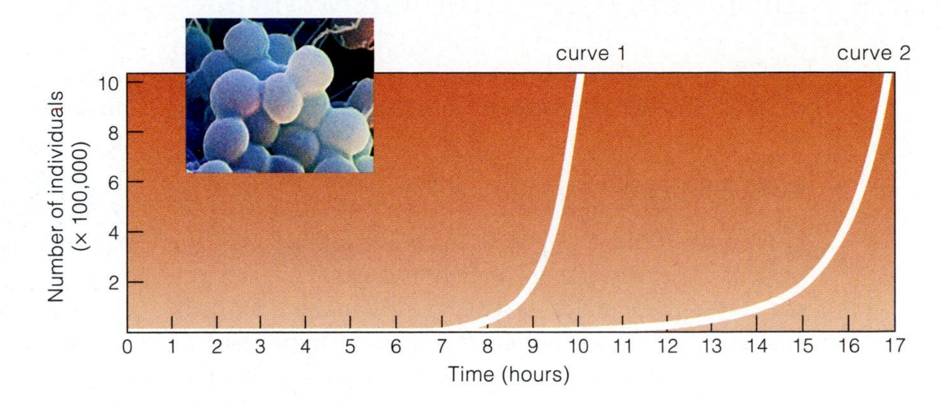

Figure 45.6 Effect of deaths on the rate of increase for two hypothetical populations of bacteria. Plot the population growth for bacterial cells that reproduce every half hour and you get growth curve 1. Next, plot the population growth of bacterial cells that divide every half hour, with 25 percent dying between divisions, and you get growth curve 2. Deaths slow the rate of increase, but as long as the birth rate exceeds the death rate and is constant, exponential growth will continue.

month. We can calculate the population growth (*G*) for each interval based on the per capita growth rate (*r*) and the number of individuals (*N*):

$$\underset{\text{(population growth per unit time)}}{G} = \underset{\text{(per capita growth rate)}}{r} \times \underset{\text{(number of individuals)}}{N}$$

After one month, 2,800 mice are scurrying about in the field (Figure 45.5*a*). A net increase of 800 fertile mice has made the reproductive base larger. They all reproduce, so the population size expands, for a net increase of 0.4 × 2,800 = 1,120. Population size is now 3,920. At this growth rate, the number of mice would rise from 2,000 to more than 1 million in two years! Plot the increases against time and you end up with a J-shaped curve that is characteristic of exponential growth (Figure 45.5*b*).

With exponential growth, a population grows faster and faster, although the per capita growth rate stays the same. It is like the compounding of interest on a bank account. The annual interest rate remains fixed, yet every year the amount of interest paid increases. Why? The annual interest paid into the account adds to the size of the balance, and the next interest payment will be calculated based on that larger balance.

In exponentially growing populations, *r* is like the interest rate. Although *r* remains constant, population growth accelerates as the population size increases. When 6,000 individuals reproduce, population growth is three times higher than it was when there were only 2,000 reproducers.

As another example, think of a single bacterium in a culture flask. After thirty minutes, the cell divides in two. Those two cells divide, and so on every thirty minutes. If no cells die between divisions, then the population size will double in every interval—from 1 to 2, then 4, 8, 16, 32, and so on. The time it takes for a population to double in size is its **doubling time**.

Consider how doubling time works in our flask of bacteria. After 9–1/2 hours, or nineteen doublings, there are more than 500,000 bacterial cells. After ten hours, or twenty doublings, there are more than one million. Curve 1 in Figure 45.6 is a plot of this change over time.

The size of *r* affects the speed of exponential growth. Suppose 25 percent of the bacteria in our hypothetical flask die every 30 minutes. Under these conditions, it would take 17 hours, rather than 10, for the population to reach 1 million (curve 2 in Figure 45.6). The higher death rate decreases *r*, so exponential growth occurs more slowly. However, as long as *r* is greater than zero and constant, growth plots out as a J-shaped curve.

What Is the Biotic Potential?

Now imagine a population living in an ideal habitat, free of all threats such as predators and pathogens. Every individual has plenty of shelter, food, and other vital resources. Under such conditions, a population would reach its **biotic potential**: the maximum possible per capita rate of increase for its species.

All species have a characteristic biotic potential. For many bacteria, it is 100 percent every half hour or so. For humans, it is about 2 to 5 percent per year.

The actual growth rate depends on many factors. A population's age distribution, how often its individuals reproduce, and how many offspring an individual can produce are examples. The human population has not reached its biotic potential, but it is growing exponentially. We will return to the topic of the human population later in the chapter.

Take-Home Message

What determines the size of a population and its growth rate?

- The size of a population is influenced by its rates of births, deaths, immigration, and emigration.
- Subtract the per capita death rate from the per capita birth rate to get *r*, the per capita growth rate of a population. As long as *r* is constant and greater than zero, a population will grow exponentially. With exponential growth, the number of individuals increases faster and faster over time.
- The biotic potential of a species is its maximum possible population growth rate under optimal conditions.

45.4 Limits on Population Growth

- Natural populations seldom continue to grow unchecked.
- Competition and crowding can slow growth.

Environmental Limits on Growth

Most of the time, a population cannot fulfill its biotic potential because of environmental limits. That is why sea stars—the females of which could make 2,500,000 eggs each year—do not fill the oceans with sea stars.

Any essential resource that is in short supply is a **limiting factor** on population growth. Food, mineral ions, refuge from predators, and safe nesting sites are examples (Figure 45.7). Many factors can potentially limit population growth. Which specific factor is the first to be in short supply and thus limit growth varies from one environment to another.

To get a sense of the limits on growth, start again with a bacterial cell in a culture flask, where you can control the variables. First, enrich the culture medium with glucose and other nutrients bacteria require for growth. Next, let the cells reproduce.

Initially, growth may be exponential. Then it slows, and population size remains relatively stable. After a brief stable period, population size plummets until all the bacterial cells are dead. What happened? The larger population required more nutrients. Over time, nutrient levels declined, and the cells could no longer divide. Even after cell division stopped, existing cells kept taking up and using nutrients. When the nutrient supply was exhausted, the last cells died out.

Suppose you continued adding nutrients to the flask. Population growth would still slow and then halt. As before, the bacteria would eventually die. Why? Like other organisms, bacteria generate metabolic wastes. Over time, this waste would accumulate and poison the habitat preventing further growth. No population can grow exponentially forever. Remove one limiting factor and another one becomes limiting.

Figure 45.7 One example of a limiting factor. (**a**) Wood ducks build nests only inside cavities of specific dimensions. With the clearing of old growth forests, the access to natural cavities of the correct size and position is now a limiting factor on wood duck population size. (**b**) Artificial nesting boxes are being placed in preserves to help ensure the health of wood duck populations.

Carrying Capacity and Logistic Growth

Carrying capacity refers to the maximum number of individuals of a population that a given environment can sustain indefinitely. Ultimately, it means that the sustainable supply of resources determines population size. We can use the pattern of **logistic growth**, shown in Figure 45.8, to reinforce this point. By this pattern, a small population starts growing slowly in size, then it grows rapidly, then its size levels off as the carrying capacity is reached.

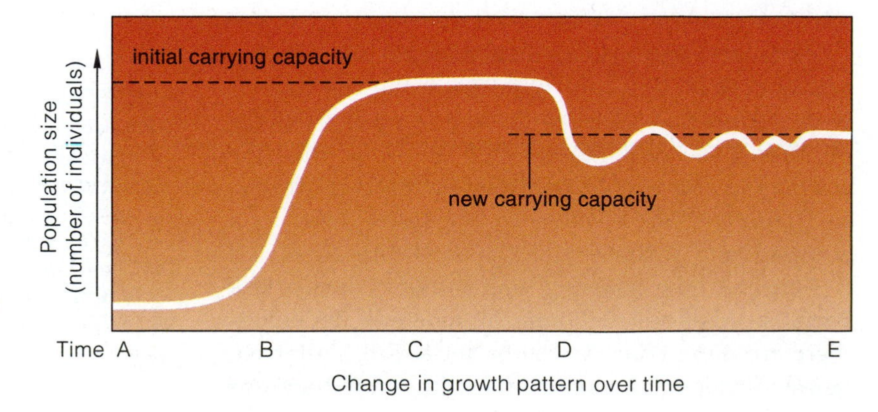

Figure 45.8 Animated Idealized S-shaped curve characteristic of logistic growth. After a rapid growth phase (time B to C), growth slows and the curve flattens as carrying capacity is reached (time C to D).

In the real world, population size often declines when a change in the environment lowers carrying capacity (time D to E). That happened to the human population of Ireland in the mid-1800s. Late blight, a disease caused by a water mold, destroyed the potato crop that was the mainstay of Irish diets (Section 22.8).

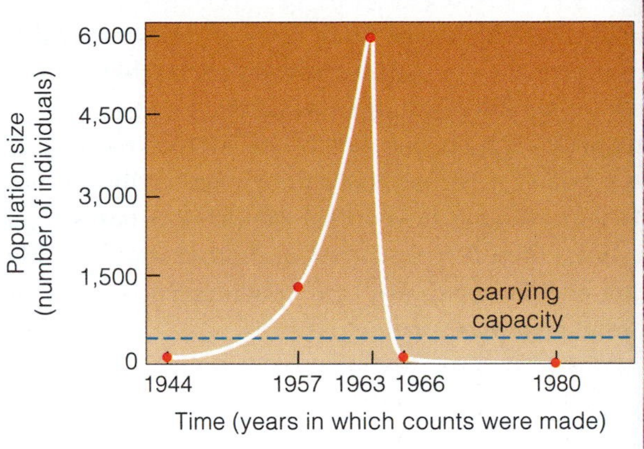

Figure 45.9 Graph of changes in a reindeer population that exceeded its habitat's carrying capacity (*blue* dashed line) and did not recover.

In 1944, during World War II, a United States Coast Guard crew established a station on St. Matthew, an island 320 kilometers (200 miles) west of Alaska in the Bering Sea. They brought in 29 reindeer as a backup food source. Reindeer eat lichens. Thick mats of lichens cloaked the island, which is no more than 51 kilometers long and 6.4 kilometers (32 miles by 4 miles) across. World War II drew to a close before any reindeer were shot. The Coast Guard pulled out, leaving behind seabirds, arctic foxes, voles—and a herd of healthy reindeer with no predators big enough to hunt them.

In 1957, biologist David Klein visited St. Matthew. On a hike from one end of the island to the other, he counted 1,350 well-fed reindeer and saw trampled and overgrazed lichens. In 1963, Klein and three other biologists returned to the island. They counted 6,000 reindeer. They could not help but notice the profusion of reindeer tracks and feces, and a lot of trampled, dead lichens.

Klein returned to St. Matthew in 1966. Bleached-out reindeer bones littered the island. Forty-two reindeer were still alive. Only one was a male; it had abnormal antlers, which made it unlikely to reproduce. There were no fawns. Klein figured out that thousands of reindeer had starved to death during the unusually harsh winter of 1963–1964. By the 1980s, there were no reindeer on the island at all.

Graphing logistic growth yields an S-shaped curve, as shown in Figure 45.8 (A to C). In equation form,

$$\text{population growth per unit time} = \text{maximum per capita population growth rate} \times \text{number of individuals} \times \text{proportion of resources not yet used}$$

An S-shaped curve is simply an approximation of what takes place in nature. Often a population that is growing fast overshoots its carrying capacity. Figure 45.9 shows what happened to a small population of reindeer. As the population size increased, more and more individuals competed for resources such as food and shelter, so each reindeer received a smaller share. More individuals died of starvation and fewer young were born. Deaths began to outnumber births. Finally, the death rate soared and the birth rate plummeted.

Two Categories of Limiting Factors

Density-dependent factors lower reproductive success and appear or worsen with crowding. Competition for limited resources leads to density-dependent effects, as does disease. Pathogens and parasites can spread more easily when hosts are crowded. As one example, human populations in cities support huge numbers of rats that can carry bubonic plague, typhus, and other deadly infectious diseases.

Density-dependent factors control population size through negative feedback. High density causes these factors to come into play, then their effects act to lower population density. A logistic growth pattern results from this feedback effect.

Density-independent factors decrease reproductive success too, but their likelihood of occurring and their magnitude of effect are unaffected by crowding. Fires, snow storms, earthquakes, and other natural disasters affect crowded and uncrowded populations alike. For example, in December of 2004, a powerful tsunami (a giant wave caused by an earthquake) hit Indonesia. It killed about 250,000 people. The degree of crowding did not make the tsunami any more or less likely to happen, or to strike any particular island. The logistic growth equation cannot be used to predict effects of density-independent factors.

Take-Home Message

How do limiting factors affect population growth?

- Carrying capacity is the maximum number of individuals of a population that can be sustained indefinitely by the resources in a given environment.
- With logistic growth, population growth is fastest when density is low, slows as the population approaches carrying capacity, and then levels off.
- Density-dependent factors such as disease result in a pattern of logistic growth. Density-independent factors such as natural disasters also affect population size.

45.5 Life History Patterns

■ Life span, age at maturity, and the number of offspring produced vary widely among organisms. Natural selection influences these life history traits.

So far, you have looked at populations as if all of their members are identical with regard to age. For most species, however, individuals that make up a group are at many different stages of development. Often, those stages require different resources, as when caterpillars that eat leaves later develop into butterflies that sip nectar. In addition, individuals might be more or less vulnerable to danger at different stages.

In short, each species has a **life history pattern**. It has a set of adaptations that affect when an individual starts reproducing, how many offspring it has at one time, how often it reproduces, and other traits. In this section and the next, we will consider variables that underlie these age-specific patterns.

Table 45.1 Life Table for an Annual Plant Cohort*

Age Interval (days)	Survivorship (number surviving at start of interval)	Number Dying During Interval	Death Rate (number dying/ number surviving)	"Birth" Rate During Interval (number of seeds from each plant)
0–63	996	328	0.329	0
63–124	668	373	0.558	0
124–184	295	105	0.356	0
184–215	190	14	0.074	0
215–264	176	4	0.023	0
264–278	172	5	0.029	0
278–292	167	8	0.048	0
292–306	159	5	0.031	0.33
306–320	154	7	0.045	3.13
320–334	147	42	0.286	5.42
334–348	105	83	0.790	9.26
348–362	22	22	1.000	4.31
362–	0	0	0	0
		996		

* *Phlox drummondii*; data from W. J. Leverich and D. A. Levin, 1979.

Table 45.2 Life Table for Humans in the United States (based on 2003 conditions)

Age Interval	Number at Start of Interval	Number Dying During Age Interval	Life Expectancy (Years Remaining) at Start of Interval	Reported Live Births
0–1	100,000	687	77.5	
1–5	99,313	124	77.0	
5–10	99,189	73	73.1	
10–15	99,116	95	68.2	6,781
15–20	99,022	328	63.2	415,262
20–25	98,693	474	58.4	1,034,454
25–30	98,219	467	53.7	1,104,485
30–35	97,752	542	48.9	965,633
35–44	97,210	767	45.2	475,606
44–45	96,444	1,157	39.5	103,679
45–50	95,287	1,702	35.0	5,748
50–55	93,585	2,441	30.6	374
55–60	91,185	3,425	26.3	
60–65	87,760	5,092	22.2	
65–70	82,668	7,133	18.4	
70–75	75,535	9,825	14.9	
75–80	65,710	12,969	11.8	
80–85	52,741	15,753	9.0	
85–90	36,988	15,648	6.8	
90–95	21,344	12,363	5.0	
95–100	8,977	6,614	3.6	
100+	2,363	2,363	2.6	

Life Tables

Each species has a characteristic life span, but only a few individuals survive to the maximum age possible. Death is more likely at some ages. Individuals tend to reproduce during an expected age interval and to be most likely to die during another interval.

Age-specific patterns in populations are useful to life insurance and health insurance companies as well as ecologists. Such investigators focus on a **cohort**—a group of individuals born during the same interval—from their time of birth until the last one dies.

Ecologists often divide a natural population into age classes and record the age-specific birth rates and mortality. The resulting data is summarized in a life table (Table 45.1). Such tables inform decisions about how changes, such as harvesting a species or altering its environment, might affect the species' numbers. Birth and death schedules for the northern spotted owl are one case in point. They were cited in federal court rulings that halted mechanized logging in the owl's habitat—old-growth forests of the Pacific Northwest.

Human life tables are usually not based on a real cohort. Instead, information about current conditions is used to predict the births and deaths for a hypothetical group. Table 45.2 is such a life table for humans based on conditions in the United States during 2003.

Survivorship Curves

A **survivorship curve** is a graph line that emerges when you plot a cohort's age-specific survival in its habitat. Each species has a characteristic survivorship curve. Three types are common in nature.

A type I curve indicates survivorship is high until late in life. Populations of large animals that bear one or, at most, a few offspring at a time and give these young extended parental care show this pattern (Figure 45.10*a*). For example, a female elephant has one calf at a time and cares for it for several years. Type I curves are typical of human populations when individuals have access to good health care.

A type II curve indicates that death rates do not vary much with age (Figure 45.10*b*). In lizards, small mammals, and big birds, old individuals are about as likely to die of disease or predation as young ones.

A type III curve indicates that the death rate for a population peaks early in life. It is typical of species that produce many small offspring and provide little or no parental care. Figure 45.10*c* shows how the curve plummets for sea urchins, which release great numbers of eggs. Sea urchin larvae are soft and tiny, so fish, snails, and sea slugs devour most of them before protective hard parts can develop. A type III curve is common for marine invertebrates, insects, fishes, fungi, and for annual plants such as phlox (Table 45.1).

Reproductive Strategies

Some organisms such as bamboo and Pacific salmon reproduce just once, then die. Others such as oak trees, mice, and humans reproduce repeatedly. A one-shot strategy is favored when an individual is unlikely to have a second chance to reproduce. For Pacific salmon, reproduction requires a life-threatening journey from the sea to a stream. For bamboo, environmental conditions that favor reproduction occur only sporadically.

Population density may also influence the optimal reproductive strategy. At low density, there will be little competition for resources, so individuals who turn resources into offspring fast are at an advantage. Such individuals reproduce while still young, produce many small offspring, and invest very little in parental care. Selection that favors traits that maximize number of offspring is called ***r*-selection**. When population density nears the carrying capacity, outcompeting others for resources becomes more important. Big individuals that reproduce later in life and produce fewer, higher quality offspring have the advantage in this scenario. Selection for traits that improve offspring quality is ***K*-selection**. Some organisms have traits associated mainly with *r*-selection or with *K*-selection, but most have a mixture of these traits.

Take-Home Message

How do researchers study and describe life history patterns?

- Tracking a cohort (a group of individuals) from their birth until the last one dies reveals patterns of reproduction, death, and migrations.
- Survivorship curves reveal differences in age-specific survival among species or among populations of the same species.
- Different environments and population densities can favor different reproductive strategies.

a Elephants have type I surviorship, with low mortality until advanced age.

b Snowy egrets are type II populations, with a fairly constant death rate.

c Sea urchins are type III populations. Spines protect this adult, but the larvae are tiny, soft-bodied, and vulnerable to predation.

Figure 45.10 Three generalized survivorship curves and examples.

45.6 Natural Selection and Life Histories

- Predation can serve as a selection pressure that shapes life history patterns.
- Links to Directional selection 18.4, Gene flow 18.8

Predation on Guppies in Trinidad Several years ago, two evolutionary biologists drenched with sweat and clutching fishnets were wading through a stream. John Endler and David Reznick were in the mountains of Trinidad, an island in the southern Caribbean Sea. They wanted to capture guppies (*Poecilia reticulata*), small fishes that live in the shallow freshwater streams (Figure 45.11). The biologists were beginning what would become a long-term study of guppy traits, including life history patterns.

Male guppies are usually smaller and more colorful than female guppies of the same age. A male's colors serve as visual signals during courtship rituals. The drabber females are less conspicuous to predators and, unlike males they continue to grow after reaching sexual maturity.

Reznick and Endler were interested in how predators influence the life history of guppies. For their study sites, they decided on streams with many small waterfalls. These waterfalls are barriers that prevent guppies in one part of a stream from moving easily to another. As a result, each stream holds several populations of guppies, and very little gene flow occurs among those populations (Section 18.8).

The waterfalls also keep guppy predators from moving into different parts of the stream. In this habitat, the main

Figure 45.11 (**a**,**b**) Guppies and two guppy eaters, a killifish and a cichlid. (**c**) Biologist David Reznick contemplating interactions among guppies and their predators in a freshwater stream in Trinidad.

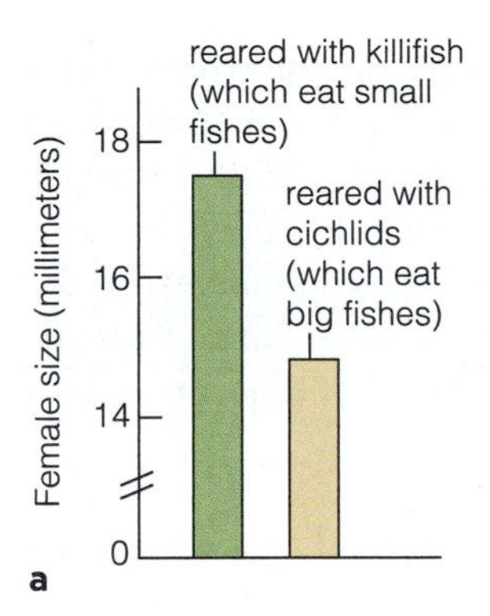

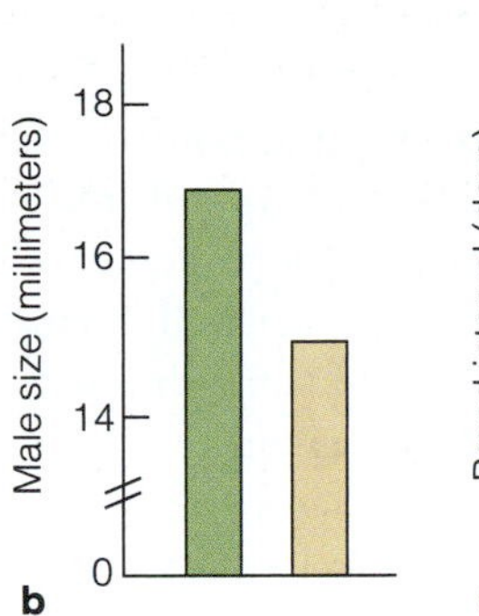

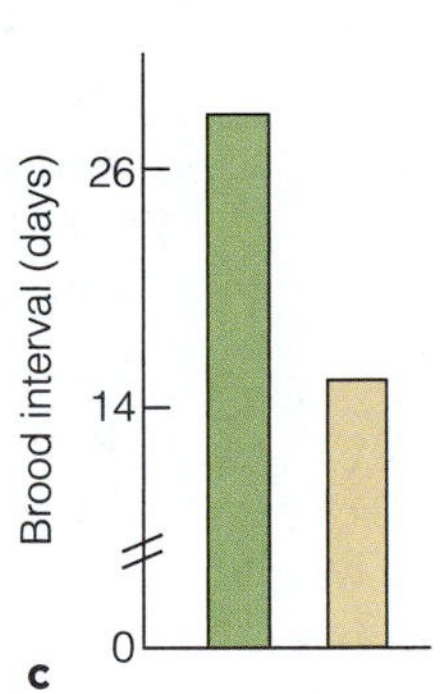

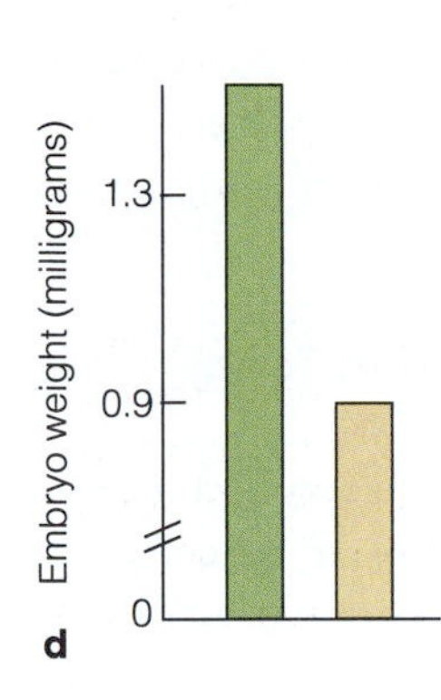

Figure 45.12 Experimental evidence of natural selection among guppy populations subject to different predation pressures. Compared to the guppies raised with killifish (*green* bars), guppies raised with cichlids (*tan* bars) differed in body size and in the length of time between broods.

guppy predators are killifish and cichlids. These two types of predatory fish differ in size and prey preferences. The killifish is relatively small and preys mostly on immature guppies. It ignores the larger adults. The cichlids are large fish. They tend to pursue mature guppies and ignore the small ones. Some parts of the streams hold one type of predator but not the other, so different guppy populations face different predation pressures.

As Reznick and Endler discovered, guppies in streams with cichlids grow faster and are smaller at maturity than those in streams with killifish (Figure 45.12). Also, guppies hunted by cichlids reproduce earlier, have more offspring at a time, and breed more frequently.

Were the differences in life history traits genetic, or did environmental differences cause them? To find out, the scientists collected guppies from cichlid-dominated and killifish-dominated streams. They reared these two groups in separate aquariums under identical conditions, with no predators present. Two generations later, the life history traits of these groups still differed, as they had in natural populations. Apparently, the differences in life history traits observed in the wild do have a genetic basis.

Reznick and Endler hypothesized that predators serve as selective agents that influence guppy life history traits. The scientists made a prediction: If life history traits are adaptive responses to predation, then these traits will change when a population is exposed to a new predator.

To test their prediction, Reznick and Endler found a stream region above a waterfall that had killifish but no guppies or cichlids. They brought in some guppies from a region below the waterfall where there were cichlids but no killifish. At the experimental site, the guppies that had previously lived only with cichlids were now exposed to killifish. The control site was the downstream region below the waterfall, where relatives of the transplanted guppies still coexisted with cichlids.

Reznik and Endler revisited the stream over the course of eleven years and thirty-six generations of guppies. They monitored traits of guppies above and below the waterfall. Their data showed that guppies at the upstream experimental site were evolving. Exposure to a novel predator had caused big changes in their rate of growth, age at first reproduction, and other life history traits. By contrast, guppies at the control site showed no such changes. As Reznick and Endler concluded, life history traits in guppies can evolve rapidly in response to the selective pressure exerted by predation.

Overfishing and the Atlantic Cod The evolution of life history traits in response to predation pressure is not merely interesting. It has commercial importance. Just as guppies evolved in response to predators, the North Atlantic codfish (*Gadus morhua*) evolved in response to fishing pressure. North Atlantic codfish can be big (*below*). From the mid-1980s to early 1990s, the number of fisherman pursuing codfish rose. Fishermen kept the largest fish, and threw smaller ones back. This human behavior put codfish that became sexually mature when they were still small at an advantage, and such fish became increasingly common. As codfish numbers declined, smaller and smaller fish were kept.

Looking back, a rapid decline in age at first reproduction was a sign that the cod population was under great pressure. In 1992, Canada banned cod fishing in some areas. That ban, and later restrictions, came too late to stop the Atlantic cod population from plummeting. The population still has not recovered from this decline.

Had biologists recognized the life history changes as a warning sign, they might have been able to save this fishery and protect the livelihood of thousands of workers. Monitoring the life history data for other economically important fishes may help prevent over-fishing of other species in the future.

45.7 Human Population Growth

- The size of the human population is at its highest level ever and is expected to continue to increase.
- Links to Infectious disease 21.8, Human dispersal 26.15

The Human Population Today

In 2008, the estimated average rate of increase for the human population was 1.16 percent per year. As long as birth rates continue to exceed death rates, annual additions will drive a larger absolute increase each year into the foreseeable future.

Although many people enjoy abundant resources, about a fifth of the human population lives in severe poverty, and more than 800 million are malnourished (Figure 45.13). More than 1 billion people lack access to clean drinking water. More than 2 billion people face a shortage in fuelwood, which they depend on to heat their homes and cook their food. Rising populations will only increase pressure on limited resources.

Extraordinary Foundations for Growth

How did we get into this predicament? For most of its history, the human population grew very slowly. The growth rate began to increase about 10,000 years ago, and during the past two centuries, growth rates soared (Figure 45.14). Three trends promoted the large increases. First, humans were able to migrate into new habitats and expand into new climate zones. Second, humans developed new technologies that increased the carrying capacity of existing habitats. Third, humans sidestepped some limiting factors that tend to restrain the growth of other species.

Geographic Expansion Early humans evolved in the dry woodlands of Africa, then moved into the savannas. We assume they subsisted mainly on plant foods, but they probably also scavenged bits of meat. Bands of hunter–gatherers moved out of Africa about 2 million years ago. By 44,000 years ago, their descendants were established in much of the world (Section 26.15).

Few species can expand into such a broad range of habitats, but the early humans had large brains that allowed them to develop the necessary skills. They learned how to start fires, build shelters, make clothing, manufacture tools, and cooperate in hunts. With the advent of language, knowledge of such skills did not die with the individual. Compared to most species, humans displayed a greater capacity to disperse fast over long distances and to become established in physically challenging new environments.

Banks of corn silos in Wisconsin

Figure 45.13 Far from well-fed humans in highly developed countries, an Ethiopian child shows the effects of starvation. Ethiopia is one of the poorest developing countries, with an annual per capita income of $120. Average caloric intake is more than 25 percent below the minimum necessary to maintain good health. Malnutrition stunts the growth, weakens the body, and impairs the brain development of about half of Ethiopia's children. Despite ongoing food shortages, Ethiopia's population has one of the highest annual rates of increase in the world. If growth continues at its current rate, the population of 75 million will double in less than 25 years.

Increased Carrying Capacity Beginning about 11,000 years ago, bands of hunter–gatherers were shifting to agriculture. Instead of counting on the migratory game herds, they were settling in fertile valleys and other regions that favored seasonal harvesting of fruits and grains. They developed a more dependable basis for life. A pivotal factor was the domestication of wild grasses, including species ancestral to modern wheat and rice. Now people harvested, stored, and planted seeds all in one place. They domesticated animals as sources of food and to pull plows. They dug irrigation ditches and diverted water to croplands.

Agricultural productivity became a basis for increases in population growth rates. Towns and cities formed. Later, food supplies increased yet again. Farmers started to use chemical fertilizers, herbicides, and pesticides to protect their crops. Transportation and food distribution improved. Even at its simplest, the management of food supplies through agricultural practices increased the carrying capacity for the human population.

Sidestepped Limiting Factors Until about 300 years ago, malnutrition and infectious diseases

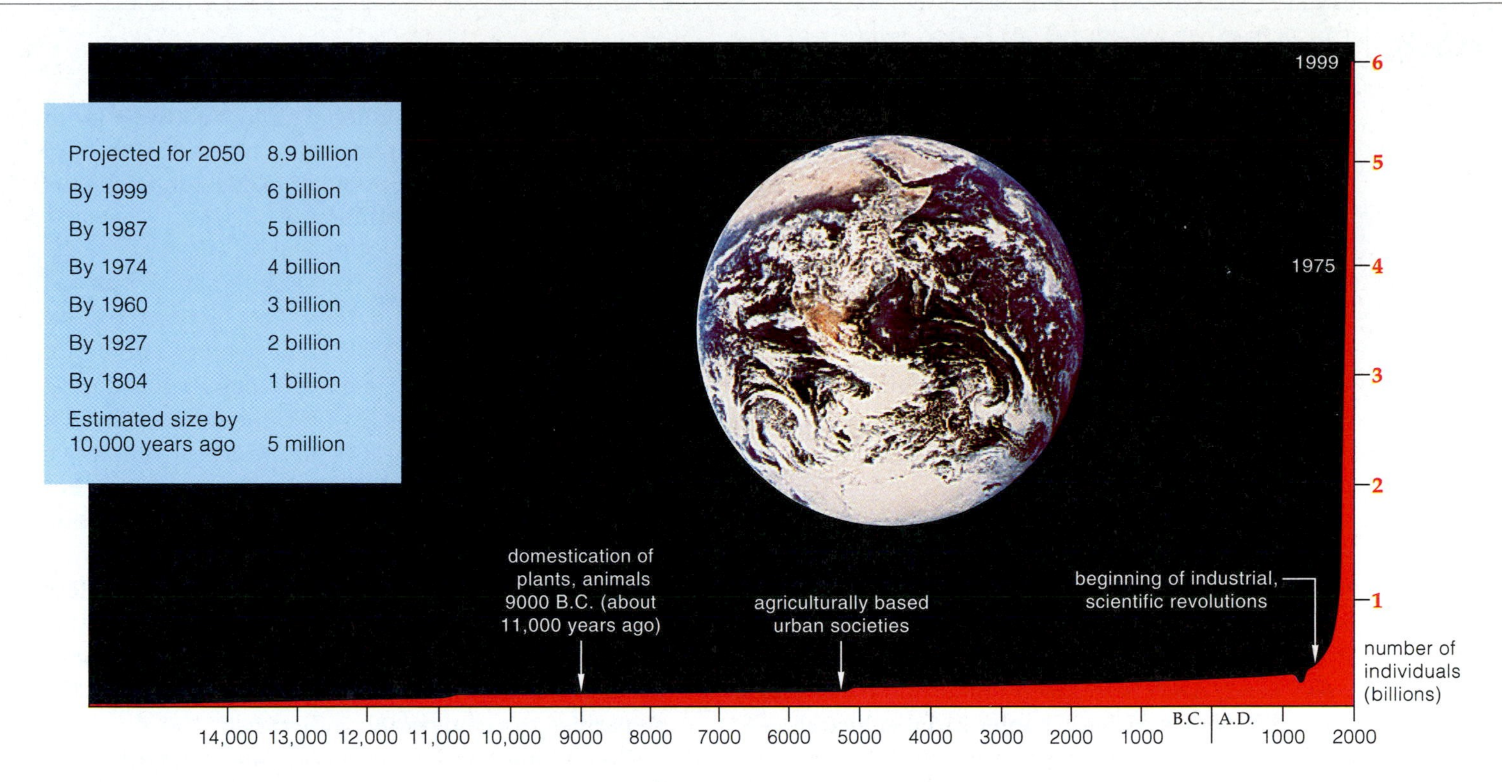

Figure 45.14 Growth curve (*red*) for the world human population. The *blue* box indicates how long it took for the human population to increase from 5 million to 6 billion. The dip between years 1347 and 1351 marks the time when 60 million people died during a pandemic that may have been a bubonic plague.

kept death rates high enough to more or less balance birth rates. Infectious diseases are density-dependent controls. Plagues swept through crowded cities. In the mid-1300s, one third of Europe's population was lost to a pandemic known as the Black Death. Waterborne diseases such as cholera that are associated with poor sanitation ran rampant. Then plumbing improved and vaccines and medications began to cut the death toll from disease. Births increasingly outpaced deaths—r became larger and exponential growth accelerated.

The industrial revolution took off in the middle of the eighteenth century. People had discovered how to harness the energy of fossil fuels, starting with coal. Within decades, cities of western Europe and North America became industrialized. World War I sparked the development of more technologies. After the war, factories turned to mass production of cars, tractors, and other affordable goods. Advances in agricultural practices meant that fewer farmers were required to support a larger population.

In sum, by controlling disease agents and tapping into fossil fuels—a concentrated source of energy—the human population sidestepped many factors that had previously limited its rate of increase.

Where have the far-flung dispersals and ongoing advances in technology and infrastructure gotten us? It took more than 100,000 years for the human population size to reach 1 billion. As Figure 45.14 shows, it took just 123 years to reach 2 billion, 33 more to reach 3 billion, 14 more to reach 4 billion, and then 13 more to get to 5 billion. It took only 12 more years to arrive at 6 billion! No doubt new technology will continue to increase Earth's human carrying capacity, but growth cannot be sustained indefinitely.

Why not? Ongoing increases in population size will cause density-dependent controls to exert their effects. For instance, globe-hopping travelers can carry pathogens to dense urban areas all around the world in a matter of weeks (Section 21.8). Also, limited resources cause economic hardship and civil strife.

Take-Home Message

Why have human populations grown so much, and what can we expect?

- Through expansion into new habitats, cultural interventions, and technological innovations, the human population has temporarily skirted environmental resistance to growth.
- Without technological breakthroughs, density-dependent controls will kick in and slow human population growth.

45.8 Fertility Rates and Age Structure

- Acknowledgment of the risks posed by rising populations has led to increased family planning in almost every region.
- Links to AIDS Chapter 21 introduction, Contraception 42.9

Some Projections

Most governments recognize that population growth, resource depletion, pollution, and quality of life are interconnected. Many offer family planning programs, and the United Nations Population Division estimates that about 60 percent of the world's married women now use some sort of contraception.

An increase in contraceptive use is contributing to a global decline in birth rate. Death rates are also falling in most regions. Improved diet and health care are lowering the infant mortality rate (the number of infants per 1,000 who die in their first year). On the other hand, AIDS has caused the death rate to soar in some African countries (Chapter 21 introduction).

World population is expected to peak at 8.9 billion by 2050, and possibly to decline as the century ends. Think of all the resources that will be required. We will have to boost food production, and find more energy and fresh water to meet even the most basic needs of billions more people. Utilizing natural resources on a larger scale will intensify pollution.

We expect to see the most growth in India, China, Pakistan, Nigeria, Bangladesh, and Indonesia, in that order. China (with 1.3 billion people) and India (with 1.09 billion) dwarf other countries; together, they hold 38 percent of the world population. Next in line is the United States, with 294 million.

Shifting Fertility Rates

The **total fertility rate** (TFR) is the average number of children born to the women of a population during their reproductive years. In 1950, the worldwide TFR averaged 6.5. Currently it is 2.7, which is still above the replacement level of 2.1—or the average number of children a couple must bear to keep the population at a constant level, given current death rates.

TFRs vary among countries. TFRs are at or below replacement levels in many developed countries; the developing countries in western Asia and Africa have the highest. Figure 45.15 has some examples of the disparities in demographic indicators.

Comparing age structure diagrams is revealing. In Figure 45.16, focus on the reproductive age category for the next fifteen years. Women generally bear children when they are 15 to 35 years old. We can expect populations that have a broad base to grow faster. The United States population has a relatively narrow base below a wide area that represents the 78 million baby-boomers (Figure 45.16*c*). This cohort began forming in 1946 when American soldiers came home after World War II and started to raise families.

Global increases in population seem certain. Even if every couple from this time forward has no more than two children, population growth cannot slow for sixty years. About 1.9 billion are about to enter their reproductive years. More than one-third of the world population is in the broad pre-reproductive base.

China has the most wide-reaching family planning program. Its government discourages premarital sex. It urges people to delay marriage and limit families to one or two children. It offers abortions, contraceptives, and sterilization at no cost to married couples, which mobile units and paramedics provide even in remote areas. Couples who follow guidelines get more food, free medical care, better housing, and salary bonuses.

Population in 2006	298 million	188 million	132 million
Population in 2025 (projected)	349 million	211 million	206 million
Population under age 15	20%	26%	42%
Population above age 65	13%	6%	3%
Total fertility rate (TFR)	2.1	1.9	5.5
Infant mortality rate	6 per 1,000 live births	29 per 1,000 live births	97 per 1,000 live births
Life expectancy	78 years	72 years	47 years
Per capita income	$43,740	$3,460	$560

Figure 45.15 Key demographic indicators for three countries, mainly in 2006. The United States (*brown* bar) is highly developed, Brazil (*red* bar) is moderately developed, and Nigeria (*beige* bar) is less developed. **Figure It Out:** What is the difference in life expectancy between the United States and Nigeria?

Answer: 31 years

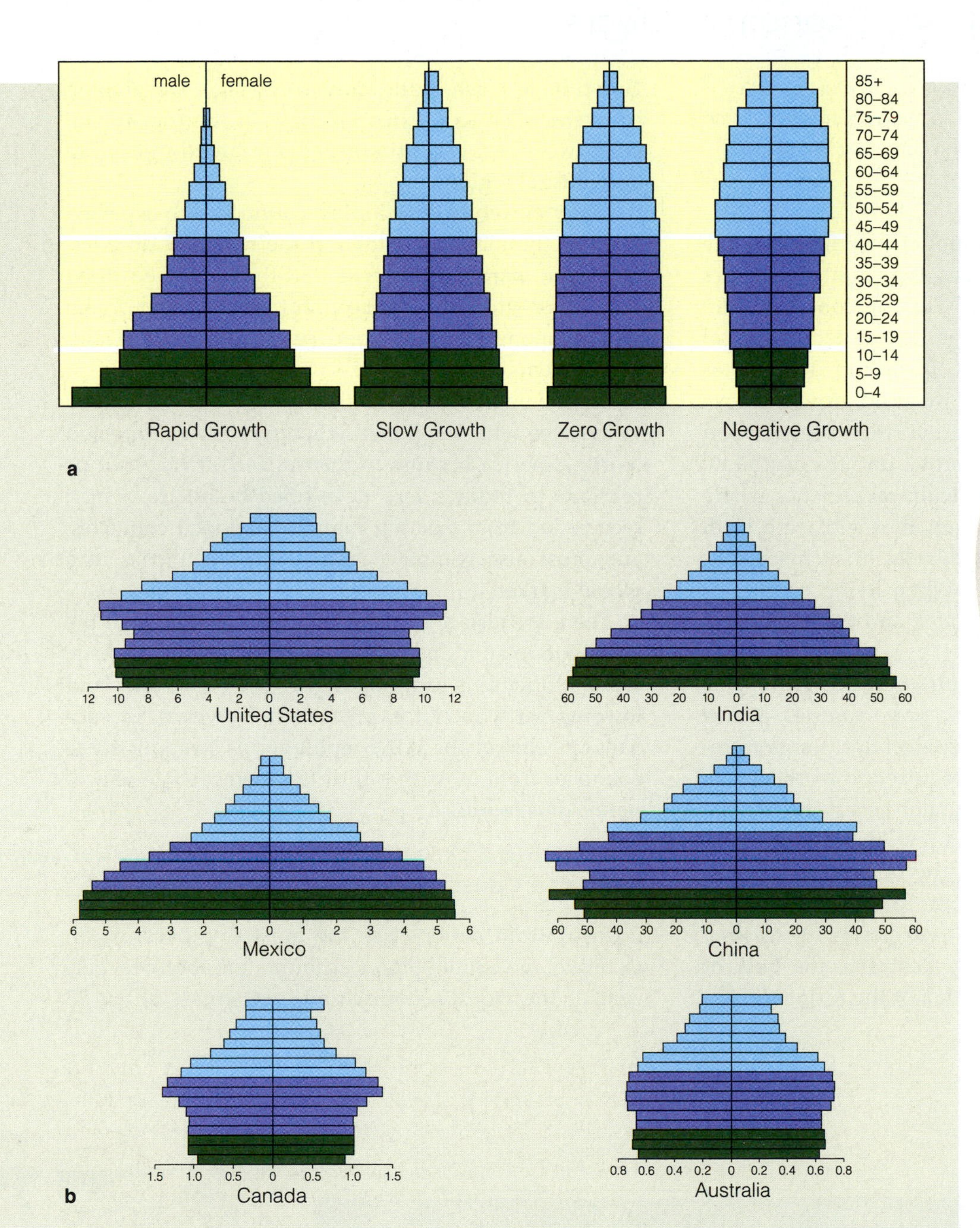

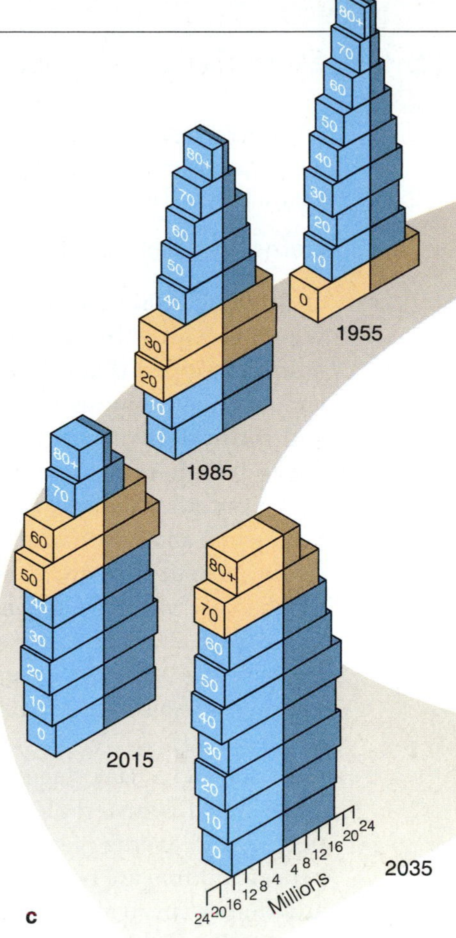

Figure 45.16 Animated (**a**) General age structure diagrams for countries with rapid, slow, zero, and negative rates of population growth. The pre-reproductive years are the *green* bars; reproductive years, *purple;* post-reproductive years, *light blue.* A vertical axis divides each graph into males (*left*) and females (*right*). Bar widths correspond to the proportions of individuals in each age group.

(**b**) 1997 age structure diagrams for six nations. Population sizes are measured in millions.

(**c**) Sequential age structure diagrams for the United States population. *Gold* bars track the baby-boomer generation.

Their offspring get free tuition and special treatment when they enter the job market. Parents having more than two children lose benefits and pay more taxes.

Since 1972, China's TFR has fallen sharply, from 5.7 to 1.75. An unintended consequence has been a shift in the country's sex ratio. Traditional cultural preference for sons, especially in rural areas, led some parents to abort female fetuses or commit infanticide. Worldwide, 1.06 boys are born for every girl. However, among those under age 15 in China, there are 1.134 boys for every girl. More than 100,000 girls are abandoned each year. The government is offering additional cash and tax incentives to the parents of girls. In the meantime, the population time bomb keeps on ticking in China. About 150 million of its young females now make up the pre-reproductive age category.

Take-Home Message

How has the human fertility rate changed and what can we expect?

- The worldwide total fertility rate has been declining but it is still above the replacement level.
- Even if total fertility rate declines to the replacement level worldwide, the population will continue to increase; more than one-third of the population is in a broad pre-reproductive base.

45.9 Population Growth and Economic Effects

■ The most developed countries have the slowest growth rates and use the most resources. As more countries become industrialized, pressure on Earth's resources will increase.

Demographic Transitions

The **demographic transition model** describes how the population growth rate changes as a country becomes more developed (Figure 45.17). Living conditions are harsh in the preindustrial stage, before technological and medical advances spread. Birth and death rates are both high, so the rate of population growth is low. In the transitional stage, industrialization begins. Food production and health care improve, and the death rate slows. Not surprisingly, in agricultural societies where families are expected to help in the fields, the birth rate is high. The annual growth rates in such societies are between 2.5 to 3 percent. When living conditions improve, the birth rate starts to fall and the population size levels off.

In the industrial stage, population growth slows. Cities filled with employment opportunities attract people, and average family size declines. Large numbers of children are no longer required to work a farm, and higher survival means it is not necessary to have many offspring to ensure that a few live.

In the postindustrial stage, the population growth rate becomes negative. The birth rate falls below the death rate, and the population size slowly decreases.

The United States, Canada, Australia, the bulk of western Europe, Japan, and much of the former Soviet Union have reached the industrial stage. Developing countries such as Mexico are now in the transitional stage, with people continuing to migrate to cities from agricultural regions.

Many currently developing countries are expected to enter the industrial stage in the next few decades. However, there are concerns that the continued rapid population growth in these countries will overwhelm their economic growth, food production, and health care systems.

The demographic transition model was developed to describe what happened when western Europe and North America became industrialized. It may not be relevant to today's less developed countries, which receive aid from existing highly developed countries, and must also compete against these countries in a global market.

There are also regional differences in how well the transition to an industrial stage is proceeding. In Asia, rising affluence is bringing higher life expectancy and lowered birth rates, as predicted. However, in sub-Saharan Africa, the AIDS epidemic is keeping some countries from moving out of the lowest stage of economic development.

Resource Consumption

Industrialized nations use the most resources. As an example, the United States accounts for about 4.6 percent of the world's population, yet it uses about 25

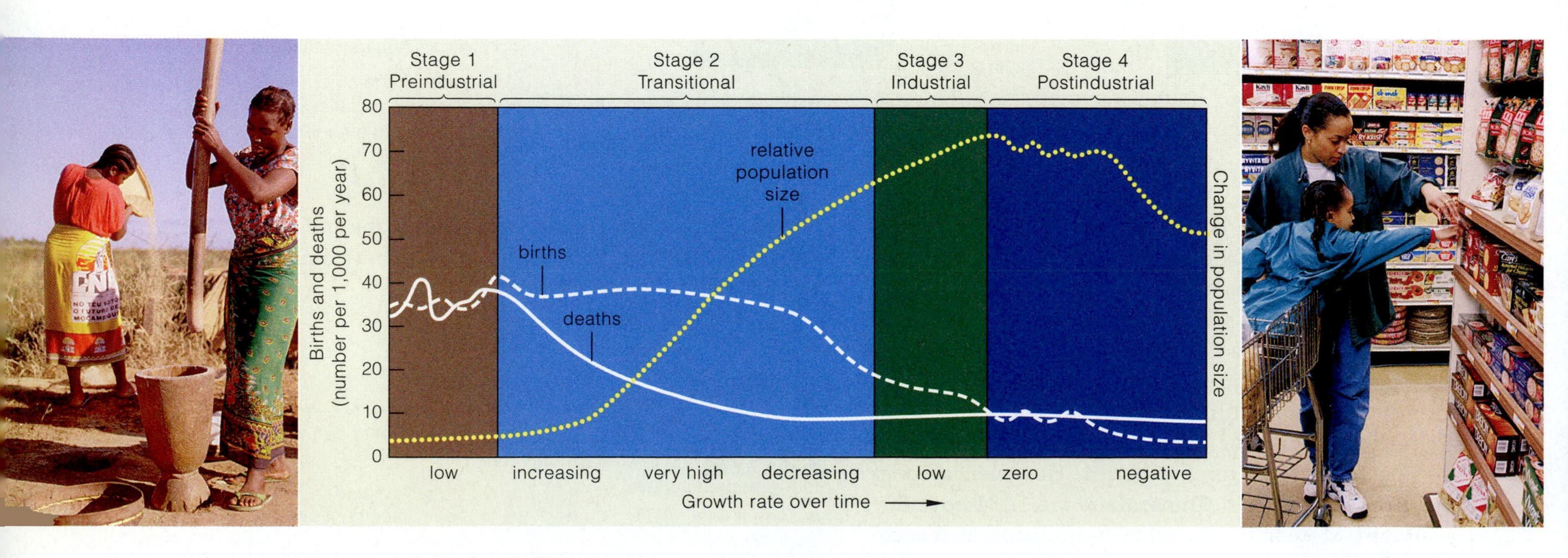

Figure 45.17 **Animated** Demographic transition model for changes in population growth rates and sizes, correlated with long-term changes in the economy.

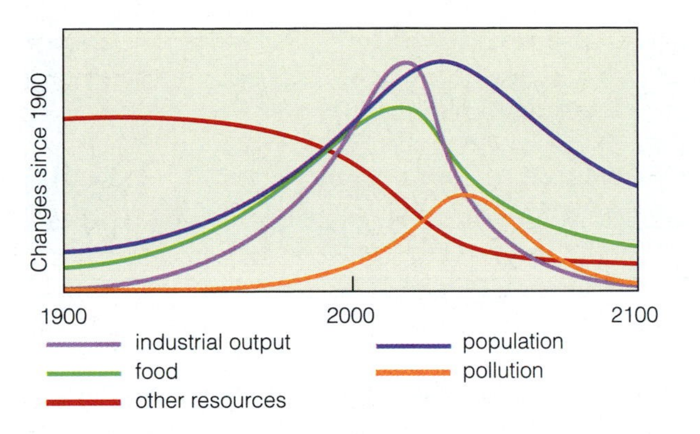

Figure 45.18 Computer-based projection of what might happen if human population size continues to skyrocket without dramatic policy changes and technological innovation. The assumptions were that the population has already overshot the carrying capacity and current trends will continue unchanged.

percent of the world's minerals and energy supplies. Billions of people living in India, China, and other less developed nations dream of owning the same kinds of consumer goods as people in developed countries. Earth does not have enough resources to make that possible. For everyone now alive to have a lifestyle like an average American would require four times the resources present on Earth.

What will happen if the human population keeps on increasing as predicted? How will we find the food, energy, water, and other basic resources needed to sustain so many people? Can we provide the necessary education, housing, medical care, and other social services? Some models suggest not (Figure 45.18). Other analysts claim we can adapt to a more crowded world if innovative technologies improve crop yields, if people rely less on meat for protein, and if resources are shared more equitably among regions. We have made great strides in increasing our agricultural output, but have been less successful in getting food to the people who need it.

Take-Home Message

How does industrialization affect population growth and resource consumption?

- Differences in population growth and resource consumption among countries can be correlated with levels of economic development. Growth rates are typically greatest during the transition to industrialization.
- Global conditions have changed so that the demographic transition model may no longer apply to modern nations.
- An average person living in a highly developed nation uses far more resources than a person in a less-developed nation.

45.10 Rise of the Seniors

- While some countries face overpopulation, others have declining birth rates and an increasing average age.

In some developed countries, the decreasing total fertility rate and increasing life expectancy have resulted in a high proportion of older adults. In Japan, people over 65 currently make up about 20 percent of the population. In the United States, the proportion of people over 65 is projected to reach this level by 2030 (Figure 45.19). In 2050, there could be as many as 31 million Americans over age 85.

The aging of a population has social implications. Older individuals have traditionally been supported by a younger workforce. In the United States, most older people receive social security payments and government-subsidized medical care. As a result of inflation and increases in life expectancy, the benefits being distributed to current seniors exceed the contributions these people paid into the program. When baby boomers begin to receive benefits, the deficit will skyrocket. Keeping the system going will require ever greater contributions from the younger, still-working population. Increasing numbers of debilitated seniors will also challenge the health care system. Thus, finding ways to keep people healthy later in life is both a social and an economic priority.

Figure 45.19 Two of the 37 million Americans over age 65.

Take-Home Message

How does slowing population growth affect age distribution?

- When population growth slows, the proportion of older individuals rises.

IMPACTS, ISSUES REVISITED The Numbers Game

Many states are struggling to control rising numbers of white-tailed deer. In Ohio, the number has risen from 17,000 deer in 1970 to more than 700,000. In West Virginia, deer are overbrowsing plants that grow on the forest floor, including wild ginseng, which is an important export crop. Biologist James McGraw argues that controlling deer and saving West Virginia's forests will require either reintroducing big predators or increasing deer hunting.

How would you vote?
Without natural predators, deer numbers are soaring. Is encouraging deer hunting the best solution? See CengageNOW for details, then vote online.

Summary

Sections 45.1, 45.2 Each population is a group of individuals of the same species. Its growth is affected by its **demographics**. These include **population size** and **age structure**, such as the size of the **reproductive base**. They also include **population density** and **population distribution**. Most populations in nature have a clumped distribution pattern.

Counting the number of individuals in **quadrats** is a way to estimate the density of a population in a specified area. **Capture–recapture methods** can be used to estimate the population density for mobile animals.

- *Use the interaction on CengageNOW to learn how to estimate population size.*

Section 45.3 **Immigration** and **emigration** permanently affect population size, but **migration** does not. The per capita birth rate minus the per capita death rate gives us r, the population's **per capita growth rate**. When births equal deaths we have **zero population growth**.

In cases of **exponential growth**, a population's growth is proportional to its size. The population size increases at a fixed rate in any given interval. The time required for a population to double is the **doubling time**. The maximum possible rate of increase is a species' **biotic potential**.

- *View the animation on CengageNOW to observe a pattern of exponential growth.*

Section 45.4 **Limiting factors** constrain population increases. With **logistic growth**, a small population starts growing slowly, then grows rapidly, then levels off once **carrying capacity** is reached. **Density-dependent factors** are conditions or events that lower reproductive success and have an increasing effect with crowding. **Density-independent factors** are conditions or events that can lower reproductive success, but their effect does not vary with crowding.

- *Watch the animation on CengageNOW to learn about logistic growth.*

Sections 45.5, 45.6 The time to maturity, number of reproductive events, number of offspring per event, and life span are aspects of a **life history pattern**. A **cohort** is a group of individuals that were born at the same time. Three types of **survivorship curves** are common: a high death rate late in life, a constant rate at all ages, or a high rate early in life. Life histories have a genetic basis and are subject to natural selection. At low population density, ***r*-selection** favors quickly producing as many offspring as possible. At a higher population density, ***K*-selection** favors investing more time and energy in fewer, higher quality offspring. Most populations have a mixture of both *r*-selected and *K*-selected traits.

Section 45.7 The human population has surpassed 6.6 billion. Expansion into new habitats and agriculture allowed early increases. Later, medical and technological innovations raised the carrying capacity and sidestepped many limiting factors.

Section 45.8 A population's **total fertility rate** (TFR) is the average number of children born to women during their reproductive years. The global TFR is declining and most countries have family planning programs of some sort. Even so, the pre-reproductive base of the world population is so large that population size will continue to increase for at least sixty years.

- *Use the interaction on CengageNOW to compare age structure diagrams.*

Section 45.9 The **demographic transition model** predicts how human population growth rates will change with industrialization. Generally, the death rate and birth rate both fall with rising industrialization, but conditions in countries can vary in ways that affect this trend.

Developed nations have a much higher per capita consumption of resources than developing nations. Earth does not have enough resources to support the current population in the style of the developed nations.

- *Use the interaction on CengageNOW to learn about the demographic transition model.*

Section 45.10 Slowing population growth leads to an increase in the proportion of elderly in the population.

Self-Quiz

Answers in Appendix III

1. Most commonly, individuals of a population show a ______ distribution through their habitat.
   a. clumped
   b. random
   c. nearly uniform
   d. none of the above

2. The rate at which population size grows or declines depends on the rate of ______.
   a. births
   b. deaths
   c. immigration
   d. emigration
   e. a and b
   f. all of the above

Data Analysis Exercise

In 1989, Martin Wikelski started a long-term study of marine iguana populations in the Galápagos Islands (Section 17.2). He marked the iguanas on two islands— Genovesa and Santa Fe—and collected data on how their body size, survival, and reproductive rates varied over time. The iguanas eat algae and have no predators, so deaths are usually the result of food shortages, disease, or old age. His studies showed that numbers decline during El Niño events, when the surrounding waters heat up.

In January 2001, an oil tanker ran aground and leaked a small amount of oil into the waters near Santa Fe—Figure 45.20 shows the number of marked iguanas that Wikelski and his team counted in their census of study populations just before the spill and about a year later.

1. Which island had more marked iguanas at the time of the first census?

2. How much did the population size on each island change between the first and second census?

3. Wikelski concluded that changes on Santa Fe were the result of the oil spill, rather than sea temperature or other climate factors common to both islands. How would the census numbers be different from those he observed if an adverse event had affected both islands?

Figure 45.20 Shifting numbers of marked marine iguanas on two Galápagos islands. An oil spill occurred near Santa Fe just before the January 2001 census (*green* bars). A second census was carried out in December 2001 (*tan* bars).

3. Suppose 200 fish are marked and released in a pond. The following week, 200 fish are caught and 100 of them have marks. There are about ______ fish in this pond.
   a. 200 b. 300 c. 400 d. 2,000

4. A population of worms is growing exponentially in a compost heap. Thirty days ago there were 400 worms and now there are 800. How many worms will there be thirty days from now, assuming conditions remain constant?
   a. 1,200 b. 1,600 c. 3,200 d. 6,400

5. For a given species, the maximum rate of increase per individual under ideal conditions is its ______ .
   a. biotic potential
   b. carrying capacity
   c. environmental resistance
   d. density control

6. ______ is a density-independent factor that influences population growth.
   a. Resource competition
   b. Infectious disease
   c. Predation
   d. Harsh weather

7. A life history pattern for a population is a set of adaptations that influence the individual's ______ .
   a. longevity
   b. fertility
   c. age at reproductive maturity
   d. all of the above

8. The human population is now over 6.6 billion. It was about half that in ______ .
   a. 2004 b. 1960 c. 1802 d. 1350

9. Compared to the less developed countries, the highly developed ones have a higher ______ .
   a. death rate
   b. birth rate
   c. total fertility rate
   d. resource consumption rate

10. ______ population growth increases the proportion of older individuals in a population.
   a. Slowing b. Accelerating

11. Match each term with its most suitable description.

___carrying capacity	a. maximum rate of increase per individual under ideal conditions
___exponential growth	b. population growth plots out as an S-shaped curve
___biotic potential	c. maximum number of individuals sustainable by the resources in a given environment
___limiting factor	d. population growth plots out as a J-shaped curve
___logistic growth	e. essential resource that restricts population growth when scarce

■ *Visit CengageNOW for additional questions.*

Critical Thinking

1. Think back to Section 45.6. When researchers moved guppies from populations preyed on by cichlids to a habitat with killifish, the life histories of the transplanted guppies evolved. They came to resemble those of guppy populations preyed on by killifish. Males became gaudier; some scales formed larger, more colorful spots. How might a decrease in predation pressure on sexually mature fish favor this change?

2. The age structure diagrams for two hypothetical populations are shown at right. Describe the growth rate of each population and discuss the current and future social and economic problems that each is likely to face.

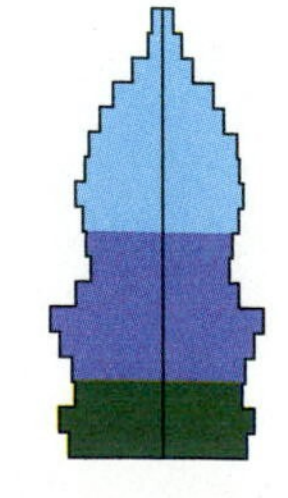

46 Community Structure and Biodiversity

IMPACTS, ISSUES Fire Ants in the Pants

Step on a nest of red imported fire ants, *Solenopsis invicta* (Figure 46.1*a*), and you will be sorry. The ants are quick to defend their nest. Ants stream out from the ground and inflict a series of stings. Venom injected by the stinger causes burning pain and results in the formation of a pus-filled bump that is slow to heal. Multiple stings can cause nausea, dizziness, and—rarely—death.

S. invicta arrived in the United States from South America in the 1930s, probably as stowaways on a ship. The ants spread out from the Southeast and have been found as far west as California and as far north as Kansas and Delaware.

Like many introduced species, the ants disrupt natural communities. They attack livestock, pets, and wildlife. They also outcompete native ants and may be contributing to the decline of other native wildlife. For example, the Texas horned lizard vanished from most of its home range when *S. invicta* moved in and displaced the native ants—the lizard's food of choice. The horned lizard cannot tolerate eating the imported fire ants.

Invicta means "invincible" in Latin and *S. invicta* is living up to its species name. Pesticides have not managed to halt the foreign ant's spread. The chemicals might even be facilitating dispersal by preferentially wiping out native ant populations.

Ecologists are enlisting biological controls. Phorid flies control *S. invicta* in its native habitat (Figure 46.1*b*). The flies are parasitoids, a type of parasite that kills its host in a rather gruesome way. A female fly pierces the cuticle of an adult ant, then lays an egg in the ant's soft tissues. The egg hatches into a larva, which grows and eats its way through tissues to the ant's head. After the larva gets big enough, it causes the ant's head to fall off (Figure 46.1*c*). The larva develops into an adult within the detached head.

Several phorid fly species have now been introduced in various southern states. The flies are surviving, reproducing, and increasing their range. They probably will never kill off all *S. invicta* in affected areas, but they are expected to reduce the density of colonies.

This example introduces community structure: patterns in the number of species and their relative abundances. As you will see, species interactions and disturbances to the habitat can shift community structure in small and large ways—some predictable, others unexpected.

See the video! Figure 46.1 (**a**) Red imported fire ant (*S. invicta*) mounds. (**b**) A phorid fly that lays its eggs on the ants. (**c**) An ant that lost its head after the larva of a phorid fly moved into it.

Key Concepts

Community characteristics

A community consists of all species in a habitat. Each species has a niche—the sum of its activities and relationships. A habitat's history, its biological and physical characteristics, and the interactions among species in the habitat affect community structure. **Section 46.1**

Types of species interactions

Commensalism, mutualism, competition, predation, and parasitism are types of interspecific interactions. They influence the population size of participating species, which in turn influences the community's structure. **Sections 46.2–46.7**

Community stability and change

Communities have certain elements of stability, as when some species persist in a habitat. Communities also change, as when new species move into the habitat and others disappear. Physical characteristics of the habitat, species interactions, disturbances, and chance events affect how a community changes over time. **Sections 46.8–46.10**

Global patterns in community structure

Biogeographers identify regional patterns in species distribution. They have shown that tropical regions hold the greatest number of species, and also that characteristics of islands can be used to predict how many species an island will hold. **Section 46.11**

Links to Earlier Concepts

- In this chapter, you will see how natural selection (Section 17.3) and coevolution (18.12) shape traits of species in communities.
- You will revisit examples of interspecific interactions such as bacteria that live inside protists (20.4), plant–pollinator interactions (23.8, 30.2), lichens (24.6), and root nodules and mycorrhizae (29.2).
- You will consider again the evolution of prey defenses such as ricin (Chapter 14 introduction), nematocysts (25.5), and the way that evolution affects pathogens (21.8).
- Knowledge of biogeography (17.1) will help you understand how communities in different regions differ.

How would you vote? Currently, only a fraction of the crates imported into the United States are inspected for the inadvertent or deliberate presence of exotic species. Would added inspections that better protect native communities be worth the cost? See CengageNOW for details, then vote online.

46.1 Which Factors Shape Community Structure?

- Community structure refers to the number and relative abundances of species in a habitat. It changes over time.
- Link to Coevolution 18.12

The type of place where a species normally lives is its **habitat**, and all species living in a habitat represent a **community**. A community has a dynamic structure. It shows shifts in its species diversity—the number and relative abundances of species.

Many factors influence community structure. First, climate and topography influence a habitat's features, including temperature, soil, and moisture. Second, a habitat has only certain kinds and amounts of food and other resources. Third, species themselves have traits that adapt them to certain habitat conditions, as in Figure 46.2. Fourth, the species interact in ways that cause shifts in their numbers and abundances. Finally, the timing and history of disturbances, both natural and human-induced, affect community structure.

The Niche

All species of a community share the same habitat—the same "address"—but each also has a "profession," or unique ecological role, that sets it apart. This role is the species' **niche**, which we describe in terms of the conditions, resources, and interactions necessary for survival and reproduction. Aspects of an animal's niche include temperatures it can tolerate, the kinds of foods it can eat, and the types of places it can breed or hide. A description of a plant's niche would include its soil, water, light, and pollinator requirements.

Figure 46.2 Three of twelve fruit-eating pigeon species in Papua New Guinea's tropical rain forests: (**a**) pied imperial pigeon, (**b**) superb crowned fruit pigeon, and (**c**) the turkey-sized Victoria crowned pigeon. The forest's trees differ in the size of fruit and fruit-bearing branches. The big pigeons eat big fruit. Smaller ones, with smaller bills, cannot peck open big, thick-skinned fruit. They eat the small, soft fruit on branches too spindly to hold big pigeons.

Trees feed the birds, which help the trees. Seeds in fruit resist digestion in the bird gut. Flying pigeons disperse seed-rich droppings, often some distance from mature trees that would outcompete new seedlings for water, minerals, and sunlight. With dispersal, seedlings have a better chance of surviving.

Table 46.1 Direct Two-Species Interactions

Type of Interaction	Effect on Species 1	Effect on Species 2
Commensalism	Helpful	None
Mutualism	Helpful	Helpful
Interspecific competition	Harmful	Harmful
Predation	Helpful	Harmful
Parasitism	Helpful	Harmful

Categories of Species Interactions

Species in a community interact in a variety of ways (Table 46.1) **Commensalism** benefits one species and does not affect the other. Most bacteria in your gut are commensal. They benefit by living inside you, but do not help or harm you. **Mutualism** provides benefits to both species. **Interspecific competition** hurts both species. **Predation** and **parasitism** help one species at another's expense. Predators are free-living organisms that kill their prey. Parasites live on or in a host and usually do not kill it.

Parasitism, commensalism, and mutualism can all be types of **symbiosis**, which means "living together." Symbiotic species, or symbionts, spend most or all of their life cycle in close association with each other. An endosymbiont is a species that lives inside its partner.

Regardless of whether one species helps or hurts another, two species that interact closely for extended periods may coevolve. With **coevolution**, each species is a selective agent that shifts the range of variation in the other (Section 18.12).

Take-Home Message

What is a biological community?

- A community consists of all species in a habitat, each with a unique niche, or ecological role.
- Species in a community interact and may benefit, harm, or have no net effect on one another. Some are symbionts; they associate closely for most or all of their life cycle.

46.2 Mutualism

- A mutualistic interaction benefits both partners.
- Links to Endosymbiosis and organelles 20.4, Pollination 23.8 and 30.2, Lichens 24.6, Plant mutualisms 29.2

Mutualists are common in nature. For example, birds, insects, bats, and other animals serve as pollinators of flowering plants (Sections 23.8 and 30.2). Pollinators feed on energy-rich nectar and pollen. In return, they transfer pollen between plants, facilitating pollination. Similarly, pigeons take food from rain forest trees but disperse their seeds to new sites (Figure 46.2).

In some mutualisms, neither species can complete its life cycle without the other. Yucca plants and the moths that pollinate them show such interdependence (Figure 46.3). In other cases, the mutualism is helpful but not a life-or-death requirement. Most plants, for example, use more than one pollinator.

Mutualists help most plants take up mineral ions (Section 29.2). Nitrogen-fixing bacteria living on roots of legumes such as peas provide the plant with extra nitrogen. Mycorrhizal fungi living in or on plant roots enhance the plant's mineral uptake.

Other fungi partner with photosynthetic bacteria or algae, thus forming lichens (Section 24.6). In all mutualisms, there is some conflict between partners. In a lichen, the fungus would do best by obtaining as much sugar as possible from its photosynthetic partner. That partner would do best by keeping as much sugar as possible for its own use.

Some mutualists defend one another. For example, most fishes avoid sea anemones, which have stinging cells called nematocysts in their tentacles. However, an anemone fish can nestle among those tentacles (Figure 46.4). A mucus layer shields the anemone fish from stings, and the tentacles keep it safe from predatory fish. The anemone fish repays its partner by chasing off the few fishes that feed on sea anemone tentacles.

Finally, reflect on a theory outlined in Section 20.4, whereby certain aerobic bacteria became mutualistic endosymbionts of early eukaryotic cells. The bacteria received nutrients and shelter. In time, they evolved into mitochondria and provided the "host" with ATP. Cyanobacteria living inside eukaryotic cells evolved into chloroplasts by a similar process.

Take-Home Message

What is mutualism?

- Mutualism is a species interaction in which each species benefits by associating with the other.
- In some cases the mutualism is necessary for both species; more often it is not essential for one or both partners.

Figure 46.3 Mutualism in the high desert of Colorado.

Each species of *Yucca* plant is pollinated by one species of yucca moth, which cannot complete its life cycle with any other plant. The moth matures when yucca plants flower. A female moth collects yucca pollen and rolls it into a ball. She flies to another flower and pierces the floral ovary, and lays eggs inside. As she crawls out, she pushes a ball of pollen onto the flower's pollen-receiving platform.

After pollen grains germinate, they give rise to pollen tubes, which grow through the ovary tissues and deliver sperm to the plant's eggs. Seeds develop after fertilization.

Meanwhile, moth eggs develop into larvae that eat a few seeds, then gnaw their way out of the ovary. Seeds that larvae do not eat give rise to new yucca plants.

Figure 46.4 The sea anemone *Heteractis magnifica*, which shelters about a dozen fish species. It has a mutualistic association with the pink anemone fish (*Amphiprion perideraion*). This tiny but aggressive fish chases away predatory butterfly fishes that would bite off tips of anemone tentacles. The fish cannot survive and reproduce without the protection of an anemone. The anemone does not need a fish to protect it, but it does better with one.

46.3 Competitive Interactions

- Resources are limited and individuals of different species often compete for access to them.
- Links to Natural selection 17.3, Limiting factor 45.4

As Charles Darwin understood, intense competition for resources among individuals of the same species leads to evolution by natural selection (Section 17.3). Competitive interactions between different species—interspecific competition—is not usually as intense. Why not? The requirements of two species might be similar, but they can never be as close as they are for individuals of the same species.

With interference competition, one species actively prevents another from accessing some resource. As an example, one species of scavenger will often chase another away from a carcass (Figure 46.5). As another example, some plants use chemical weapons against potential competition. Aromatic chemicals that ooze from tissues of sagebrush plants, black walnut trees, and eucalyptus trees seep into the soil around these plants. The chemicals prevent other kinds of plants from germinating or growing.

In exploitative competition, species do not interact directly; each reduces the amount of resources available to the other by using that resource. For example, deer and blue jays both eat acorns in oak forests. The more acorns the birds eat, the fewer there are for the deer.

Figure 46.5 Interspecific competition among scavengers. (**a**) A golden eagle and a red fox face off over a moose carcass. (**b**) In a dramatic demonstration of interference competition, the eagle attacks the fox with its talons. After this attack, the fox retreated, leaving the eagle to exploit the carcass.

Effects of Competition

Deer and blue jays share a fondness for acorns, but each also has other sources of food. Any two species differ in their resource requirements. Species compete most intently when the supply of a shared resource is the main limiting factor for both (Section 45.4).

In the 1930s, G. Gause conducted experiments with two species of ciliated protists (*Paramecium*) that compete for bacterial prey. When cultured separately, the growth curves for these species were about the same. When grown together, growth of one species outpaced the other, and drove it to extinction (Figure 46.6).

Experiments by Gause and others are the basis for the concept of **competitive exclusion**: Whenever two species require the same limited resource to survive or reproduce, the better competitor will drive the less competitive species to extinction in that habitat.

Competitors can coexist when their resource needs are not exactly the same, however, competition generally supresses population growth of both species. For instance, Gause also studied two *Paramecium* species with differing food preferences. When grown together, one fed on bacteria suspended in culture tube liquid. The other ate yeast cells near the bottom of the tube. When grown together, population growth rates fell for both species, but they continued to coexist.

Experiments by Nelson Hairston showed the effects of competition between slimy salamanders (*Plethodon glutinosus*) and Jordan's salamanders (*P. jordani*). The salamanders coexist in wooded habitats (Figure 46.7). Hairston removed all slimy salamanders from certain test plots and Jordan's salamanders from others. He left a final group of plots unaltered as controls.

After five years, the numbers and abundances of the two species had not changed in the control plots. In the plots with slimy salamanders alone, population density had soared. Numbers also increased in plots with Jordan's salamanders alone. Hairston concluded

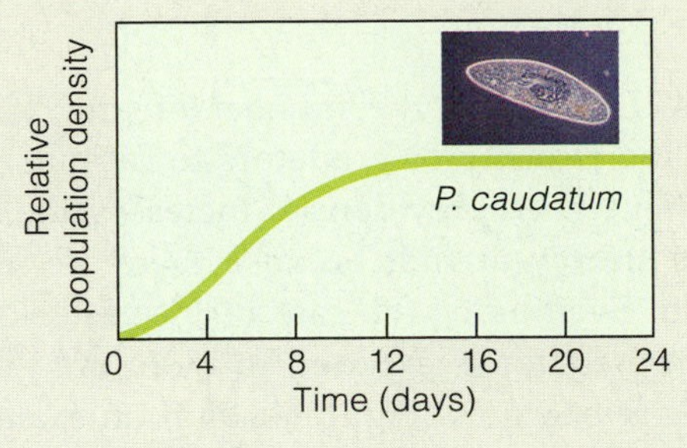

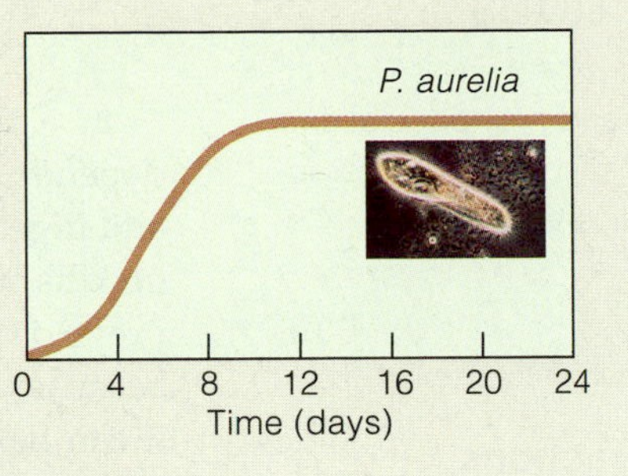

A *Paramecium caudatum* and *P. aurelia* grown in separate culture flasks established stable populations. The S-shaped graph curves indicate logistic growth and stability.

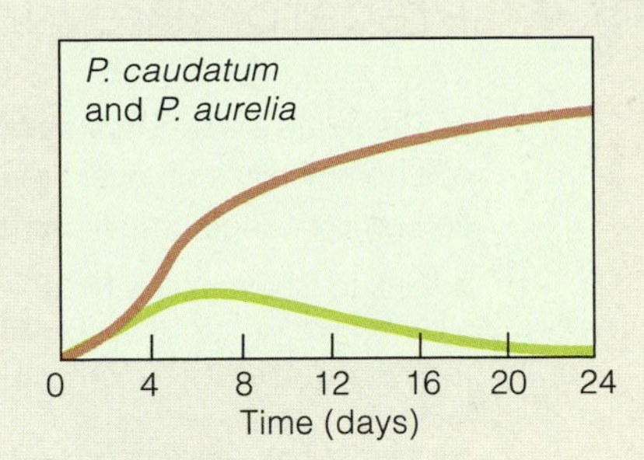

B For this experiment, the two species were grown together. *P. aurelia* (*brown* curve) drove *P. caudatum* toward extinction (*green* curve).

Figure 46.6 Animated Results of competitive exclusion between two related species that compete for the same food. Two species cannot coexist indefinitely in the same habitat *when they require identical resources.*

that whenever these salamanders coexist, competitive interactions suppress the population growth of both.

Resource Partitioning

Think back on those fruit-eating pigeon species. They all require fruit, but each eats fruits of a certain size. Their preferences are a case of **resource partitioning**: a subdividing of an essential resource, which reduces the competition among species that require it.

Similarly, three annual plant species live in the same field. They all require minerals and water, but their roots take them up at different depths (Figure 46.8).

When species with very similar requirements share a habitat, competition puts selective pressure on them. In each species, individuals who differ most from the competing species are favored. The outcome may be **character displacement**: Over the generations, a trait of one species diverges in a way that lowers the intensity of competition with the other species. Modification of the trait promotes partitioning of a resource.

Figure 46.7 Two species of salamanders, *Plethodon glutinosus* (*top*) and *P. jordani* (*bottom*), that compete in areas where their habitats overlap.

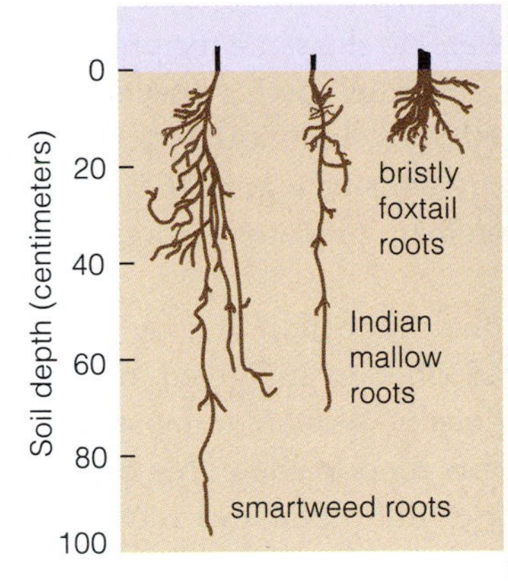

Figure 46.8 A case of resource partitioning among three annual plant species in a plowed but abandoned field. Roots of each species take up water and mineral ions from a different soil depth. This reduces competition among them and allows them to coexist.

For example, researchers Peter and Rosemary Grant demonstrated a change in beak size in the Galápagos finch *Geospiza fortis*. It occurred after a larger finch, *G. magnirostris*, moved onto the island where *G. fortis* had previously been alone. Arrival of *G. magnirostris* put big-beaked *G. fortis* individuals at a disadvantage. They now had to compete with G. *magnirostris* for big seeds. Small-beaked *G. fortis* had no such competition, and enjoyed higher reproductive success. As a result, the average beak size of *G. fortis* declined over time.

Take-Home Message

What happens when species compete for resources?

- In some interactions, one species actively blocks another's access to a resource. In other interactions, one species is simply better than another at exploiting a shared resource.
- When two species compete, selection favors individuals whose needs are least like those of the competing species.

46.4 Predator-Prey Interactions

- The relative abundances of predator and prey populations of a community shift over time in response to species interactions and changing environmental conditions.
- Link to Coevolution 18.12

Models for Predator-Prey Interactions

Predators are consumers that get energy and nutrients from **prey**, which are living organisms that predators capture, kill, and eat. The quantity and types of prey species affect predator diversity and abundance, and predator types and numbers do the same for prey.

The extent to which a predator species affects prey numbers depends in part on how individual predators respond to changes in prey density. Figure 46.9*a* compares models for the three main predator responses to increases in density.

In a type I response, the proportion of prey killed is constant, so the number killed in any given interval depends solely on prey density. Web-spinning spiders and other passive predators tend to show this type of response. As the number of flies in an area increases, more and more become caught in each spider's web. Filter-feeding predators also show a type I response.

In a type II response, the number of prey killed depends on the capacity of predators to capture, eat, and digest prey. When prey density increases, the rate of kills rises steeply at first because there are more prey to catch. Eventually, the rate of increase slows, because each predator is exposed to more prey than it can handle at one time. Figure 46.9*b* is an example of this type of response, which is common in nature. A wolf that just killed a caribou will not hunt another until it has eaten and digested the first one.

In a type III response, the number of kills increases slowly until prey density exceeds a certain level, then rises rapidly, and finally levels off. This response is common in nature in three situations. In some cases, the predator switches among prey, concentrating its efforts on the species that is most abundant. In other cases, the predators need to learn how to best capture each prey species; they get more lessons when more prey are around. In still other cases, the number of hiding places for prey is limited. Only after prey density rises and some individual prey have no place to hide, does the number of kills increase.

Knowing which type of response a predator makes to prey helps ecologists predict long-term effects of predation on a prey population.

The Canadian Lynx and Snowshoe Hare

In some cases, a time lag in the predator's response to prey density leads to cyclic changes in abundance of predators and prey. When prey density becomes low, the number of predators declines. As a result, prey are safer and their number increases. This increase allows predators to increase. Then predation causes another prey decline, and the cycle begins again.

Consider a ten-year oscillation in populations of a predator, the Canadian lynx, and the snowshoe hare

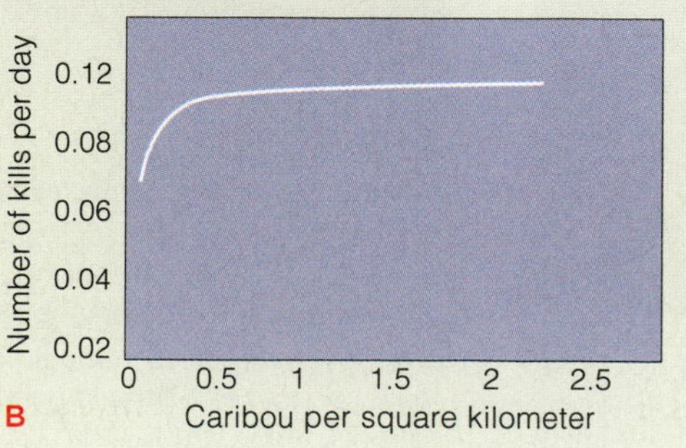

Figure 46.9 Animated (**a**) Three models for responses of predators to prey density. Type I: Prey consumption rises linearly as prey density rises. Type II: Prey consumption is high at first, then levels off as predator bellies stay full. Type III: When prey density is low, it takes longer to hunt prey, so the predator response is low. (**b**) A type II response in nature. For one winter month in Alaska, B. W. Dale and his coworkers observed four wolf packs (*Canis lupus*) feeding on caribou (*Rangifer tarandus*). The interaction fit the type II model for the functional response of predators to the prey density.

Figure 46.10 Graph of the abundances of Canadian lynx (*dashed* line) and snowshoe hares (*solid* line), based on counts of pelts sold by trappers to Hudson's Bay Company during a ninety-year period. Charles Krebs observed that predation causes heightened alertness among snowshoe hares, which continually look over their shoulders during the declining phase of each cycle. The photograph at *right* supports the Krebs hypothesis that there is a three-level interaction going on, one that involves plants.

The graph may be a good test of whether you tend to accept someone else's conclusions without questioning their basis in science. Remember those sections in Chapter 1 that introduced the nature of scientific methods?

What other factors may have had an impact on the cycle? Did the weather vary, with more severe winters imposing greater demand for hares (to keep lynxes warmer) and higher death rates? Did the lynx compete with other predators, such as owls? Did the predators turn to alternative prey during low points of the hare cycle?

that is its main prey (Figure 46.10). To determine the causes of this pattern, Charles Krebs and coworkers tracked hare population densities for ten years in the Yukon River Valley of Alaska. They set up one-square-kilometer control plots and experimental plots. They used fences to keep predatory mammals out of some plots. Extra food or fertilizers that helped plants grow were used in other plots. The researchers captured and put radio collars on more than 1,000 snowshoe hares, lynx, and other animals, and then released them.

In predator-free plots, the hare density doubled. In plots with extra food, it tripled. In plots having extra food and fewer predators, it increased elevenfold.

The experimental manipulations delayed the cyclic declines in population density but did not stop them. Why not? Owls and other raptors flew over the fences. Only 9 percent of the collared hares starved to death; predators killed some of the others. Krebs concluded that a simple predator–prey or plant–herbivore model did not fully explain his results. Other variables were at work, in a multilevel interaction.

Coevolution of Predators and Prey

Interactions among predators and prey can influence characteristic species traits. If a certain genetic trait in a prey species helps it escape predation, that trait will increase in frequency. If some predator characteristic helps overcome a prey defense, it too will be favored. Each defensive improvement selects for a countering improvement in predators, which selects for another defensive improvement, and so on, in a never-ending arms race. The next section describes some outcomes.

Take-Home Message

How do predator and prey populations change over time?

- Predator populations show three general patterns of response to changes in prey density. Population levels of prey may show recurring oscillations.
- The numbers in predator and prey populations often vary in complex ways that reflect the multiple levels of interaction in a community.
- Predator and prey populations exert selective pressures on one another.

46.5 An Evolutionary Arms Race

- Predators select for better prey defenses, and prey select for more efficient predators.
- Links to Ricin Chapter 14 introduction, Coevolution 18.12, Nematocysts 25.5

Figure 46.11 Prey camouflage. (**a**) What bird? When a predator approaches its nest, the least bittern stretches its neck (which is colored like the surrounding withered reeds), points its bill upward, and sways like reeds in the wind. (**b**) An inedible bird dropping? No. This caterpillar's body coloration and its capacity to hold its body in a rigid position help camouflage it from predatory birds. (**c**) Find the plants (*Lithops*) hiding in the open from herbivores with the help of their stonelike form, pattern, and coloration.

Prey Defenses

Earlier chapters, including Chapter 25, introduced some examples of prey defenses. Many species have hard parts that make them difficult to eat. Spikes in a sponge body, clam and snail shells, lobster and crab exoskeletons, sea urchin spines—all of these traits help deter predators and thereby contribute to evolutionary success.

Also, many heritable traits function in **camouflage**: body shape, color pattern, behavior, or a combination of factors make an individual blend with its surroundings. Predators cannot eat prey they cannot find. Section 18.4 explains how alleles that improved the camouflage of a prey species, the desert pocket mouse, were adaptive in particular habitats.

Camouflage is widespread. Marsh birds called bitterns live among tall reeds. When threatened, a bittern points its beak skyward and blends with the reeds (Figure 46.11*a*). On a breezy day, the bird enhances the effect by swaying slightly. A caterpillar with mottled color patterns appears to be a bird dropping (Figure 46.11*b*). Desert plants of the genus *Lithops* usually look like rocks (Figure 46.11*c*). They flower only during a brief rainy season, when plenty of other plants tempt herbivores.

Many prey species contain chemicals that taste bad or sicken predators. Some produce toxins through metabolic processes. Others use chemical or physical weapons that they get from their prey. For instance, after sea slugs dine on a sea anemone or a jellyfish, they can store its stinging nematocysts in their own tissues (Figure 25.24*c*).

Leaves, stems, and seeds of many plants contain bitter, hard-to-digest, or toxic chemicals. Remember the Chapter 14 introduction? It explains how ricin acts to kill or sicken animals. Ricin evolved in castor bean seeds as a defense against herbivores. Caffeine in coffee beans and nicotine in tobacco leaves evolved as defenses against insects.

Many prey species advertise their bad-tasting or toxin-laden properties by **warning coloration**. They have flashy patterns and colors that predators learn to recognize and avoid. For instance, a toad might catch a yellow jacket once. But a painful sting from this wasp teaches the toad that black and yellow stripes mean *AVOID ME!*

Mimicry is an evolutionary convergence in body form; species come to resemble one another. In some cases, two or more well-defended organisms end up looking alike.

a A dangerous model

b One of its edible mimics

c Another edible mimic

d And another edible mimic

Figure 46.12 Examples of mimicry. Edible insect species often resemble toxic or unpalatable species that are not at all closely related. (**a**) A yellow jacket can deliver a painful sting. It might be the model for nonstinging wasps (**b**), beetles (**c**), and flies (**d**) of strikingly similar appearance.

In others, a tasty, harmless prey species evolves the same warning coloration as an unpalatable or well-defended one (Figure 46.12). Predators may avoid the mimic after experiencing the disgusting taste, irritating secretion, or painful sting of the species it resembles.

When an animal is cornered or under attack, survival may depend on a last-chance trick. Opossums "play dead," Other animals startle predators. Section 1.7 describes an experiment that tested the peacock butterfly defenses—a show of eye-like spots and hissing. Other species puff up, bare sharp teeth, or flare neck ruffs (Figure 26.19*d*). When cornered, many animals, including skunks, some snakes, many toads, and certain insects, secrete or squirt stinky or irritating repellents (Figure 46.13*a*).

Adaptive Responses of Predators

A predator's evolutionary success hinges on eating prey. Stealth, camouflage, and ways of avoiding repellents are countermeasures to prey defenses. For example, some edible beetles spray noxious chemicals at their attackers. A grasshopper mouse grabs the beetle and plunges the sprayer end into the ground, and then chews on the tasty, unprotected head (Figure 46.13*b*). Some evolved traits in herbivores are responses to plant defenses. The digestive tract of koalas can handle tough, aromatic eucalyptus leaves that would sicken other herbivorous mammals.

Also, a speedier predator catches more prey. Consider the cheetah, the world's fastest animal on land. One was clocked at 114 kilometers (70 miles) per hour. Compared with other big cats, a cheetah has longer legs relative to body size and nonretractable claws that act like cleats to increase traction. Thomson's gazelle, its main prey, can run longer but not as fast (80 kilometers per hour). Without a head start, the gazelle is likely to be outrun.

Camouflaging helps predators as well as prey. Think of white polar bears stalking seals on ice, striped tigers crouched in tall-stalked, golden grasses, and scorpionfish on the sea floor (Figure 46.13*c*). Camouflage can be quite stunning among predatory insects (Figure 46.13*d*). Even so, with each new, improved camouflaging trait, predators select for enhanced predator-detecting ability in prey.

Figure 46.13 Predator responses to prey defenses. (**a**) Some beetles spray noxious chemicals at attackers, which deters them some of the time. (**b**) Grasshopper mice plunge the chemical-spraying tail end of their beetle prey into the ground and feast on the head end. (**c**) This leaf scorpionfish, is a venomous predator with camouflaging fleshy flaps, multiple colors, and many spines. (**d**) Where do the pink flowers end and the pink praying mantis begin?

46.6 Parasite-Host Interactions

- Predators have only a brief interaction with prey, but parasites live on or in their hosts.
- Link to Evolution and disease 21.8

Parasites and Parasitoids

Parasites spend all or part of their life living in or on other organisms, from which they steal nutrients. Although most parasites are small, they can have a major impact on populations of their hosts. Many parasites are pathogens; they cause disease in their hosts. For example, *Myxobolus cerebralis* is a parasite of trout, salmon and related fishes. Following infection, a host fish develops deadly whirling disease (Figure 46.14).

Even when a parasite does not cause such dramatic symptoms, infection can weaken the host so it is more vulnerable to predation or less attractive to potential mates. Some parasitic infections cause sterility. Others shift the sex ratio of their host species. Parasites affect host numbers by altering birth and death rates. They also indirectly affect species that compete with their host. The decline in trout caused by whirling disease allows competing fish populations to increase.

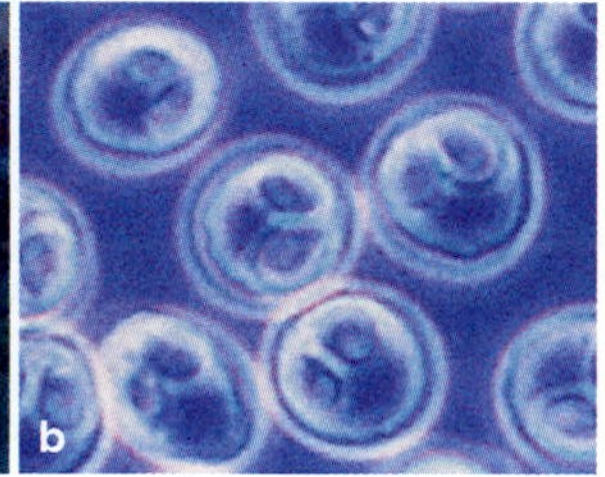

Figure 46.14 (**a**) A young trout with a twisted spine and darkened tail caused by whirling disease, which damages cartilage and nerves. Jaw deformities and whirling movements are other symptoms. (**b**) Spores of *Myxobolus cerebralis*, the parasite that causes the disease. The disease now occurs in many lakes and streams in western and northeastern states.

Sometimes the gradual drain of nutrients during a parasitic infection indirectly leads to death. The host is so weak that it cannot fight off secondary infections. A rapid death is rare. Usually death happens only after a parasite attacks a novel host—one with no coevolved defenses—or after the body is overwhelmed by a huge population of parasites.

In evolutionary terms, killing the host too quickly is bad for the parasite. Ideally, a host will live long enough to give the parasite time to produce plenty of offspring. The longer the host survives, the more offspring the parasite can produce. That is why we can predict that natural selection will favor parasites with less-than-fatal effects on hosts (Section 21.8).

Unit Four describes many parasites. Some spend their entire life in or on a single host species. Others have different hosts during different stages of the life cycle. Insects and other arthropods can act as **vectors**: organisms that convey a parasite from host to host.

Even a few plants are parasitic. Nonphotosynthetic species such as dodders obtain energy and nutrients from a host plant (Figure 46.15). Other species carry out photosynthesis but steal nutrients and water from their host. Most mistletoe are like this; their modified roots tap into the vascular tissues of host trees.

Many tapeworms, flukes, and certain roundworms are parasitic invertebrates (Figure 46.16). So are ticks, many insects, and some crustaceans.

Parasitoids are insects that lay eggs in other insects. Larvae hatch, develop in the host's body, eat its tissue, and eventually kill it. The fire ant–killing phorid flies described in this chapter's introduction do this. As many as 15 percent of all insects may be parasitoids.

Social parasites are animals that take advantage of the behavior of a host to complete their life cycle. Cuckoos and North American cowbirds, as explained shortly, are social parasites.

Figure 46.15 Dodder (*Cuscuta*), also known as strangleweed or devil's hair. This parasitic flowering plant has almost no chlorophyll. Leafless stems twine around a host plant during growth. Modified roots penetrate the host's vascular tissues and absorb water and nutrients from them.

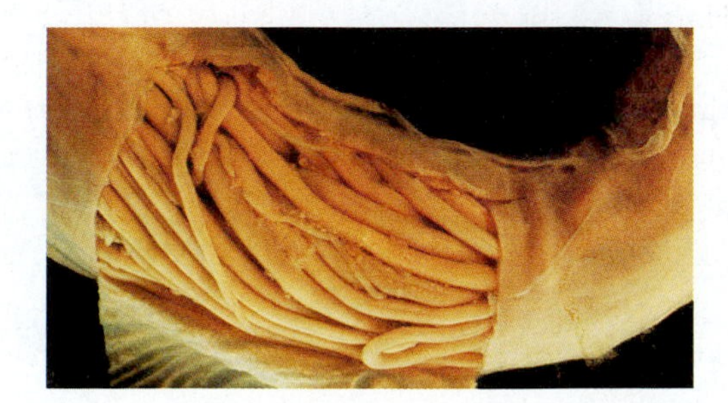

Figure 46.16 Adult roundworms (*Ascaris*), an endoparasite, inside the small intestine of a host pig. Sections 25.6 and 25.11 show more examples of parasitic worms.

Figure 46.17 Biological control agent: a commercially raised parasitoid wasp about to deposit an egg in an aphid. After the egg it laid hatches, a wasp larva will devour the aphid from the inside.

Biological Control Agents

Some parasites and parasitoids are now raised commercially for use as biological control agents. Use of such agents is promoted as an alternative to pesticides. For example, some parasitoid wasps attack aphids, which are widespread plant pests (Figure 46.17).

Effective biological control agents are adapted to a specific host species and to its habitat. They are good at finding the hosts. Their population growth rate is high compared to the host's. Their offspring are good at dispersing. Also, they make a type III response to changes in prey density (Section 46.4), without much lag time after the prey or host population size shifts.

Biological control is not without risks of its own. Releasing multiple species of biological control agent in an area may allow competition among them, and lower their effectiveness against an intended target. Also, introduced parasites sometimes go after nontargeted species in addition to, or instead of, those species they were introduced to control.

For example, parasitoids deliberately introduced to the Hawaiian Islands attacked the wrong target. They were brought in to control stinkbugs that are pests of Hawaii's crops. Instead, the parasitoids decimated the population of koa bugs, Hawaii's largest native bug. Introduced parasitoids also have been implicated in ongoing declines of many native Hawaiian butterfly and moth populations.

Take-Home Message

What are parasites, parasitoids, and social parasites?

- Parasitic species feed on another species but generally do not kill their host.
- Parasitoids are insects that eat other insects from inside out.
- Social parasites manipulate the social behavior of another species to their own benefit.

46.7 Strangers in the Nest

■ The brown-headed cowbird's genus name (*Molothrus*) means intruder in Latin. They intrude into other birds' nests and lay their eggs there.

Brown-headed cowbirds (*Molothrus ater*) evolved in the Great Plains of North America and they were commensal with bison. Great herds of these hefty ungulates stirred up plenty of tasty insects as they migrated through the grasslands, and, being insect-eaters, cowbirds wandered around with them (Figure 46.18*a*).

Cowbirds are social parasites that lay their eggs in the nests constructed by other birds, so young cowbirds are reared by foster parents. Many species became "hosts" to cowbirds; they did not have the capacity to recognize the differences between cowbird eggs and their own eggs. Concurrently, cowbird hatchlings became innately wired for hostile takeovers. They demand to be fed by unwitting, and often smaller, foster parents (Figure 46.18*b*). For thousands of years, cowbirds have perpetuated their genes at the expense of hosts.

When American pioneers moved west, many cleared swaths of woodlands for pastures. Cowbirds now moved in the other direction. They adapted easily to a life with new ungulates—cattle—in the man-made grasslands; hence their name. They started to penetrate adjacent woodlands and exploit novel species. Today, brown-headed cowbirds parasitize at least fifteen kinds of native North American birds. Some of those birds are threatened or endangered.

Besides being successful opportunists, cowbirds are big-time reproducers. A female can lay an egg a day for ten days, give her ovaries a rest, do the same again, and then again in one season. As many as thirty eggs in thirty nests—that is a lot of cowbirds.

Figure 46.18 (**a**) Brown-headed cowbirds (*Molothrus ater*) originally evolved as commensalists with bison of the North American Great Plains. (**b**) Cowbirds are social parasites. The large nestling at the *left* is a cowbird. The smaller foster parent is rearing the cowbird in place of its own offspring.

46.8 Ecological Succession

- Which species are present in a community depends on physical factors such as climate, biotic factors such as which species arrived earlier, and the frequency of disturbances.
- Links to Mosses 23.3, Lichens 24.6, Nitrogen-fixing bacteria 29.2

Figure 46.19 One observed pathway of primary succession in Alaska's Glacier Bay region. (**a**) As a glacier retreats, meltwater leaches minerals from the rocks and gravel left behind. (**b**) Pioneer species include lichens, mosses, and some flowering plants such as mountain avens (*Dryas*), which associate with nitrogen-fixing bacteria. Within 20 years, alder, cottonwood, and willow seedlings take hold. Alders also have nitrogen-fixing symbionts. (**c**) Within 50 years, alders form dense, mature thickets in which cottonwood, hemlock, and a few evergreen spruce grow. (**d**) After 80 years, western hemlock and spruce crowd out alders. (**e**) In areas deglaciated for more than a century, tall Sitka spruce are the predominant species.

Successional Change

Species composition of a community can change over time. Species often alter the habitat in ways that allow other species to come in and replace them. We call this type of change ecological succession.

The process of succession starts with the arrival of **pioneer species**, which are opportunistic colonizers of new or newly vacated habitats. Pioneers species have high dispersal rates, grow and mature fast, and produce many offspring. Later, other species replace the pioneers. Then replacements are replaced, and so on.

Primary succession is a process that begins when pioneer species colonize a barren habitat with no soil, such as a new volcanic island or land exposed by the retreat of a glacier (Figure 46.19). The earliest pioneers to colonize a new habitat are often mosses and lichens (Sections 23.3 and 24.6). They are small, have a brief life cycle, and can tolerate intense sunlight, extreme temperature changes, and little or no soil. Some hardy, annual flowering plants with wind-dispersed seeds are also among the pioneers.

Pioneers help build and improve the soil. In doing so, they may set the stage for their own replacement. Many pioneer species partner with nitrogen-fixing bacteria, so they can grow in nitrogen-poor habitats. Seeds of later species find shelter inside mats of the pioneers. Organic wastes and remains accumulate and, by adding volume and nutrients to soil, this material helps other species take hold. Later successional species often shade and eventually displace earlier ones.

In **secondary succession**, a disturbed area within a community recovers. If improved soil is still present, secondary succession can be fast. It commonly occurs in abandoned fields, burned forests, and tracts of land where plants were killed by volcanic eruptions.

Factors Affecting Succession

When the concept of ecological succession was first developed in the late 1800s, it was thought to be a predictable and directional process. Physical factors such as climate, altitude, and soil type were considered to be the main determinants of which species appeared in what order during succession. Also by this view, succession culminates in a "climax community," an array of species that will persist over time and will be reconstituted in the event of a disturbance.

Ecologists now realize that the species composition of a community changes frequently, in unpredictable ways. Communities do not journey along a well-worn path to some predetermined climax state.

Figure 46.20 A natural laboratory for succession after the 1980 Mount Saint Helens eruption (**a**). The community at the base of this Cascade volcano was destroyed. (**b**) In less than a decade, pioneer species came in. (**c**) Twelve years later, seedlings of a dominant species, Douglas firs, took hold.

Random events can determine the order in which species arrive in a habitat and thus affect the course of succession. Arrival of a certain species may make it easier or more difficult for others to take hold. As an example, surf grass can only grow along a shoreline if algae have already colonized that area. The algae act as an anchoring site for the grass. In contrast, when sagebrush gets established in a dry habitat, chemicals it secretes into the soil keep most other plants out.

Ecologists had an opportunity to investigate these factors after the 1980 eruption of Mount Saint Helens leveled about 600 square kilometers (235 square miles) of forest in Washington State (Figure 46.20). Ecologists recorded the natural pattern of colonization. They also carried out experiments in plots inside the blast zone. They added seeds of certain pioneer species to some plots and left other plots seedless. The results showed that some pioneers helped other later arriving plants become established. Different pioneers kept the same late arrivals out.

Disturbances also can influence the species composition in communities. According to the **intermediate disturbance hypothesis**, species richness is greatest in communities where disturbances are moderate in their intensity or frequency. In such habitats, there is enough time for new colonists to arrive and become established but not enough for competitive exclusion to cause extinctions:

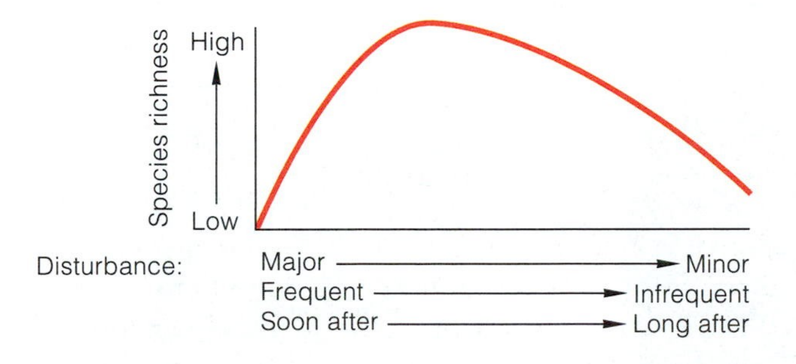

In short, the modern view of succession holds that the species composition of a community is affected by (1) physical factors such as soil and climate, (2) chance events such as the order in which species arrive, and (3) the extent of disturbances in a habitat. Because the second and third factors may vary even between two geographically close regions, it is generally difficult to predict exactly what any given community will look like at any point in the future.

Take-Home Message

What is succession?

- Succession, a process in which one array of species replaces another over time. It can occur in a barren habitat (primary succession), or a region in which a community previously existed (secondary succession).
- Chance events make successional changes difficult to predict.

46.9 Species Interactions and Community Instability

- The loss or addition of even one species may destabilize the number and abundances of species in a community.
- Link to Sudden oak death 22.8

The Role of Keystone Species

As you read earlier, short-term physical disturbances can influence the species composition of a community. Long-term changes in climate or some other environmental variable also have an effect. In addition, a shift in species interactions can dramatically alter the community by favoring some species and harming others.

The uneasy balance of forces in a community comes into focus when we observe the effects of a keystone species. A **keystone species** has a disproportionately large effect on a community relative to its abundance. Robert Paine was the first to describe the effect of a keystone species after his experiments on the rocky shores of California's coast. Species living in the rocky intertidal zone withstand pounding surf by clinging to rocks. A rock to cling to is a limiting factor. Paine set up control plots with the sea star *Pisaster ochraceus* and its main prey—chitons, limpets, barnacles, and mussels. In experimental plots he removed all sea stars.

Mussels (*Mytilus*) happen to be the prey of choice for sea stars. In the absence of sea stars, they took over Paine's experimental plots; they became the strongest competitors and crowded out seven other species of invertebrates. In this intertidal zone, predation by sea stars normally keeps the number of prey species high because it restricts competitive exclusion by mussels. Remove all the sea stars, and the community shrinks from fifteen species to eight.

The impact of a keystone species can vary between habitats that differ in their species arrays. Periwinkles (*Littorina littorea*) are alga-eating snails that live in the intertidal zone. Jane Lubchenco found removing them can increase *or* decrease the diversity of algal species, depending on the habitat (Figure 46.21).

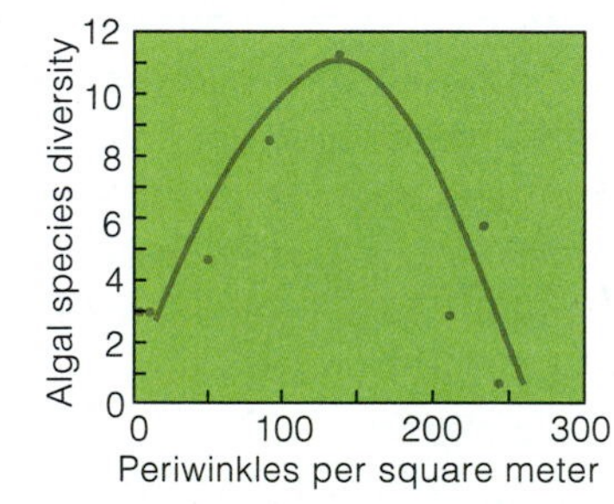

d Algal diversity in tidepools

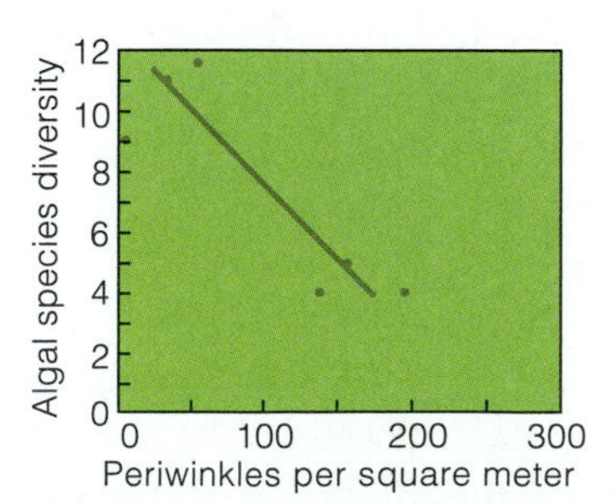

e Algal diversity on rocks that become exposed at high tide

Figure 46.21 Effect of competition and predation in an intertidal zone. (**a**) Grazing periwinkles (*Littorina littorea*) affect the number of algal species in different ways in different marine habitats. (**b**) *Chondrus* and (**c**) *Enteromorpha*, two kinds of algae in their natural habitats. (**d**) By grazing on the dominant alga in tidepools (*Enteromorpha*), the periwinkles promote the survival of less competitive algal species that would otherwise be overgrown. (**e**) *Enteromorpha* does not grow on rocks. Here, *Chondrus* is dominant. Periwinkles find *Chondrus* tough and dine instead on less competitive algal species. By doing so, periwinkles decrease the algal diversity on the rocks.

Table 46.2 Outcomes of Some Species Introductions Into the United States

Species Introduced	Origin	Mode of Introduction	Outcome
Water hyacinth	South America	Intentionally introduced (1884)	Clogged waterways; other plants shaded out
Dutch elm disease:			
Ophiostoma ulmi (fungus)	Asia (by way of Europe)	Accidental; on infected elm timber (1930)	Millions of mature elms destroyed
Bark beetle (vector)		Accidental; on unbarked elm timber (1909)	
Chestnut blight fungus	Asia	Accidental; on nursery plants (1900)	Nearly all eastern American chestnuts killed
Zebra mussel	Russia	Accidental; in ballast water of ship (1985)	Clogged pipes and water intake valves of power plants; displaced native bivalves in Great Lakes
Japanese beetle	Japan	Accidental; on irises or azaleas (1911)	Close to 300 plant species (e.g., citrus) defoliated
Sea lamprey	North Atlantic	Accidental; on ship hulls (1860s)	Trout, other fish species destroyed in Great Lakes
European starling	Europe	Intentional release, New York City (1890)	Outcompetes native cavity-nesting birds; crop damage; swine disease vector
Nutria	South America	Accidental release of captive animals being raised for fur (1930)	Crop damage, destruction of levees, overgrazing of marsh habitat

In tidepools, periwinkles prefer to eat a certain alga (*Enteromorpha*) which can outgrow other algal species. By keeping that alga in check, periwinkles help other, less competitive algal species survive. On rocks of the lower intertidal zone, *Chondrus* and other tough, red algae dominate. Here, periwinkles preferentially graze on competitively weaker algae. Periwinkles promote species richness in tidepools but reduce it on rocks.

Not all keystone species are predators. For example, beavers can be a keystone species. These large rodents cut down trees by gnawing through their trunks. Some of the felled trees are used to build dams that create a pool where only a shallow stream would otherwise exist. Thus the presence of beavers affects which types of fish and aquatic invertebrates are present.

Species Introductions Can Tip the Balance

Instabilities are also set in motion when residents of an established community move out from their home range, then successfully take up residence elsewhere. This type of directional movement, called geographic dispersal, happens in three ways.

First, over a number of generations, a population might expand its home range by slowly moving into any outlying regions that prove hospitable. Second, a population might be moved away from a home range by continental drift, at an almost imperceptibly slow pace over long spans of time. Third, some individuals might be rapidly transported across great distances, an event called jump dispersal. Birds that travel long distances facilitate such jumps by carrying seeds of plants. For some time now, humans have been a major cause of jump dispersal. They have introduced species that benefit them, as by bringing crop plants from the Americas to Europe. They have also unknowingly transported stowaways, as when Asian long-horned beetles were imported along with wood products.

When you hear someone speaking enthusiastically about exotic species, you can safely bet the speaker is not an ecologist. An **exotic species** is a resident of an established community that dispersed from its home range and became established elsewhere. Unlike most imports, which never do take hold outside the home range, an exotic species permanently insinuates itself into a new community.

In its new locale, the exotic species is often untroubled by competitors, predators, parasites, and diseases that kept it in check back home. Freed from its usual constraints, the exotic species can often outcompete similar species native to its new habitat.

You have already learned how some imports are affecting community structure. The chapter introduction described how red imported fire ants that arrived from South America outcompete North American ant species. Sudden oak death, described in Section 22.8, is caused by a protist from Asia. A parasite from Europe is the cause of whirling disease in trout. The list of detrimental exotic species is depressingly long. Table 46.2 lists some well-known imports, and the next section describes four others in some detail.

Take-Home Message

How can a single species affect community structure?

- A keystone species is one that has a major effect on species richness and relative abundances in a habitat.
- Removal of a keystone species or introductions of an exotic species can affect the types and abundances of species in a community.

46.10 Exotic Invaders

- Nonnative species introduced by human activities are affecting native communities on every continent.
- Link to Green algae 22.9

Battling Algae

The long, green, feathery branches of *Caulerpa taxifolia* look great in saltwater aquariums, so researchers at the Stuttgart Aquarium in Germany developed a sterile strain of this green alga and shared it with other marine institutions. Was it from Monaco's Oceanographic Museum that the hybrid strain escaped into the wild? Some say yes, Monaco says no.

In any case, a small patch of the aquarium strain was found growing in the Mediterranean near Monaco in 1984. Boat propellers and fishing nets dispersed the alga, and it now blankets tens of thousands of acres of sea floor in the Mediterranean and Adriatic (Figure 46.22*a*).

Just how bad is *C. taxifolia?* The aquarium strain can thrive on sandy or rocky shores and in mud. It can live ten days after being discarded in meadows. Unlike its tropical parents, it can also survive in cool water and polluted water. It has the potential to displace endemic algae, overgrow reefs, and destroy marine food webs. Its success is due in part to production of a toxin (Caulerpenyne) that poisons invertebrates and fishes, including algae eaters that keep other algae in check.

In 2000, scuba divers discovered *C. taxifolia* growing near the southern California coast. Someone might have drained water from a home aquarium into a storm drain or into the lagoon itself. The government and private groups quickly sprang into action. So far, eradication and surveillance programs have worked, but at a cost of more than $3.4 million.

Importing *C. taxifolia* or any closely related species of *Caulerpa* into the United States is now illegal. To protect native aquatic communities, aquarium water should never be dumped into storm drains or waterways. It should be discarded into a sink or toilet so wastewater treatment can kill any algal spores (Section 22.9).

The Plants That Overran Georgia

In 1876, kudzu (*Pueraria montana*) was introduced to the United States from Japan. In its native habitat, this perennial vine is a well-behaved legume with an extensive root system. It *seemed* like a good idea to use it for forage and to control erosion on slopes. But kudzu grew faster in the American Southeast. No native herbivores or pathogens were adapted to attack it. Competing plant species posed no serious threat to it.

With nothing to stop it, kudzu can grow 60 meters (200 feet) per year. Its vines now blanket streambanks, trees, telephone poles, houses, and almost anything else in their path (Figure 46.22*b*). Kudzu withstands burning, and grows back from its deep roots. Grazing goats and herbicides help. But goats eat most other plants along with it, and herbicides taint freshwater supplies. Kudzu invasions now stretch from Connecticut down to Florida and are reported in Arkansas. It crossed the Mississippi River into Texas. Thanks to jump dispersal, it is now an invasive species in Oregon.

Figure 46.22 (**a**) Aquarium strain of *Caulerpa taxifolia* suffocating yet another richly diverse marine ecosystem.

(**b**) Kudzu (*Pueraria montana*) taking over part of Lyman, South Carolina. This vine has become invasive in many states from coast to coast. Ruth Duncan of Alabama (*above*), who makes 200 kudzu vine baskets a year, can't keep up.

Figure 46.23 Rabbit-proof fence? Not quite. This photo shows part of a fence built in 1907 to hold back rabbits that were wreaking havoc with the vegetation in Australia. The fence did not solve the rabbit problem, but it did restrict movements of native wildlife such as kangaroos and emus.

On the bright side, Asians use a starch extracted from kudzu in drinks, herbal medicines, and candy. A kudzu processing plant in Alabama may export this starch to Asia, where the demand currently exceeds the supply. Also, kudzu may help save forests; it can be an alternative source for paper and other wood products. Today, about 90 percent of Asian wallpaper is kudzu-based.

The Rabbits That Ate Australia

During the 1800s, British settlers in Australia just could not bond with koalas and kangaroos, and so they imported familiar animals from home. In 1859, in what would be the start of a major ecological disaster, a landowner in northern Australia imported and released two dozen European rabbits (*Oryctolagus cuniculus*). Good food and great sport hunting—that was the idea. An ideal rabbit habitat with no natural predators—that was the reality.

Six years later, the landowner had killed 20,000 rabbits and was besieged by 20,000 more. The rabbits displaced livestock and caused the decline of native wildlife. Now 200 million to 300 million are hippity-hopping through the southern half of the country. They graze on grasses in good times and strip bark from shrubs and trees during droughts. Thumping hordes turn shrublands as well as grasslands into eroded deserts. Their burrows undermine the soil and set the stage for widespread erosion.

Rabbits have been shot and their warrens fumigated, plowed under, and dynamited. The first assaults killed 70 percent of them, but the rabbits rebounded in less than a year. When a fence 2,000 miles long was built to protect western Australia, rabbits made it from one side to the other before workers could finish the job (Figure 46.23).

In 1951, the government introduced a myxoma virus that normally infects South American rabbits. The virus causes myxomatosis. This disease has mild effects on its coevolved host but nearly always kills *O. cuniculus*. Fleas and mosquitoes transmit the virus to new hosts. With no coevolved defenses against the import, European rabbits died in droves. But natural selection has since favored a rise in rabbit populations resistant to the imported virus.

In 1991, on an uninhabited island in Australia's Spencer Gulf, researchers released rabbits that were injected with a calicivirus. The rabbits died from blood clots in their lungs, heart, and kidneys. Then, in 1995, the test virus escaped from the island to the mainland, perhaps on insect vectors.

The combination of the two imported viruses, along with traditional control methods has brought the rabbit population under control. There still are some rabbits, but vegetation is growing back and native herbivores are increasing in numbers.

Gray Squirrels Versus Red Squirrels

The eastern gray squirrel (*Sciurus carolinensis*) is native to eastern North America, where it is a welcome sight in forests, yards, and parks. It has become similarly common throughout Britain and parts of Italy where it has been introduced. Here, the squirrel is considered an exotic pest that has thrived at the expense of Europe's native red squirrel (*Sciurus vulgaris*). In Britain, the imported grays now outnumber the native reds 66 to 1.

The gray squirrels are at an advantage over their European cousins because they excel at detecting and stealing nuts that red squirrels stored for the winter. In addition, gray squirrels carry and spread a virus that kills Britain's red squirrels, but are not themselves affected by the virus.

To protect the remaining red squirrels, the British have begun trapping and killing gray squirrels. Efforts are also under way to develop a contraceptive drug that would be effective against grays, but not the native reds.

46.11 Biogeographic Patterns in Community Structure

- The richness and relative abundances of species differ from one habitat or region of the world to another.
- Link to Biogeography 17.1

Biogeography is the scientific study of how species are distributed in the natural world (Section 17.1). We see patterns that correspond with differences in sunlight, temperature, rainfall, and other factors that vary with latitude, elevation, or water depth. Still other patterns relate to the history of a habitat and the species in it. Each species has its own unique physiology, capacity for dispersal, resource requirements, and interactions with other species.

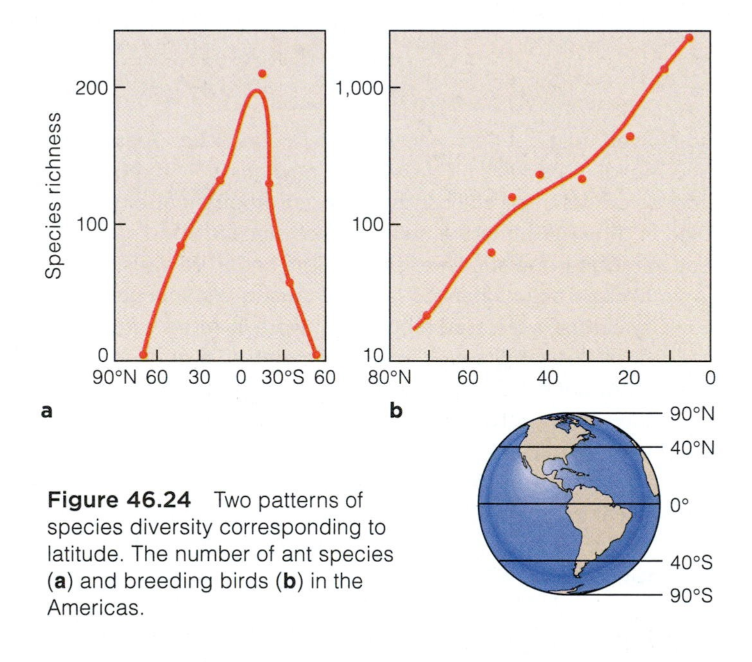

Figure 46.24 Two patterns of species diversity corresponding to latitude. The number of ant species (**a**) and breeding birds (**b**) in the Americas.

Mainland and Marine Patterns

Perhaps the most striking pattern of species richness corresponds with distance from the equator. For most major plants and animal groups, the number of species is greatest in the tropics and declines from the equator to the poles. Figure 46.24 illustrates two examples of this pattern. Consider just a few factors that help bring about such a pattern and maintain it.

First, for reasons explained in Section 48.1, tropical latitudes intercept more intense sunlight and receive more rainfall, and their growing season is longer. As one outcome, resource availability tends to be greater and more reliable in the tropics than elsewhere. One result is a degree of specialized interrelationships not possible where species are active for shorter periods.

Second, tropical communities have been evolving for a long time. Some temperate communities did not start forming until the end of the last ice age.

Third, species richness may be self-reinforcing. The number of species of trees in tropical forests is much greater than in comparable forests at higher latitudes. Where more plant species compete and coexist, more species of herbivores also coexist, partly because no single herbivore species can overcome all the chemical defenses of all plants. In addition, more predators and parasites can evolve in response to more kinds of prey and hosts. The same principles apply to tropical reefs.

Island Patterns

As you saw in Section 45.4, islands are laboratories for population studies. They have also been laboratories for community studies. For instance, in the mid-1960s volcanic eruptions formed a new island 33 kilome-

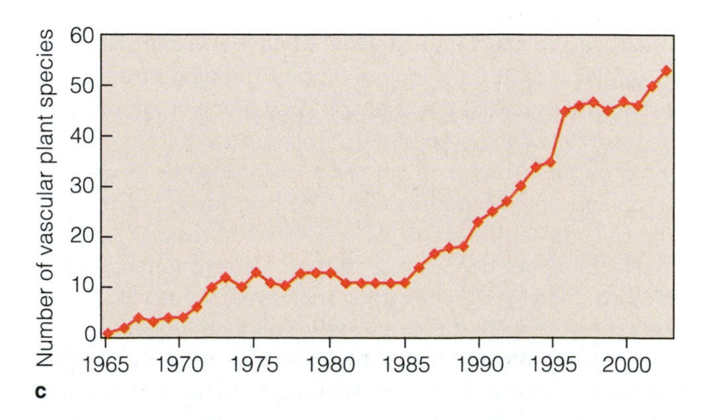

Figure 46.25 Surtsey, a volcanic island, at the time of its formation (**a**) and in 1983 (**b**). The graph (**c**) shows the number of vascular plant species found in yearly surveys. Sea gulls began nesting on the island in 1986.

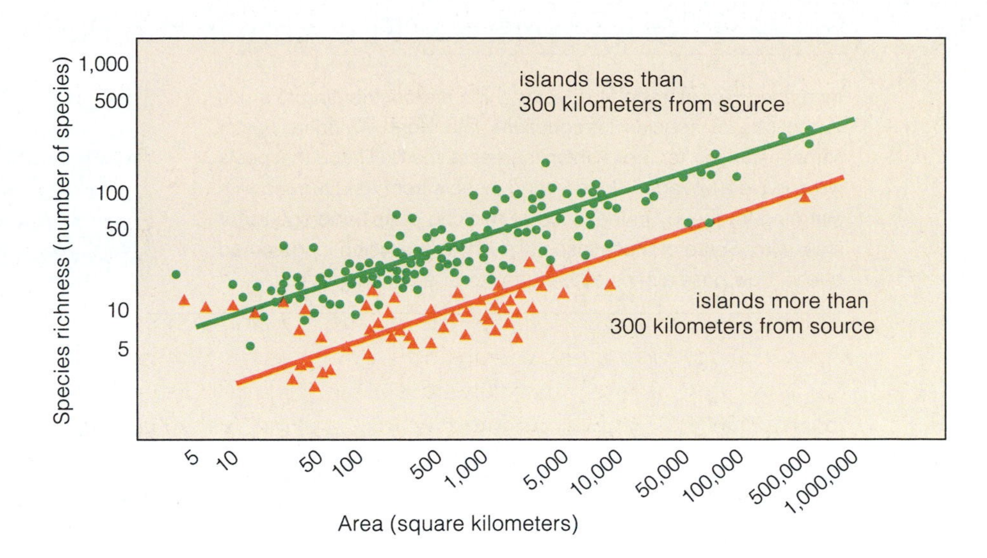

Figure 46.26 Island biodiversity patterns.

Distance effect: Species richness on islands of a given size declines as distance from a source of colonists rises. *Green* circles are values for islands less than 300 kilometers from the colonizing source. *Orange* triangles are values for islands more than 300 kilometers (190 miles) from a source of colonists.

Area effect: Among islands the same distance from a source of colonists, larger islands tend to support more species than smaller ones.

Figure It Out: Which is likely to have more species, a 100-km^2 island more than 300 km from a colonizing source or a 500-km^2 island less than 300 km from a colonist source? *Answer: The 500-km^2 island*

ters (21 miles) from the coast of Iceland. The island was named Surtsey (Figure 46.25). Bacteria and fungi were early colonists. The first vascular plant became established on the island in 1965. Mosses appeared two years later and thrived (Figure 46.25*b*). The first lichens were found five years after that. The rate of arrivals of new vascular plants picked up considerably after a seagull colony became established in 1986 (Figure 46.25*c*). This example illustrates the important role birds play in introducing species to islands.

The number of species on Surtsey will not continue increasing forever. Can we estimate how many species there will be when the number levels off? The **equilibrium model of island biogeography** addresses this question. According to this model, the number of species living on any island reflects a balance between immigration rates for new species and extinction rates for established ones. The distance between an island and a mainland source of colonists affects immigration rates. An island's size affects both immigration rates and extinction rates.

Consider first the **distance effect**: Islands far from a source of colonists receive fewer immigrants than those closer to a source. Most species cannot disperse very far, so they will not turn up far from a mainland.

Species richness also is shaped by the **area effect**: Big islands tend to support more species than small ones. More colonists will happen upon a larger island simply by virtue of its size. Also, big islands are more likely to offer a variety of habitats, such as high and low elevations. These options make it more likely that a new arrival will find a suitable habitat. Finally, big islands can support larger populations of species than small islands. The larger a population, the less likely it is to become locally extinct as the result of some random event.

Figure 46.26 illustrates how interactions between the distance effect and the area effect can influence the number of species on islands.

Robert H. MacArthur and Edward O. Wilson first developed the equilibrium model of island biogeography in the late 1960s. Since then it has been modified and its use has been expanded to help scientists think about habitat islands—natural settings surrounded by a "sea" of degraded habitat. Many parks and wildlife preserves fit this description. Island-based models can help estimate the size of an area that must be set aside as a protected reserve to ensure survival of a species.

One more note about island communities: An island often differs from its source of colonists in physical aspects, such as rainfall and soil type. It also differs with regard to species array; not all species reach the island. As a result of these differences, a population on an island often faces different selection pressures than its same-species relatives on the mainland and evolves in a different way as a result.

In a pattern that is the opposite of character displacement, a species may find itself on an island that lacks a major competitor found on the mainland. In the absence of this competition, traits of the island population may become more like those of the competitor that it left behind.

Take-Home Message

What are some biogeographic patterns in species richness?

- Generally, species richness is highest in the tropics and lowest at the poles. Tropical habitats have conditions that more species can tolerate, and tropical communities have often been evolving for longer than temperate ones.
- When a new island forms, species richness rises over time and then levels off. The size of an island and its distance from a colonizing source influence its species richness.

IMPACTS, ISSUES REVISITED Fire Ants in the Pants

Increased global trade and faster ships are contributing to a rise in the rate of species introductions into North America. Faster ships mean shorter trips, which increases the likelihood that pests will survive a voyage. Wood-eating insects from Asia turn up with alarming frequency in the wood of packing crates and spools for steel wire. Some of these insects, such as the Asian long-horned beetle, now pose a serious threat to North America's forests.

How would you vote?

Is inspecting more imported goods to detect potentially harmful exotic species worth the added cost? See CengageNOW for details, then vote online.

Summary

Section 46.1 Each species occupies a certain **habitat** characterized by physical and chemical features and by the array of other species living in it. All populations of all species in a habitat are a **community**. Each species in a community has its own **niche**, or way of living. Species interactions between members of a community include **commensalism**, which does not help or harm either species, **mutualism**, which benefits both species, **interspecific competition**, which harms both species, and **parasitism** and **predation**, in which one species benefits at the expense of another. Commensalism, mutualism, and parasitism may be a **symbiosis**, in which species live together. Interacting species undergo **coevolution**.

Section 46.2 In a mutualism, two species interact and both benefit. Some mutualists cannot complete their life cycle without the interaction.

Section 46.3 By the process of **competitive exclusion**, one species outcompetes a rival with the same resource needs, driving it to extinction. **Character displacement** makes competing species less similar, which facilitates **resource partitioning**.

- *Use the animation on CengageNOW to learn about competitive interactions.*

Sections 46.4, 46.5 **Predators** are free-living and usually kill their **prey**. Predator and prey numbers often fluctuate in cycles. Carrying capacity, predator behavior, and availability of other prey affect these cycles. Predators and their prey exert selection pressure on one another. Evolutionary results of such selection include **warning coloration**, **camouflage**, and **mimicry**.

- *Use the interaction on CengageNOW to learn about three alternative models for predator responses to prey density.*

Sections 46.6, 46.7 **Parasites** live in or on a host and withdraw nutrients from its tissues. Hosts may or may not die as a result. An animal **vector** often carries the parasite between hosts. **Parasitoids** lay eggs on a host, then their larvae devour the host. **Social parasites** manipulate some aspect of a host's behavior.

Section 46.8 Ecological succession is the sequential replacement of one array of species by another over time. **Primary succession** happens in new habitats. **Secondary succession** occurs in disturbed ones. The first species of a community are **pioneer species**. The pioneers may help, hinder, or have no effect on later colonists.

The older idea that all communities eventually reach a predictable climax state has been replaced by models that emphasize the role of chance and disturbances. The **intermediate disturbance hypothesis** holds that disturbances of moderate intensity and frequency maximize species diversity.

Sections 46.9, 46.10 Community structure reflects an uneasy balance of forces that operate over time. Major forces are competition and predation. **Keystone species** are especially important in maintaining the composition of a community. The removal of a keystone species or introduction of an **exotic species**—one that evolved in a different community—can alter community structure in ways that may be permanent.

Section 46.11 Species richness, the number of species in a given area, varies with latitude, elevation, and other factors. Tropical regions tend to have more species than higher latitude regions. The **equilibrium model of island biogeography** helps ecologists estimate the number of species that will become established on an island. The **area effect** is the tendency of large islands to have more species than small islands. The **distance effect** is the tendency of islands near a source of colonists to have more species than distant islands.

- *Learn about the area effect and distance effect with the interaction on CengageNOW.*

Self-Quiz

Answers in Appendix III

1. A habitat ______.
   a. has distinguishing physical and chemical features
   b. is where individuals of a species normally live
   c. is occupied by various species
   d. all of the above
2. A species' niche includes its ______.
   a. habitat requirements
   b. food requirements
   c. reproductive requirements
   d. all of the above
3. Which cannot be a symbiosis?
   a. mutualism
   b. parasitism
   c. commensalism
   d. interspecific competition

Data Analysis Exercise

Ant-decapitating phorid flies are just one of the biological control agents used to battle imported fire ants. Researchers have also enlisted the help of *Thelohania solenopsae*, another natural enemy of the ants. This microsporidian is a parasite that infects ants and shrinks the ovaries of the colony's egg-producing female (the queen). As a result, a colony dwindles in numbers and eventually dies out.

Are these biological controls useful against imported fire ants? To find out, USDA scientists treated infested areas with either traditional pesticides or pesticides plus biological controls (both flies and the parasite). The scientists left some plots untreated as controls. Figure 46.27 shows the results.

1. How did population size in the control plots change during the first four months of the study?

2. How did population size in the two types of treated plots change during this same interval?

3. If this study had ended after the first year, would you conclude that biological controls had a major effect?

4. How did the two types of treatment (pesticide alone versus pesticide plus biological controls) differ in their longer-term effects?

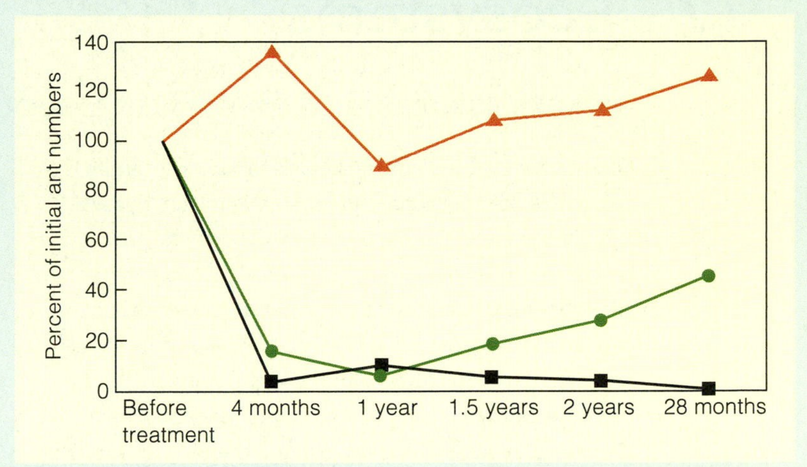

Figure 46.27 Effects of two methods of controlling red imported fire ants. The graph shows the numbers of red imported fire ants over a 28-month period. *Orange* triangles represent untreated control plots. *Green* circles are plots treated with pesticides alone. *Black* squares are plots treated with pesticide and biological control agents (phorid flies and a microsporidian parasite).

4. Lizards and songbirds that share a habitat and both eat flies are an example of ______ competition.
   a. exploitative
   b. interference
   c. intraspecific
   d. interspecific
   e. both a and d

5. With character displacement, two competing species become ______ .
   a. more alike
   b. less alike
   c. symbionts
   d. extinct

6. Predator and prey populations ______ .
   a. always coexist at relatively stable levels
   b. may undergo cyclic or irregular changes in density
   c. cannot coexist indefinitely in the same habitat
   d. both b and c

7. Match the terms with the most suitable descriptions.
   ___predation
   ___mutualism
   ___commensalism
   ___parasitism
   ___interspecific competition
   a. one free-living species feeds on another and usually kills it
   b. two species interact and both benefit by the interaction
   c. two species interact and one benefits while the other is neither helped nor harmed
   d. one species feeds on another but usually does not kill it
   e. two species attempt to utilize the same resource

8. By a currently favored hypothesis, species richness of a community is greatest between physical disturbances of ______ intensity or frequency.
   a. low
   b. intermediate
   c. high
   d. variable

9. True or false? Parasitoids usually live inside their host without killing it.

10. Match the terms with the most suitable descriptions.
   ___geographic dispersal
   ___area effect
   ___pioneer species
   ___climax community
   ___keystone species
   ___exotic species
   ___resource partitioning
   a. opportunistic colonizer of barren or disturbed habitat
   b. greatly affects other species
   c. individuals leave home range, become established elsewhere
   d. more species on large islands than small ones at same distance from the source of colonists
   e. array of species at the end of successional stages in a habitat
   f. allows competitors to coexist
   g. often outcompete, displace native species of established community

■ *Visit CengageNOW for additional questions.*

Critical Thinking

1. With antibiotic resistance rising, researchers are looking for ways to reduce use of these drugs. Some cattle once fed antibiotic-laced food now get probiotic feed that can bolster populations of helpful bacteria in the animal's gut. The idea is that if a large population of beneficial bacteria is in place, then harmful bacteria cannot become established or thrive. Which ecological principle is guiding this research?

2. Flightless birds that live on islands often have relatives on the mainland that can fly. The island species presumably evolved from fliers that, in the absences of predators, lost their ability to fly. Many flightless birds on islands are now declining because rats and other predators have been introduced to their previously isolated island. Despite the change in selective pressure, no flightless island bird has yet regained the ability to fly. Why is this unlikely to happen?

47 Ecosystems

IMPACTS, ISSUES Bye-Bye, Blue Bayou

Each Labor Day, the coastal Louisiana town of Morgan City celebrates the Louisiana Shrimp and Petroleum Festival. The state is the nation's top shrimp harvester and the third-largest producer of petroleum, which is refined into gasoline and other fossil fuels. But the petroleum industry's success may be contributing indirectly to the decline of the state's fisheries. Why? The lower atmosphere is warming up, and fossil fuel burning is one of the causes (Section 7.9). As the climate heats up, the ocean's surface waters get warmer and expand, glaciers melt, and sea level rises.

If current trends continue, some coastal lowlands will be submerged. With more than 40 percent of the nation's salt-water wetlands, Louisiana has the most to lose. This state's coastal marshes, or bayous, are already in danger. Dams and levees keep back sediments that would normally be deposited in the marshes. Since the 1940s, Louisiana has lost an area of marshland the size of Rhode Island (Figure 47.1).

Louisiana's marshes are an ecological treasure. Millions of migratory birds overwinter there. The marshes are also the source of more than $3.5 billion worth of fish, shrimp, and shellfish. If the marshes disappear, so will the revenue.

Equally troubling is what will happen to low-lying towns and cities along the coasts after the marshes are gone. Then, there will be nothing to buffer devastating storm surges that threaten the coasts during hurricanes.

In 2005, the category 5 hurricane Katrina slammed into the Gulf Coast. High winds and flooding ruined countless buildings, and more than 1,700 people died. Climate change models suggest that if temperatures continue to rise, more hurricanes are likely to reach category 5 status.

The models also indicate that warming seas will promote overgrowth of algae, which can kill fish. Warmer water can encourage growth of many types of pathogenic bacteria, so more people are expected to become sick after swimming in contaminated water, or eating shellfish harvested from it.

Inland, heat waves are becoming more intense as global temperatures rise, and more people are dying of heat stroke. Fueled by rising temperatures and extended dry seasons, wildfires are becoming more frequent and more devastating. Disease-spreading mosquitoes are now spreading into regions that were too cold for them even a few years ago.

This chapter is about the flow of energy and nutrients through ecosystems. It will give you the tools to do some of your own critical thinking about human impacts on Earth's environments. We have become major players in the global flows of energy and nutrients even before we fully understand how ecosystems work. Decisions we make today about global climate change and other environmental issues are likely to shape Earth's environments—and the quality of human life—far into the future.

See the video! Figure 47.1 *Left*, Fishing camp in Louisiana. It was built in a once-thriving marsh that has since given way to the open waters of Barataria Bay.

Above, a marsh restoration project in Louisiana's Sabine National Wildlife Refuge. In marshland that has become open water, sediments are barged in and marsh grasses are planted on them.

Key Concepts

Organization of ecosystems

An ecosystem consists of a community and its physical environment. A one-way flow of energy and a cycling of raw materials among its interacting participants maintain it. It is an open system, with inputs and outputs of energy and nutrients. Section 47.1

Food webs

Food chains are linear sequences of feeding relationships. Food chains cross-connect as food webs. Most of the energy that enters a food web returns to the environment, mainly as metabolic heat. Nutrients are recycled within the food web. Section 47.2

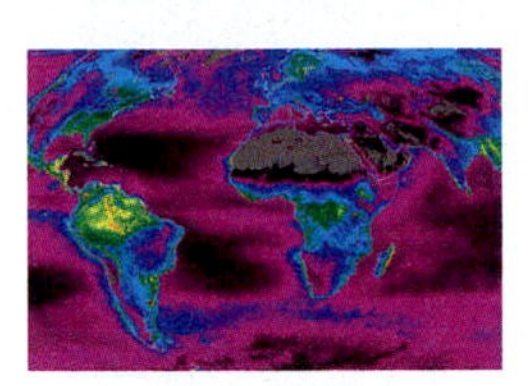

Energy and materials flow

Ecosystems differ in how much energy their producers capture and how much is stored in each trophic level. Some toxins that enter an ecosystem can become increasingly concentrated as they pass from one trophic level to another. Sections 47.3, 47.4

Cycling of water and nutrients

The availability of water, carbon, nitrogen, phosphorus, and other substances influences primary productivity. These substances move slowly in global cycles, from environmental reservoirs, into food webs, then back to reservoirs. Sections 47.5–47.10

Links to Earlier Concepts

- This chapter builds on your understanding of the laws of thermodynamics (Section 6.1). We discuss ecological roles of producers such as phytoplankton (22.7), and of decomposers (21.6 and 24.5).
- You will be reminded of the importance of water to the world of life (2.5) and how transpiration works (29.3). We also revisit the effects of acid rain (2.6) and the role of water in leaching nutrients (29.1).
- You will see how nitrogen fixation (21.6 and 29.2) plays an essential role in nutrient cycles and how excess nitrogen contributes to algal blooms (22.5).

  You will also learn more about carbon imbalances (7.9), and be reminded that carbon is stored in peat bogs (23.3) and the shells of protists such as foraminiferans (22.3). You will also hear again about attempts to control the protist-caused disease malaria (22.6).
- Discussions of nutrient cycles will also draw on your knowledge of tectonic plates (17.9).

How would you vote? Exhaust from motor vehicles contains greenhouse gases. The better mileage a vehicle gets, the fewer greenhouse gases it emits per mile. Should minimum fuel economy standards for cars and trucks be increased? See CengageNOW for details, then vote online.

47.1 The Nature of Ecosystems

- In an ecosystem, energy and nutrients from the environment flow among a community of species.
- Links to Laws of thermodynamics 6.1, Leaching 29.1

Overview of the Participants

Diverse natural systems abound on Earth's surface. In climate, soil type, array of species, and other features, prairies differ from forests, which differ from tundra and deserts. Reefs differ from the open ocean, which differs from streams and lakes. Yet, despite all these differences, all systems are alike in many aspects of their structure and function.

We define an **ecosystem** as an array of organisms and a physical environment, all interacting through a one-way flow of energy and a cycling of nutrients. It is an open system, because it requires ongoing inputs of energy and nutrients to endure (Figure 47.2).

All ecosystems run on energy captured by **primary producers**. These autotrophs, or "self-feeders," obtain energy from a nonliving source—generally sunlight—and use it to build organic compounds from carbon dioxide and water. Plants and phytoplankton are the main producers. Chapter 7 explains how they capture energy from the sun to assemble sugars from carbon dioxide and water, by the process of photosynthesis.

Consumers are heterotrophs that get energy and carbon by feeding on tissues, wastes, and remains of producers and one another. We can describe consumers by their diets. Herbivores eat plants. Carnivores eat the flesh of animals. Parasites live inside or on a living host and feed on its tissues. Omnivores devour both animal and plant materials. **Detritivores**, such as earthworms and crabs, dine on small particles of organic matter, or detritus. **Decomposers** feed on organic wastes and remains and break them down into inorganic building blocks. The main decomposers are bacteria and fungi.

Energy flows one way—into an ecosystem, through its many living components, then back to the physical environment (Section 6.1). Light energy captured by producers is converted to bond energy in organic molecules, which is then released by metabolic reactions that give off heat. This is a one-way process because heat energy cannot be recycled; producers cannot convert heat into chemical bond energy.

In contrast, many nutrients are cycled within an ecosystem. The cycle begins when producers take up hydrogen, oxygen, and carbon from inorganic sources, such as the air and water. They also take up dissolved nitrogen, phosphorus, and other minerals necessary for biosynthesis. Nutrients move from producers into the consumers who eat them. After an organism dies, decomposition returns nutrients to the environment, from which producers take them up again.

Not all nutrients remain in an ecosystem; typically there are gains and losses. Mineral ions are added to an ecosystem when weathering processes break down rocks, and when winds blow in mineral-rich dust from elsewhere. Leaching and soil erosion remove minerals (Section 29.1). Gains and losses of each mineral tend to balance out over time in a healthy ecosystem.

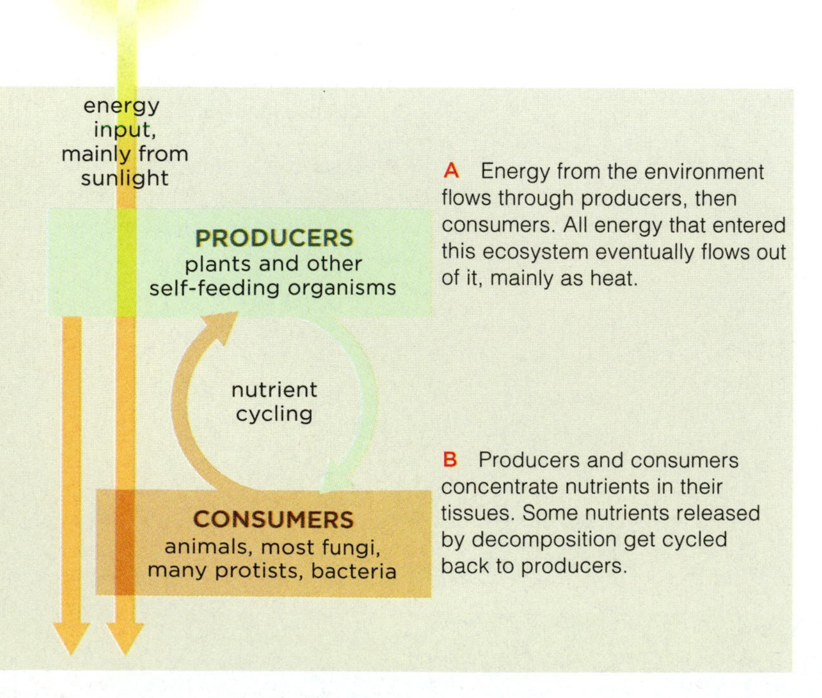

Figure 47.2 Animated Model for ecosystems on land, in which energy flow starts with autotrophs that capture energy from the sun. Energy flows one way, into and out of the ecosystem. Nutrients get cycled among producers and heterotrophs.

Trophic Structure of Ecosystems

All organisms of an ecosystem take part in a hierarchy of feeding relationships called **trophic levels** ("troph" means nourishment). When one organism eats another, energy is transferred from the eaten to the eater. All organisms at the same trophic level in an ecosystem are the same number of transfers away from the energy input into that system.

Figure 47.3 Example of a food chain and corresponding trophic levels in tallgrass prairie, Kansas.

A **food chain** is a sequence of steps by which some energy captured by primary producers is transferred to organisms at successively higher trophic levels. For example, big bluestem grass and other plants are the major primary producers in a tallgrass prairie (Figure 47.3). They are at this ecosystem's first trophic level. In one food chain, energy flows from bluestem grass to grasshoppers, to sparrows, and finally to hawks. Grasshoppers are primary consumers; they are at the second trophic level. Sparrows that eat grasshoppers are second-level consumers and at the third trophic level. Hawks are third-level consumers, and they are at the fourth trophic level.

At each trophic level, organisms interact with the same sets of predators, prey, or both. Omnivores feed at several levels, so we would partition them among different levels or assign them to a level of their own.

Identifying one food chain is a simple way to start thinking about who eats whom in ecosystems. Bear in mind, many different species usually are competing for food in complex ways. Tallgrass prairie producers (mainly flowering plants) feed grazing mammals and herbivorous insects. But many more species interact in the tallgrass prairie and in most other ecosystems, particularly at lower trophic levels. A number of food chains cross-connect with one another—as food webs—and that is the topic of the next section.

Take-Home Message

What is the trophic structure of an ecosystem?

- An ecosystem includes a community of organisms that interact with their physical environment by a one-way energy flow and a cycling of materials.
- Autotrophs tap into an environmental energy source and make their own organic compounds from inorganic raw materials. They are the ecosystem's primary producers.
- Autotrophs are at the first trophic level of a food chain, a linear sequence of feeding relationships that proceeds through one or more levels of heterotrophs, or consumers.

47.2 The Nature of Food Webs

■ All food webs consist of multiple interconnecting food chains. Ecologists who untangled the chains of many food webs discovered patterns of organization. The patterns reflect environmental constraints and the inefficiency of energy transfers from one trophic level to the next.

Interconnecting Food Chains

A **food web** diagram illustrates trophic interactions among species in one particular ecosystem. Figure 47.4 shows a small sampling of the participants in an arctic food web. Nearly all food webs include two types of food chains. In a **grazing food chain**, the energy stored

Figure 47.4 Animated A very small sampling of organisms in an arctic food web on land.

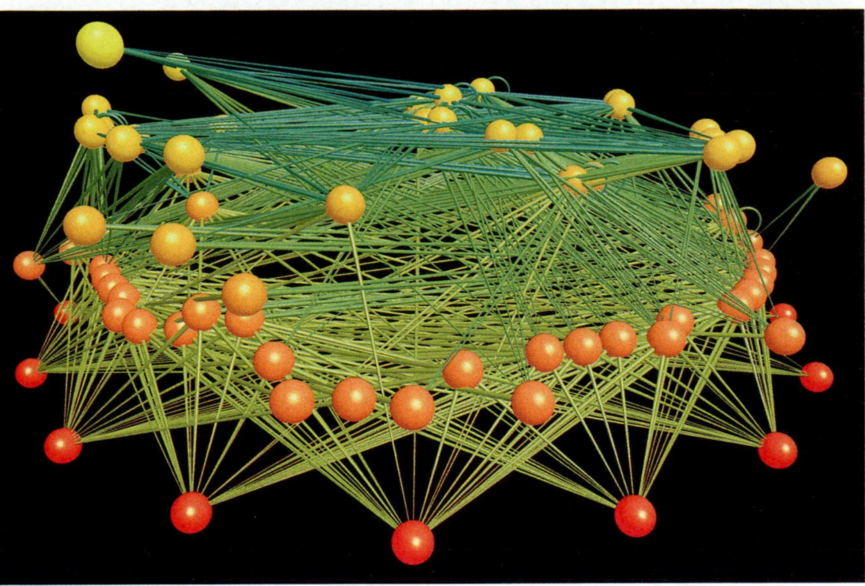

Figure 47.5 Computer model for a food web in East River Valley, Colorado. Balls signify species. Their colors identify trophic levels, with producers (coded *red*) at the bottom and predators (*yellow*) at top. The connecting lines thicken, starting from an eaten species to the eater.

in producer tissues flows to herbivores, which tend to be relatively large animals. In a **detrital food chain**, the energy in producers flows to detritivores, which tend to be smaller animals, and to decomposers.

In most land ecosystems, the bulk of the energy that becomes stored in producer tissues moves through detrital food chains. For example, in an arctic ecosystem, grazers such as voles, lemmings, and hares graze on some plant parts. However, far more plant matter becomes detritus. Bits of dead plant material sustain detritivores such as nematodes and soil-dwelling insects, and decomposers such as soil bacteria and fungi.

Grazing food chains tend to predominate in aquatic ecosystems. Zooplankton (heterotrophic protists and tiny animals that drift or swim) consume most of the phytoplankton. A smaller amount of phytoplankton ends up on the ocean floor as detritus.

Detrital food chains and grazing food chains interconnect to form the overall food web. For example, animals at higher trophic levels often eat both grazers and detritivores. Also, after grazers die, the energy in their tissues flows to detritivores and decomposers.

How Many Transfers?

When ecologists looked at food webs for a variety of ecosystems, they discovered some common patterns. For example, the energy captured by producers usually passes through no more than four or five trophic levels. Even in ecosystems with many species, the number of transfers is limited. Remember that energy transfers are not that efficient (Section 6.1). Energy losses limit the length of a food chain.

Field studies and computer simulations of aquatic and land food ecosystems reveal more patterns. Food chains tend to be shortest in habitats where conditions vary widely over time. Chains tend to be longer in stable habitats, such as the ocean depths. The most complex webs tend to have a large variety of herbivores, as in grasslands. By comparison, the food webs with fewer connections tend to have more carnivores.

Diagrams of food webs help ecologists predict how ecosystems will respond to change. Neo Martinez and his colleagues constructed the one shown in Figure 47.5. By comparing different food webs, they realized that trophic interactions connect species more closely than people thought. On average, each species in any food web was two links away from all other species. Ninety-five percent of species were within three links of one another, even in large communities with many species. As Martinez concluded in a paper discussing his findings, "Everything is linked to everything else." He cautioned that extinction of any species in a food web may have an impact on many other species.

Take-Home Message

How does energy flow affect food chains and food webs?

- Tissues of living plants and other producers are the basis for grazing food chains. Remains of producers are the basis for detrital food webs.
- Nearly all ecosystems include both grazing food chains and detrital food chains that interconnect as the system's food web.
- The cumulative energy losses from energy transfers between trophic levels limits the length of food chains.
- Even when an ecosystem has many species, trophic interactions link each species with many others.

47.3 Energy Flow Through Ecosystems

- Primary producers capture energy and take up nutrients, which then move to other trophic levels.
- Link to Phytoplankton 22.7

Capturing and Storing Energy

The flow of energy through an ecosystem begins with **primary production**: the rate at which producers (most often plants or photosynthetic protists) capture and store energy. The amount of energy captured by all producers in the ecosystem is defined as the system's gross primary production. The portion of energy that producers invest in growth and reproduction (rather than in maintenance) is net primary production.

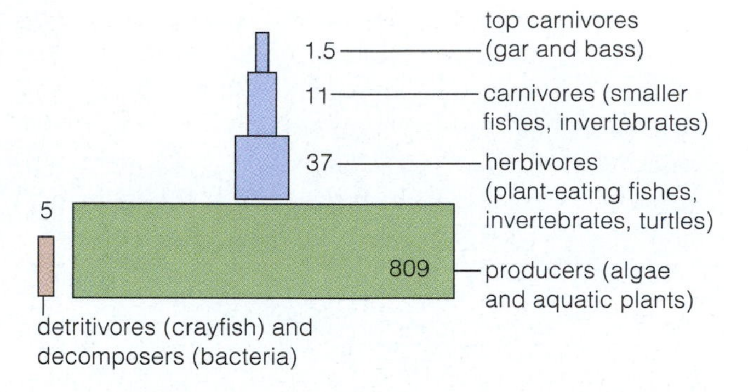

Figure 47.7 Biomass (in grams per square meter) for Silver Springs, a freshwater aquatic ecosystem in Florida. In this system, primary producers make up the bulk of the biomass.

Factors such as temperature and the availability of water and nutrients affect producer growth, and thus influence primary production. As a result, the primary production varies among habitats and may also vary seasonally (Figure 47.6). Per unit area, the net primary production on land tends to be higher than that in the oceans. However, because oceans cover about 70 percent of Earth's surface, they contribute nearly half of the global net primary productivity.

Ecological Pyramids

Ecologists often represent the trophic structure of an ecosystem in the form of ecological pyramids. In such diagrams, primary producers collectively form a base for successive tiers of consumers above them.

A **biomass pyramid** illustrates the dry weight of all organisms at each trophic level in an ecosystem. Figure 47.7 shows the biomass pyramid for Silver Springs, an aquatic ecosystem in Florida.

Most commonly, primary producers make up most of the biomass in a pyramid, and top carnivores make up very little. If you visited Silver Springs, you would see a lot of aquatic plants but very few gars (the main top predator in this ecosystem). Similarly, when you walk through a prairie, you would see more grams of grass than of hawks.

However, if producers are small and reproduce rapidly, a biomass pyramid can have its smallest tier at the bottom. For example, producers in the open ocean are

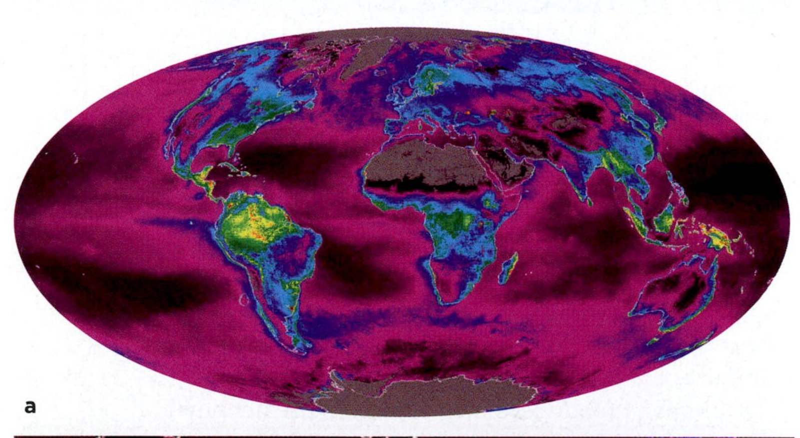

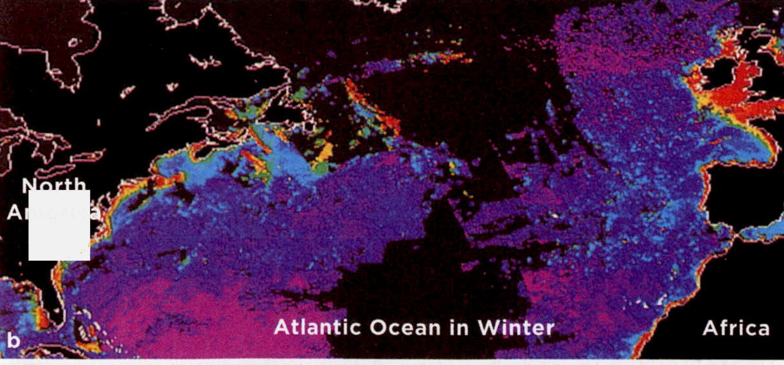

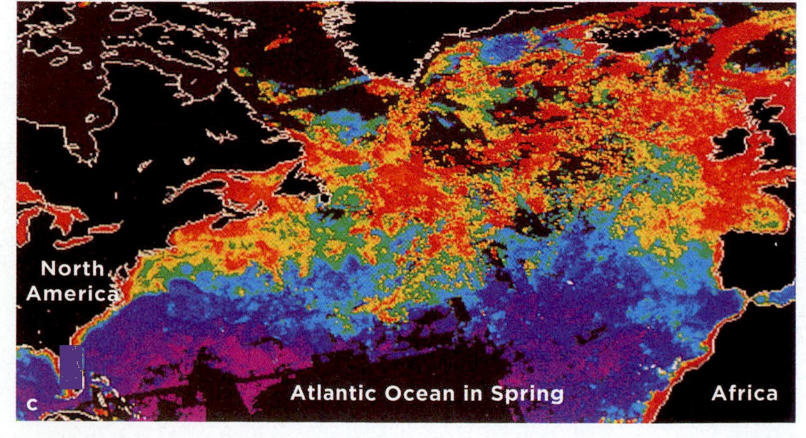

Figure 47.6 Primary productivity. (**a**) Summary of satellite data on net primary production during 2002. Productivity is coded as *red* (highest) down through *orange*, *yellow*, *green*, *blue*, and *purple* (lowest). (**b**,**c**) Satellite data showing seasonal shifts in net primary productivity for the North Atlantic Ocean.

single-celled protists that devote most energy that they harness to rapid reproduction, rather than to building a big body. They get eaten as fast as they reproduce, so a smaller biomass of phytoplankton can support a greater biomass of zooplankton and bottom feeders.

An **energy pyramid** illustrates how the amount of usable energy diminishes as it is transferred through an ecosystem. Sunlight energy is captured at the base (the primary producers) and declines with successive levels to its tip (the top carnivores). Energy pyramids are always "right-side-up," with their largest tier at the bottom. Such pyramids depict energy flow per unit of water (or land) per unit of time. Figure 47.8 shows the energy pyramid for the Silver Springs ecosystem and the energy flow that this pyramid represents.

Ecological Efficiency

Anywhere between 5 and 30 percent of the energy in the tissues of organisms at one trophic level ends up in the tissues of those at the next trophic level. Several factors influence the efficiency of transfers. First, not all energy harvested by consumers is used to build biomass. Some is lost as metabolic heat. Second, not all biomass can be digested by most consumers. Few herbivores have the ability to break down the lignin and cellulose that reinforce bodies of most land plants. Similarly, many animals have some biomass tied up in an internal or external skeleton. Hair, feathers, and fur are also part of the biomass that is difficult to digest.

The ecological efficiency of energy transfers is usually higher in aquatic ecosystems than on land. Algae lack lignin, and so are more easily digested than land plants. Also, aquatic ecosystems usually have a higher proportion of ectotherms (cold-blooded animals), such as fish, than land ecosystems do. Ectotherms lose less energy as heat than endotherms (warm-blooded animals) so more is transferred to the next level. Higher efficiencies of transfers allow for longer food chains.

Take-Home Message

How does energy flow through ecosystems?

- Primary producers capture energy and convert it into biomass. We measure this process as primary production.
- A biomass pyramid depicts dry weight of organisms at each trophic level in an ecosystem. Its largest tier is usually producers, but the pyramid for some aquatic systems is inverted.
- An energy pyramid depicts the amount of energy that enters each level. Its largest tier is always at the bottom (producers).
- Efficiency of transfers tends to be greatest in aquatic systems, where primary producers usually lack lignin and consumers tend to be ectotherms.

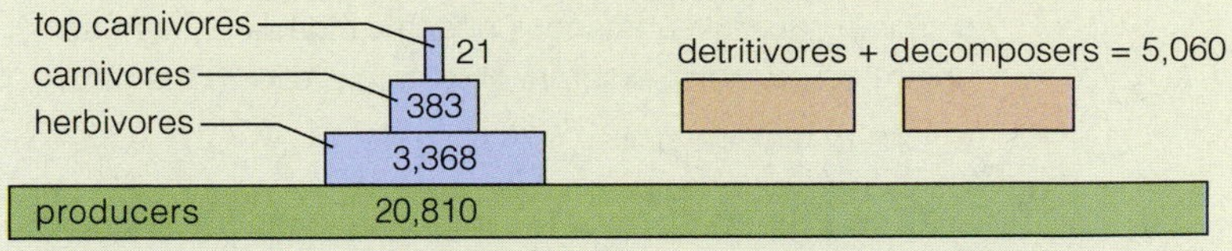

A Energy pyramid for the Silver Springs ecosystem. The size of each step in the pyramid represents the amount of energy that enters that trophic level annually, as shown in detail below.

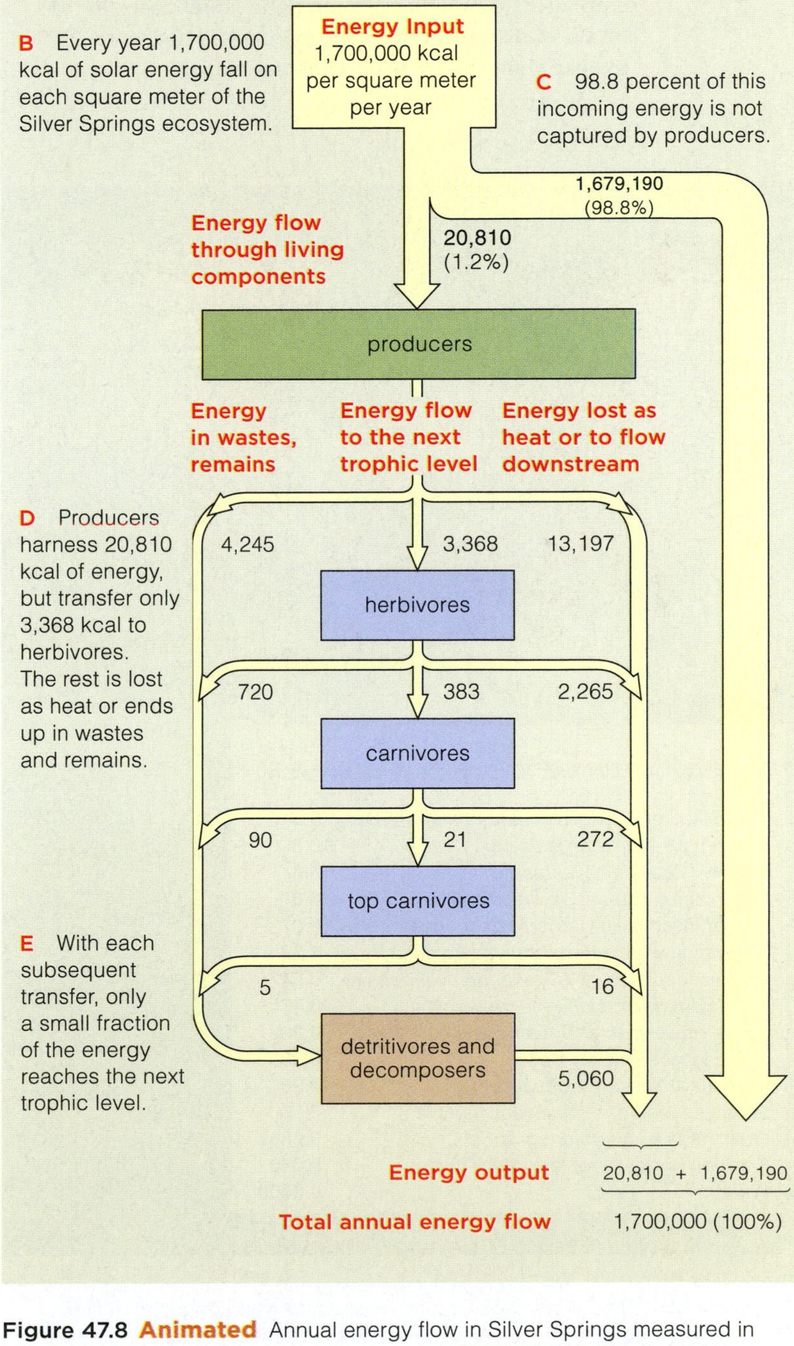

Figure 47.8 Animated Annual energy flow in Silver Springs measured in kilocalories (kcal) per square meter per year. **Figure It Out:** What percent of the energy carnivores received from herbivores was later passed on to top carnivores?

Answer: 21/383 × 100 = 5.5 percent

47.4 Biological Magnification

- Some harmful substances become more and more concentrated as they pass from one trophic level to the next.
- Link to Malaria 22.6

DDT and Silent Spring The synthetic pesticide dichlorodiphenyl-trichloroethane, or DDT, was invented in the late 1800s and came into widespread use in the 1940s. Spraying DDT saved many human lives by killing lice that spread typhus, and mosquitoes that carried malaria. Farmers also embraced this new chemical that increased crop yields by killing common agricultural pests. In the 1950s, swelling numbers of suburbanites turned to DDT to keep their shrubbery free of leaf-munching insects.

DDT Residues
(In parts per million wet weight of the whole organism)

Ring-billed gull fledgling (*Larus delawarensis*)	75.5
Herring gull (*Larus argentatus*)	18.5
Osprey (*Pandion haliaetus*)	13.8
Green heron (*Butorides virescens*)	3.57
Atlantic needlefish (*Strongylura marina*)	2.07
Summer flounder (*Paralichthys dentatus*)	1.28
Sheepshead minnow (*Cyprinodon variegatus*)	0.94
Hard clam (*Mercenaria mercenaria*)	0.47
Marsh grass shoots (*Spartina patens*)	0.33
Flying insects (mostly flies)	0.30
Mud snail (*Nassarius obsoletus*)	0.26
Shrimps (composite of several samples)	0.16
Green alga (*Cladophora gracilis*)	0.083
Plankton (mostly zooplankton)	0.040
Water	0.00005

Figure 47.9 Biological magnification in an estuary on Long Island, New York, as reported in 1967 by George Woodwell, Charles Wurster, and Peter Isaacson. Effects of DDT vary among species. Ospreys such as the one in the upper photo are highly sensitive. At 4 ppm of DDT, osprey eggs are fragile and unlikely to hatch. Gulls tolerate far higher doses of DDT without eggshell effects.

Unfortunately, DDT also affected nonpest species. Where DDT was sprayed to control Dutch elm disease, songbirds died. In forests sprayed to kill budworm larvae, DDT got into streams and killed fishes.

Rachel Carson, who had worked for the U.S. Fish and Wildlife Service, began compiling information about the harmful effects of pesticide use. She published her findings in 1962 as the book *Silent Spring*. The public embraced Carson's ideas but the pesticide industry mounted a campaign to discredit her. At the time, Carson was battling terminal breast cancer. Yet she vigorously defended her position until her death in 1964.

After Carson's death, study of DDT's impact increased. Researchers showed that DDT, like some other synthetic chemicals, undergoes **biological magnification**. By this process, a chemical that degrades slowly or not at all becomes increasingly concentrated in tissues of organisms as it moves up a food chain (Figure 47.9). In birds that are top carnivores such as ospreys, brown pelicans, bald eagles, and peregrine falcons, high DDT levels made eggs fragile, causing population sizes to plummet.

In recognition of the ecological effects of DDT, the United States has banned its use and export. Predatory bird populations in this country have largely recovered. Some countries still use DDT to fight malaria-carrying mosquitoes, but application is limited to indoor spraying. Even this use is controversial; some people would like to see a worldwide ban on the chemical. In additional to the environmental concerns, they cite studies indicating that maternal exposure to DDT during pregnancy may cause premature births and affects a child's mental development.

The Mercury Menace Birds bore the brunt of DDT's effects but fish get the spotlight when it comes to mercury pollution. Coal-burning power plants and some industrial processes put mercury into the air, then rain washes it into aquatic habitats. In some regions, runoff from abandoned or operating mines also contributes to aquatic mercury.

Like DDT, mercury accumulates as it moves up through food chains. Mercury adversely affects development of the human nervous system, so children and women who are pregnant or nursing should not eat fish that are top carnivores. Shark, swordfish, king mackerel, and tilefish are riskiest. You should also avoid these high-mercury fish if you are planning on becoming pregnant in the near future. Once mercury settles into your tissues, it can take a year for your body to get rid of it.

Everyone should avoid making fish that can have a high mercury content a major part of their diet. You can receive the health benefits of eating fish by choosing other species that are lower in mercury. For example, catfish, salmon, sardines, pollack, and canned light tuna are good choices. If you fish and plan to eat what you catch, check for local advisories about contaminants. The EPA website www.epa.gov/waterscience/fish/states.htm can link you to the appropriate agency.

47.5 Biogeochemical Cycles

- Nutrients move from nonliving environmental reservoirs into living organisms, then back into those reservoirs.
- Links to Tectonic plates 17.9, Nitrogen fixation 21.6

In a **biogeochemical cycle**, an essential element moves from one or more nonliving environmental reservoirs, through living organisms, then back to the reservoirs (Figure 47.10). As explained in the Chapter 2 introduction, oxygen, hydrogen, carbon, nitrogen, and phosphorus are some of the elements essential to all forms of life. We refer to these and other required elements as nutrients.

Depending on the element, environmental reservoirs may include Earth's rocks and sediments, waters, and atmosphere. Chemical and geologic processes move elements to and from these reservoirs. For example, elements that had been locked in rocks become part of the atmosphere as a result of volcanic activity. Uplifting elevates rocks where they are exposed to erosive forces of wind and rain. The rocks slowly dissolve; elements in them enter rivers, and eventually seas.

Elements enter the living part of an ecosystem by way of primary producers. Photosynthetic organisms take up essential ions dissolved in water. Land plants also take up carbon dioxide from the air.

Some bacteria fix nitrogen gas (Section 21.6). Their action makes this nutrient available to producers.

Nutrients move through food webs when organisms eat one another. Fungi and prokaryotes speed nutrient cycling within an ecosystem by decomposing remains and wastes of other organisms, so elements that were tied up in those materials are once again available to primary producers.

The next sections describe the four biogeochemical cycles that affect the most abundant elements in living organisms. In the water cycle, oxygen and hydrogen move on a global scale as part of molecules of water. In atmospheric cycles, a gaseous form of a nutrient such as carbon or nitrogen moves through ecosystems. A nutrient that does not often occur as a gas, such as phosphorus, moves in sedimentary cycles. Such nutrients accumulate on the ocean floor, then return to land by slow movements of Earth's crust (Section 17.9).

Take-Home Message

What are biogeochemical cycles?

- Biogeochemical cycles describe the continual flow of nutrients between nonliving environmental reservoirs and living organisms.
- Prokaryotes play a pivotal role in transfers between the living and nonliving portions of the cycle.
- Elements that occur in gases move through atmospheric cycles. Elements that do not normally occur as a gas move in sedimentary cycles.

Figure 47.10 Generalized biogeochemical cycle. In such cycles, a nutrient moves among nonliving environmental reservoirs and into and out of the living portion of an ecosystem. For all nutrients, the portion tied up in environmental reservoirs far exceeds the amount in living organisms.

47.6 The Water Cycle

- All organisms are mostly water and the cycling of this essential resource has implications for all life.
- Links to Properties of water 2.5, Leaching and erosion 29.1, Transpiration 29.3

How and Where Water Moves

The world ocean holds most of Earth's water (Table 47.1). As Figure 47.11 shows, in the **water cycle**, water moves among the atmosphere, the oceans, and environmental reservoirs on land. Sunlight energy drives evaporation, the conversion of water from liquid form to a vapor. Transpiration, explained in Section 29.3, is evaporation of water from plant parts. In cool upper layers of the atmosphere, condensation of water vapor into droplets gives rise to clouds. Later, clouds release the water as precipitation—as rain, snow, or hail.

A **watershed** is an area from which all precipitation drains into a specific waterway. It may be as small as a valley that feeds a stream, or as large as the Mississippi River Basin, which covers about 41 percent of the continental United States.

Most precipitation falling in a watershed seeps into the ground. Some collects in **aquifers**, permeable rock layers that hold water. **Groundwater** is water in soil and aquifers. When soil gets saturated, water becomes **runoff**; it flows over the ground into streams.

Table 47.1 Environmental Water Reservoirs

Main Reservoirs	Volume (10^3 cubic kilometers)
Ocean	1,370,000
Polar ice, glaciers	29,000
Groundwater	4,000
Lakes, rivers	230
Soil moisture	67
Atmosphere (water vapor)	14

Flowing water moves dissolved nutrients into and out of a watershed. Experiments in New Hampshire's Hubbard Brook watershed illustrated that vegetation helps slow nutrient losses. Experimental deforestation caused a spike in loss of mineral ions (Figure 47.12).

A Global Water Crisis

Our planet has plenty of water, but most of it is too salty to drink or use for irrigation. If all Earth's water filled a bathtub, the amount of fresh water that could be used sustainably in a year would fill a teaspoon.

Of the fresh water we use, about two-thirds goes to agriculture, but irrigation can harm soil. Piped-in water

Figure 47.11 Animated The water cycle. Arrows identify processes that move water. The numbers shown indicate the amounts moved, as measured in cubic kilometers per year.

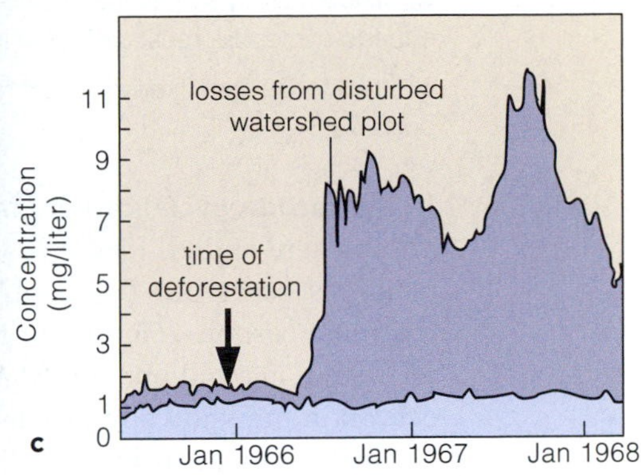

Figure 47.12 Hubbard Brook experimental watershed. (**a**) Runoff in this watershed is collected by concrete basins for easy monitoring. (**b**) This plot of land was stripped of all vegetation as an experiment. (**c**) After experimental deforestation, calcium levels in runoff increased sixfold (*medium blue*). A control plot in the same watershed showed no similar increase during this time (*light blue*).

often has high concentrations of salts. **Salinization**, the buildup of mineral salts in soil, stunts crop plants and decreases yields.

Groundwater supplies drinking water to about half of the United States population. Pollution of this water now poses a threat. Chemicals leaching from landfills, hazardous waste facilities, and underground storage tanks often contaminate it. Unlike flowing rivers and streams, which can recover fast, polluted groundwater is difficult and expensive to clean up.

Water overdrafts are also common; water is drawn from aquifers faster than natural processes replenish it. When too much fresh water is withdrawn from an aquifer near the coast, salt water moves in and replaces it. Figure 47.13 highlights regions of aquifer depletion and saltwater intrusion in the United States.

Overdrafts have now depleted half of the Ogallala aquifer, which extends from South Dakota into Texas. This aquifer supplies the irrigation water for about 20 percent of the nation's crops. For the past thirty years, withdrawals have exceeded replenishment by a factor of ten. What will happen when water runs out?

Contaminants such as sewage, animal wastes, and agricultural chemicals make water in rivers and lakes unfit to drink. In addition, pollutants disrupt aquatic ecosystems, and in some cases they drive vulnerable species to local extinction.

Desalinization, the removal of salt from seawater, may help increase freshwater supplies. However, the process requires a lot of fossil fuel. Desalinization is feasible mainly in Saudi Arabia and other places that have small populations and very large fuel reserves. In addition, the process produces mountains of waste salts that must be disposed of.

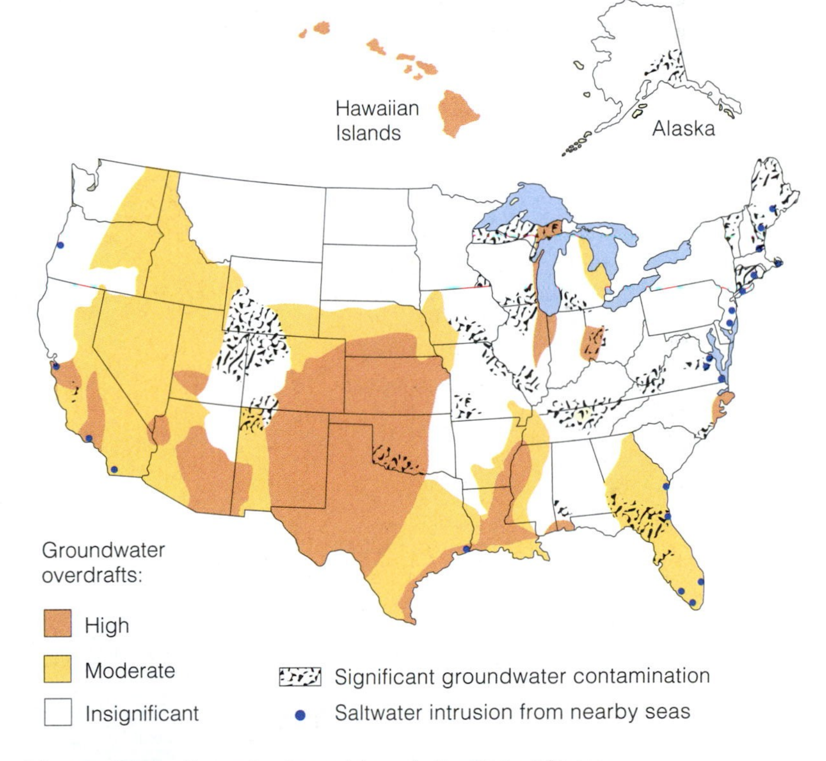

Figure 47.13 Groundwater problems in the United States.

Take-Home Message

What is the water cycle and how do humans affect it?

- In the water cycle, water moves on a global scale. It moves slowly from the world ocean—the main reservoir—through the atmosphere, onto land, then back to the ocean.
- Of the fresh water that human populations use, about two-thirds sustains agriculture.
- Aquifers that supply much of the world's drinking water are becoming polluted and depleted.

47.7 Carbon Cycle

- Carbon dioxide in air makes the carbon cycle an atmospheric cycle, but most carbon is in sediments and rocks.
- Links to Carbon fixation 7.6, Foraminiferans 22.3, Peat bogs 23.3

In the **carbon cycle**, carbon moves through the lower atmosphere and all food webs on its way to and from its largest reservoirs (Figure 47.14). Earth's crust holds the most carbon—66 million to 100 million gigatons. A gigaton is a billion tons. There are 4,000 gigatons of carbon in the known fossil fuel reserves.

Organisms contribute to Earth's carbon deposits. Single-celled protists such as foraminiferans (Section 22.3) produce shells rich in calcium carbonate. Over hundreds of millions of years, uncountable numbers of these cells died, sank, and were buried in seafloor sediments. The carbon in their remains cycles slowly, as movements of Earth's crust uplift portions of the sea floor, making it part of a land ecosystem.

Most of the annual carbon movement takes place between the ocean and atmosphere. The ocean holds 38,000–40,000 gigatons of dissolved carbon, primarily in the form of bicarbonate and carbonate ions. The air holds about 766 gigatons of carbon, mainly combined with oxygen in the form of carbon dioxide (CO_2).

On land, detritus in soil holds 1,500–1,600 gigatons of carbon. Peat bogs and the permafrost, a perpetually frozen layer of soil that underlies arctic regions, are major reservoirs. Another 540–610 gigatons is present in biomass, or tissues of organisms.

Ocean currents move carbon from upper ocean waters into deep sea reservoirs. Carbon dioxide enters warm surface waters and is converted to bicarbonate. Then, prevailing winds and regional differences in density drive the flow of bicarbonate-rich seawater in a gigantic loop from the surface of the Pacific and Atlantic oceans down to the Atlantic and Antarctic sea floors. Here, bicarbonate moves into cold, deep storage

Figure 47.14 Animated *Right*, carbon cycling in (**a**) marine ecosystems and (**b**) land ecosystems. *Gold* boxes highlight the most important carbon reservoirs. The vast majority of carbon atoms are in sediments and rocks, followed by lesser amounts in seawater, soil, the atmosphere, and biomass (in that order). Typical annual fluxes in global distribution of carbon, in gigatons, are:

From atmosphere to plants by carbon fixation	120
From atmosphere to ocean	107
To atmosphere from ocean	105
To atmosphere from plants	60
To atmosphere from soil	60
To atmosphere from fossil fuel burning	5
To atmosphere from net destruction of plants	2
To ocean from runoff	0.4
Burial in ocean sediments	0.1

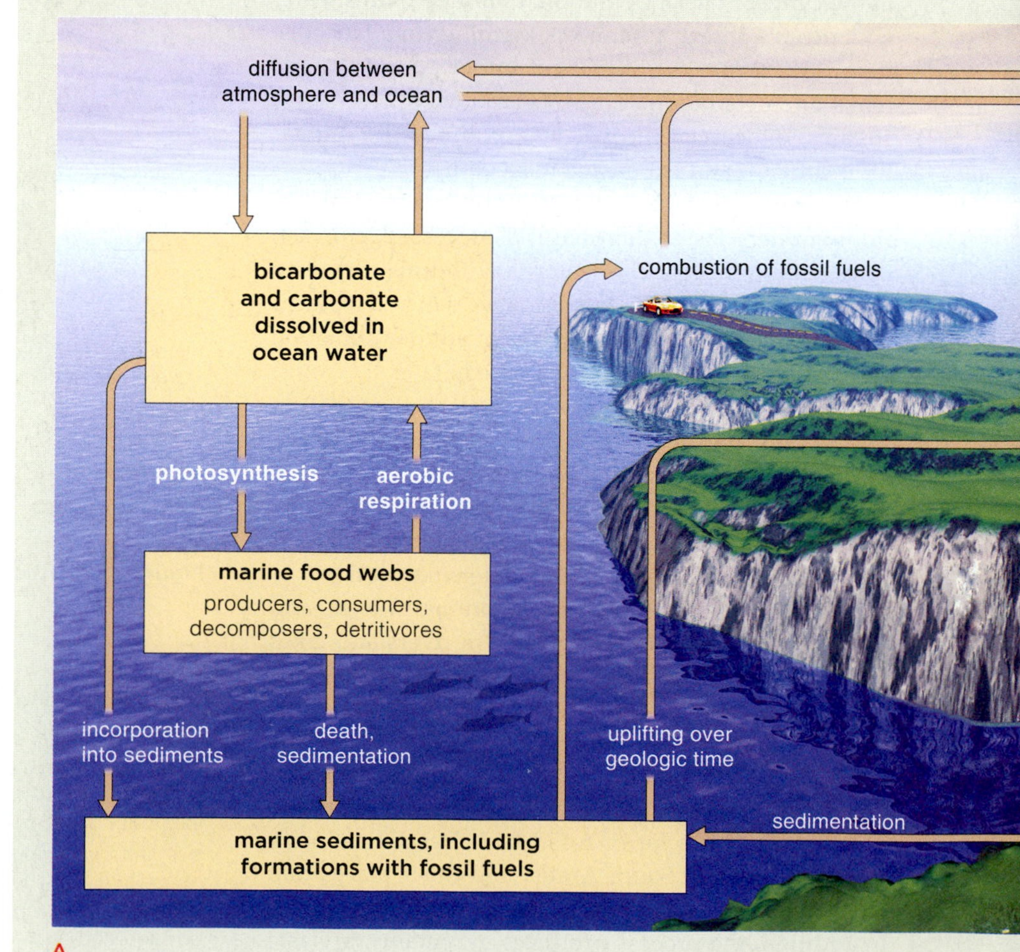

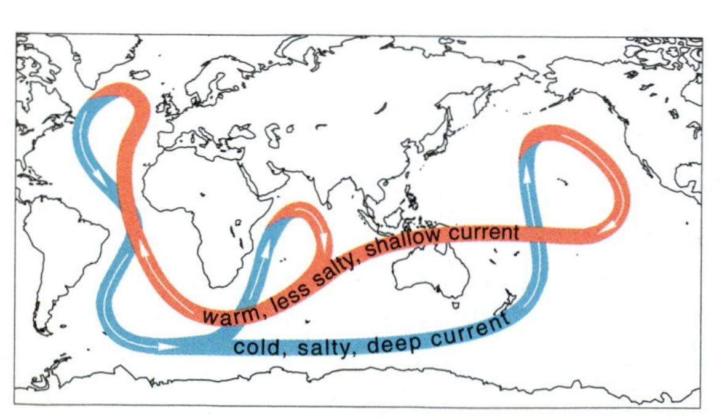

Figure 47.15 Loop that moves carbon dioxide to carbon's deep ocean reservoir. The loop sinks in the cold, salty North Atlantic. It rises in the warmer Pacific.

reservoirs before water loops back up (Figure 47.15). Storage of carbon in the deep sea helps dampen any short-term effects of increases in atmospheric carbon.

Biologists sometimes refer to the global cycling of carbon in the form of carbon dioxide and bicarbonate as a carbon–oxygen cycle. Plants, phytoplankton, and some bacteria fix carbon when they engage in photosynthesis (Section 7.6). Each year, they tie up billions of metric tons of carbon in sugars and other organic compounds. Breakdown of those compounds by aerobic respiration releases carbon dioxide into the air. More carbon dioxide escapes into the air when fossil fuels or forests burn and when volcanoes erupt.

The time that an ecosystem holds a given carbon atom varies. Organic material decomposes rapidly in tropical forests, so carbon does not build up at the soil surface. By contrast, bogs and other anaerobic habitats do not favor decomposition, so material accumulates, as in peat bogs (Section 23.3).

Humans are altering the carbon cycle. Each year, we withdraw 4 to 5 gigatons of fossil fuel from environmental reservoirs. Our activities put about 6 gigatons more carbon in the air than can be moved into ocean reservoirs by natural processes. Only about 2 percent of the excess carbon entering the atmosphere becomes dissolved in ocean water. Carbon dioxide in the air traps heat, so increased outputs of it may be a factor in global climate change. The next section looks at this possibility and some environmental implications.

Take-Home Message

What is the carbon cycle?

- In the carbon–oxygen cycle, carbon moves into and out of ecosystems mainly combined with oxygen, as in carbon dioxide, bicarbonate, and carbonate.
- Earth's crust is the largest carbon reservoir, followed by the world ocean. Most of the annual cycling of carbon occurs between the ocean and atmosphere.

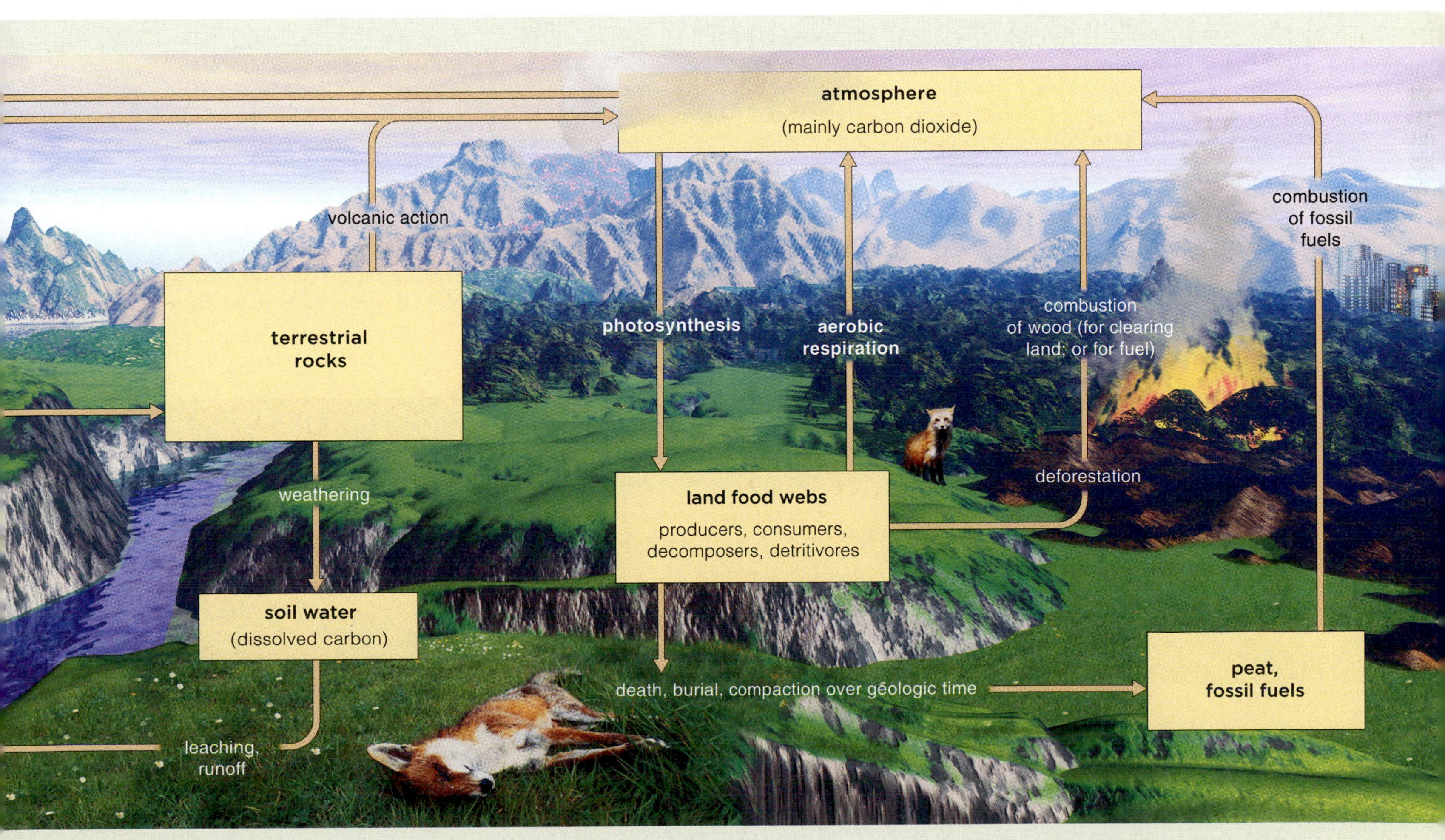

47.8 Greenhouse Gases and Climate Change

- Concentrations of gases in Earth's atmosphere help determine the temperature near Earth's surface. Human activities are altering gas concentrations and causing climate change.
- Link to Carbon imbalances 7.9

Concentrations of various gaseous molecules profoundly influence the average temperature of the atmosphere near Earth's surface. That temperature, in turn, has far-reaching effects on global and regional climates.

Atmospheric molecules of carbon dioxide, water, nitrous oxide, methane, and chlorofluorocarbons (CFCs) are among the main players in interactions that can shift global temperatures. Collectively, the gases trap heat a bit like a greenhouse does, hence the familiar name "greenhouse gases."

Radiant energy from the sun passes through the atmosphere and is absorbed by Earth's surface. The energy warms the surface, which means that the surface emits infrared radiation (heat). The infrared energy radiates back toward space, but greenhouse gases in the atmosphere interfere with its progress. How? The gases absorb some of the infrared energy, and then emit a portion of it back toward Earth's surface (Figure 47.16). Without this process, which is called the **greenhouse effect**, Earth's surface would be so cold that very little life would survive.

In the 1950s, researchers at a laboratory on Hawaii's highest volcano began to measure the atmospheric concentrations of greenhouse gases. That remote site is almost free of local airborne contamination. It also is representative of atmospheric conditions for the Northern Hemisphere. What did they find? Briefly, concentrations of CO_2 follow annual cycles of primary production. They decline in summer, when the rates of photosynthesis are highest. They rise in winter, when photosynthesis declines but aerobic respiration and fermentation continue.

Figure 47.17 *Facing page*, graphs of recent increases in four categories of atmospheric greenhouse gases. A key factor is the sheer number of gasoline-burning vehicles in large cities. *Above*, Mexico City on a smoggy morning. With 10 million residents, it is the world's largest city.

The alternating troughs and peaks along the graph line in Figure 47.17*a* are annual lows and highs of global CO_2 concentrations. For the first time, researchers saw the effects of carbon dioxide fluctuations for the entire hemisphere. Notice the midline of the troughs and peaks in the cycle. It shows that carbon dioxide concentration is steadily increasing—as are concentrations of other major greenhouse gases.

Atmospheric levels of greenhouse gases are far higher than they were for most of the past. Carbon dioxide may

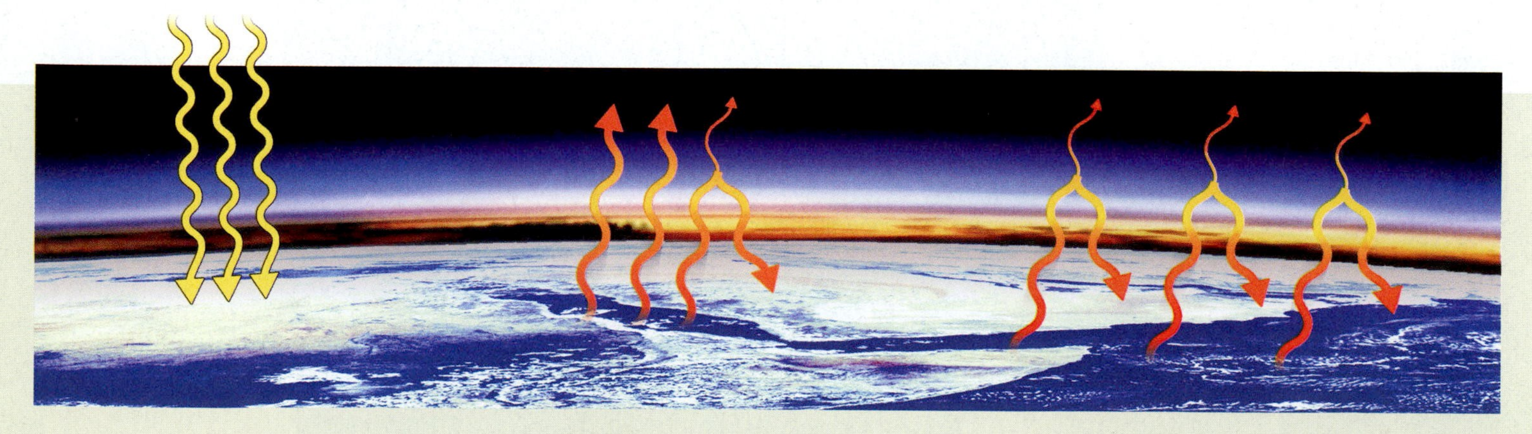

A Radiant energy from the sun penetrates the lower atmosphere, and it warms Earth's surface.

B The warmed surface radiates heat (infrared radiation) back toward space. Greenhouse gases absorb some of the infrared energy, and then emit a portion of it back toward Earth.

C Increased concentrations of greenhouse gases trap more heat near Earth's surface. Sea surface temperatures rise, so more water evaporates into the atmosphere. Earth's surface temperature rises.

Figure 47.16 Animated The greenhouse effect.

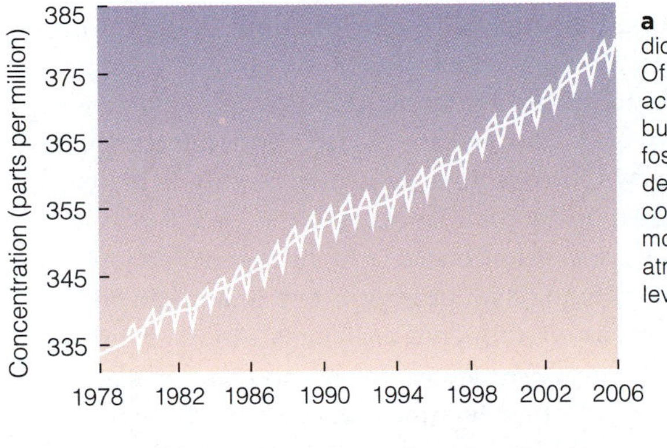

a Carbon dioxide (CO_2). Of all human activities, the burning of fossil fuels and deforestation contribute the most to rising atmospheric levels.

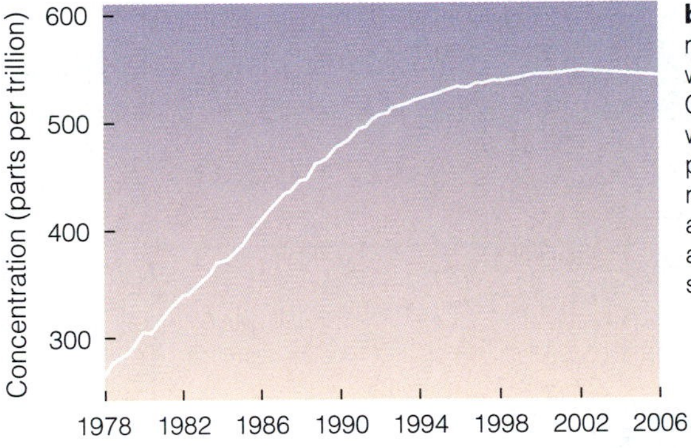

b CFCs. Until restrictions were in place, CFCs were widely used in plastic foams, refrigerators, air conditioners, and industrial solvents.

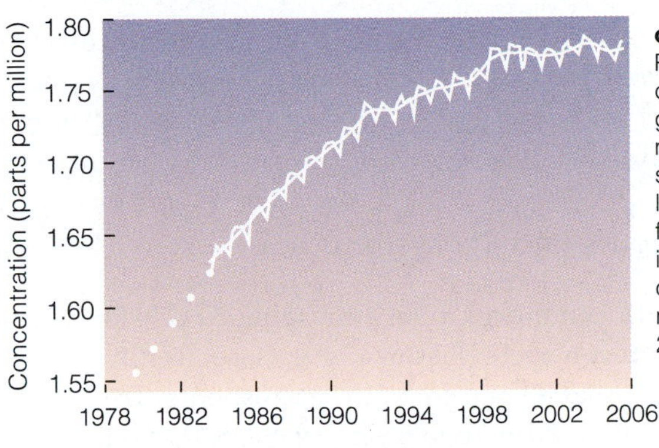

c Methane (CH_4). Production and distribution of natural gas as fuel adds to methane released by some bacteria that live in swamps, rice fields, landfills, and in the digestive tract of cattle and other ruminants (Section 21.7).

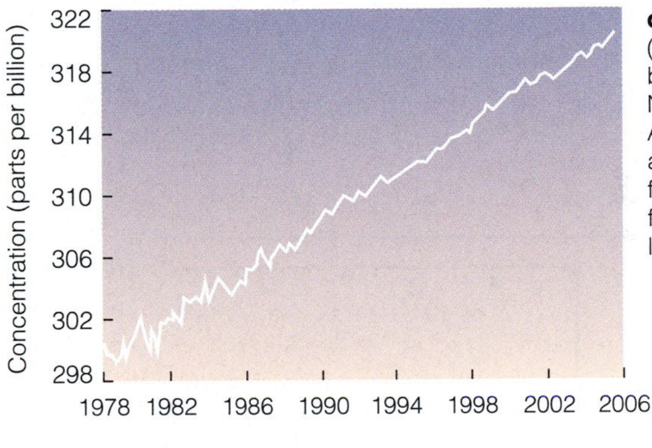

d Nitrous oxide (N_2O). Denitrifying bacteria produce N_2O in metabolism. Also fertilizers and animal waste from large-scale feedlots release large amounts.

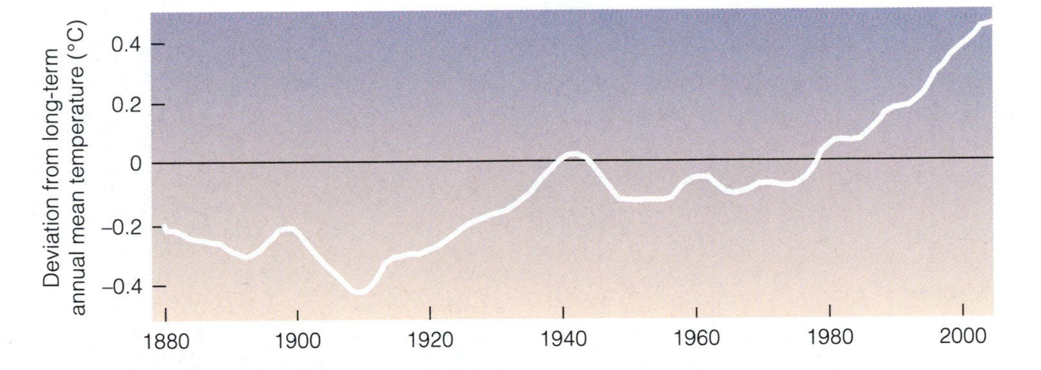

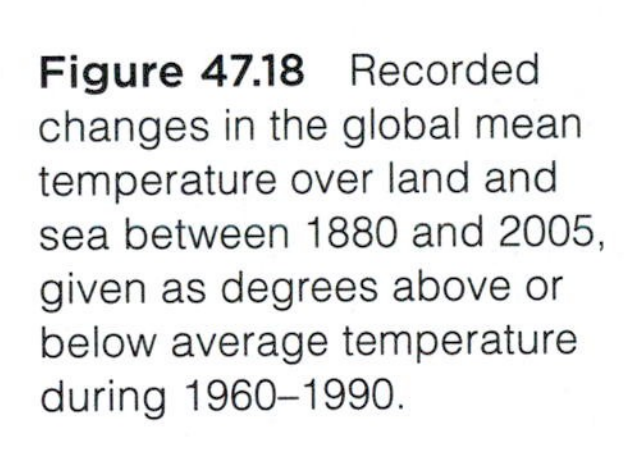

Figure 47.18 Recorded changes in the global mean temperature over land and sea between 1880 and 2005, given as degrees above or below average temperature during 1960–1990.

be at its highest level since 470,000 years ago, possibly since 20 million years ago. There is scientific consensus that human activities—mainly the burning of fossil fuels—are contributing significantly to the current increases in greenhouse gases. The big worry is that the increase may have far-reaching environmental consequences.

The increase in greenhouse gases may be a factor in **global warming**, a long-term increase in temperature near Earth's surface (Figure 47.18). In the past thirty years, the global surface temperature increased at a faster rate, to 1.8°C (3.2°F) per century. Warming is most dramatic at the upper latitudes of the Northern Hemisphere.

Data from satellites, weather stations and balloons, research ships, and computer programs suggest that some irreversible climate changes are already under way. Water expands as it is heated, and heating also melts glaciers and other ice. Together, thermal expansion and addition of meltwater will cause sea level to rise. In the past century, the sea level may have risen as much as 20 centimeters (8 inches) and the rate of rise appears to be accelerating.

Scientists expect continued temperature increases to have far-reaching effects on climate. An increased rate of evaporation will alter global rainfall patterns. Intense rains and flooding probably will become more frequent in some regions, while droughts increase in others. Hurricanes probably will become more intense.

It bears repeating: As investigations continue, a key research goal is to investigate all of the variables in play. With respect to consequences of climate change, the most crucial variable may be the one we do not know.

47.9 Nitrogen Cycle

- Gaseous nitrogen makes up about 80 percent of the lower atmosphere, but most organisms can't use this gaseous form.
- Links to Acid rain 2.6, Nitrogen fixation 21.6 and 29.2, Algal blooms 22.5, Decomposers 21.6 and 24.5, Leaching 29.1

Inputs Into Ecosystems

Nitrogen moves in an atmospheric cycle known as the **nitrogen cycle** (Figure 47.19). Gaseous nitrogen makes up about 80 percent of the atmosphere. Triple covalent bonds hold its two atoms of nitrogen together as N_2, or N≡N. Plants cannot use gaseous nitrogen, because they do not make the enzyme that can break its triple bond. Volcanic eruptions and lightning can convert some N_2 into forms that enter food webs. Far more is converted through **nitrogen fixation**. By this process, bacteria break all three bonds in N_2, then incorporate the N atoms into ammonia (NH_3). Ammonia gets converted into ammonium (NH_4^+) and nitrate (NO_3^-). These two nitrogen salts dissolve readily in water and are taken up by plant roots.

Many species of bacteria fix nitrogen (Section 21.6). Nitrogen-fixing cyanobacteria live in aquatic habitats, soil, and as components of lichens. Another nitrogen-fixing group, *Rhizobium*, forms nodules on the roots of peas and other legumes. Each year, nitrogen-fixing bacteria collectively take up about 270 million metric tons of nitrogen from the atmosphere.

The nitrogen incorporated into plant tissues moves up through trophic levels of ecosystems. It ends up in

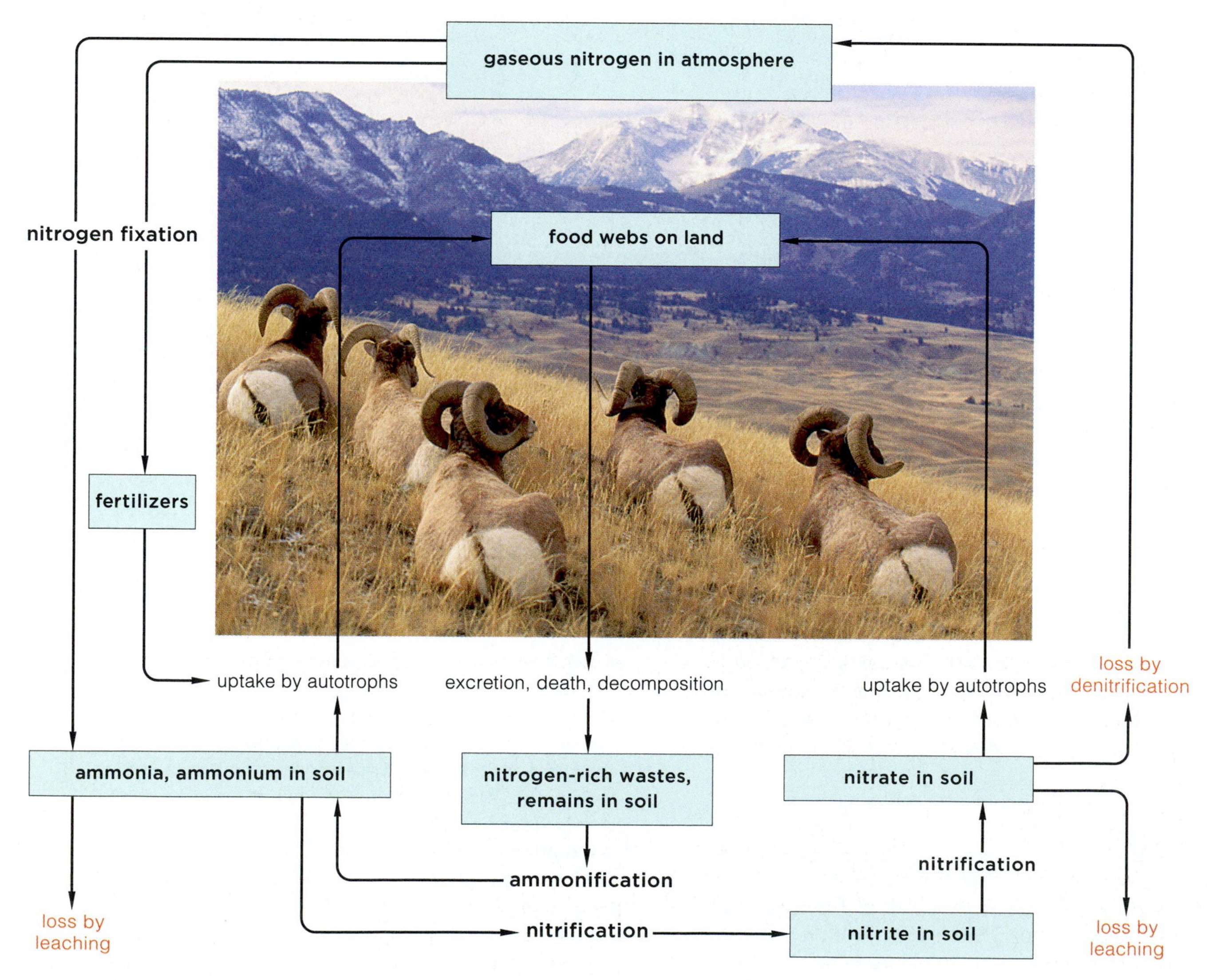

Figure 47.19 Animated Nitrogen cycle in an ecosystem on land. Nitrogen becomes available to plants through the activities of nitrogen-fixing bacteria. Other bacterial species cycle nitrogen to plants. They break down organic wastes to ammonium and nitrates.

nitrogen-rich wastes and remains, which bacteria and fungi decompose (Sections 21.6 and 24.5). By the process of **ammonification**, these organisms break apart proteins and other nitrogen-containing molecules and produce ammonium. Some of the ammonium product gets released into the soil, where plants and nitrifying bacteria take it up. **Nitrification** begins when bacteria convert ammonium to nitrite (NO_2^-). Other nitrifying bacteria then use the nitrite in reactions that end with the formation of nitrate. Nitrate, like ammonium, can be taken up by plant roots.

Natural Losses From Ecosystems

Ecosystems lose nitrogen through **denitrification**. By this process, denitrifying bacteria convert nitrate or nitrite to gaseous nitrogen or to nitrogen oxide (NO_2). Denitrifying bacteria are typically anaerobes that live in waterlogged soils and aquatic sediments.

Ammonium, nitrite, and nitrate also are lost from a land ecosystem in runoff and by leaching, the removal of some nutrients as water trickles down through the soil (Section 29.1). Nitrogen-rich runoff enters streams and other aquatic ecosystems.

Disruptions by Human Activities

Deforestation and conversion of grassland to farmland also causes nitrogen losses from an ecosystem. With each clearing and harvest of plants, nitrogen stored in plant tissues is removed. Plant removal also makes soil more vulnerable to erosion and leaching.

Farmers can counter nitrogen depletion by rotating their crops. For example, they plant corn and soybeans in the same field in alternating years. Nitrogen-fixing bacteria that associate with legumes such as soybeans add nitrogen to the soil (Section 29.2).

In developed countries, most farmers also spread synthetic nitrogen-rich fertilizers. High temperature and pressure converts nitrogen and hydrogen gases to ammonia fertilizers. Although the manufactured fertilizers improve crop yields, they also modify soil chemistry. Adding ammonium to the soil increases the concentration of hydrogen ions, as well as nitrogen. High acidity encourages ion exchange: Nutrient ions bound to particles of soil get replaced by hydrogen ions. As a result, calcium and magnesium ions needed for plant growth seep away in soil water.

Burning of fossil fuel in power plants and by vehicles releases nitrogen oxides. These gases contribute to global warming and acid rain (Section 2.6). Winds frequently carry gaseous pollutants far from their sources.

Figure 47.20 Dead and dying trees in Great Smoky Mountains National Park. Forests are among the casualties of nitrogen oxides and other forms of air pollution.

By some estimates, pollutants blowing into the Great Smoky Mountains National Park have increased the amount of nitrogen in the soil sixfold (Figure 47.20).

Nitrogen in acid rain can have the same effects as use of manufactured fertilizers. Different plant species respond in different ways to increased nitrogen level. Changes in soil nitrogen disrupt the balance among competing species in a community, causing diversity to decline. The impact can be especially pronounced in forests at high elevations or at high latitudes, where soils tend to be naturally nitrogen-poor.

Some human activities disrupt aquatic ecosystems through nitrogen enrichment. For instance, about half of the nitrogen in fertilizers applied to fields runs off into rivers, lakes, and estuaries. More nitrogen enters waters in sewage from cities and in animal wastes. As one result, nitrogen inputs promote algal blooms (Section 22.5). Phosphorus in fertilizers has the same negative effects, as explained in the next section.

Take-Home Message

What is the nitrogen cycle?

- The ecosystem phase of the nitrogen cycle starts with nitrogen fixation. Bacteria convert gaseous nitrogen in the air to ammonia and then to ammonium, which is a form that plants easily take up.
- By ammonification, bacteria and fungi make additional ammonium available to plants when they break down nitrogen-rich organic wastes and remains.
- By nitrification, bacteria convert nitrites in soil to nitrate, which also is a form that plants easily take up.
- The ecosystem loses nitrogen when denitrifying bacteria convert nitrite and nitrate back to gaseous nitrogen, and when nitrogen is leached from soil.

47.10 The Phosphorus Cycle

- Unlike carbon and nitrogen, phosphorus seldom occurs as a gas. Like nitrogen, it can be taken up by plants only in ionized form, and it, too, is often a limiting factor on plant growth.

In the **phosphorus cycle**, phosphorus passes quickly through food webs as it moves from land to ocean sediments, then slowly back to dry land. Earth's crust is the largest reservoir of phosphorus.

Phosphorus in rocks is mainly in the form of phosphate (PO_4^{3-}). Weathering and erosion put phosphate ions from rocks into streams and rivers, which deliver them to oceans (Figure 47.21). There, the phosphates accumulate as underwater deposits along the edges of continents. After millions of years, movements of Earth's crust result in uplifting of parts of the sea floor. Once uplifted, the rocky phosphate deposits on land are subject to weathering and erosion, which release phosphates from the rocks and start the phosphorus cycle over again.

Phosphates are required building blocks for ATP, phospholipids, nucleic acids, and other compounds. Plants take up dissolved phosphates from soil water. Herbivores get them by eating plants; carnivores get them by eating herbivores. Animals lose phosphate in urine and in feces. Bacterial and fungal decomposers release phosphate from organic wastes and remains, then plants take them up again.

The water cycle helps move phosphorus and other minerals through ecosystems. Water evaporates from the ocean and falls on land. As it flows back to the ocean, it transports silt and dissolved phosphates that the primary producers require for growth.

Of all minerals, phosphorus most frequently acts as the limiting factor for plant growth. Only newly weathered, young soil has an abundance of phosphorus. Many tropical and subtropical ecosystems that are already low in phosphorus are likely to be further depleted by human actions. In an undisturbed forest, decomposition releases phosphorus stored in biomass. When forest is converted to farmland, the ecosystem loses phosphorus that had been stored in trees. Crop yields soon decline. Later, after the fields are abandoned, regrowth remains sparse. Spreading finely ground, phosphate-rich rock can help restore fertility, but many developing countries lack this resource.

Many developed countries have a different problem. Phosphorous in runoff from heavily fertilized fields pollutes water. Sewage from cities and factory farms also contain phosphorus. Dissolved phosphorus that

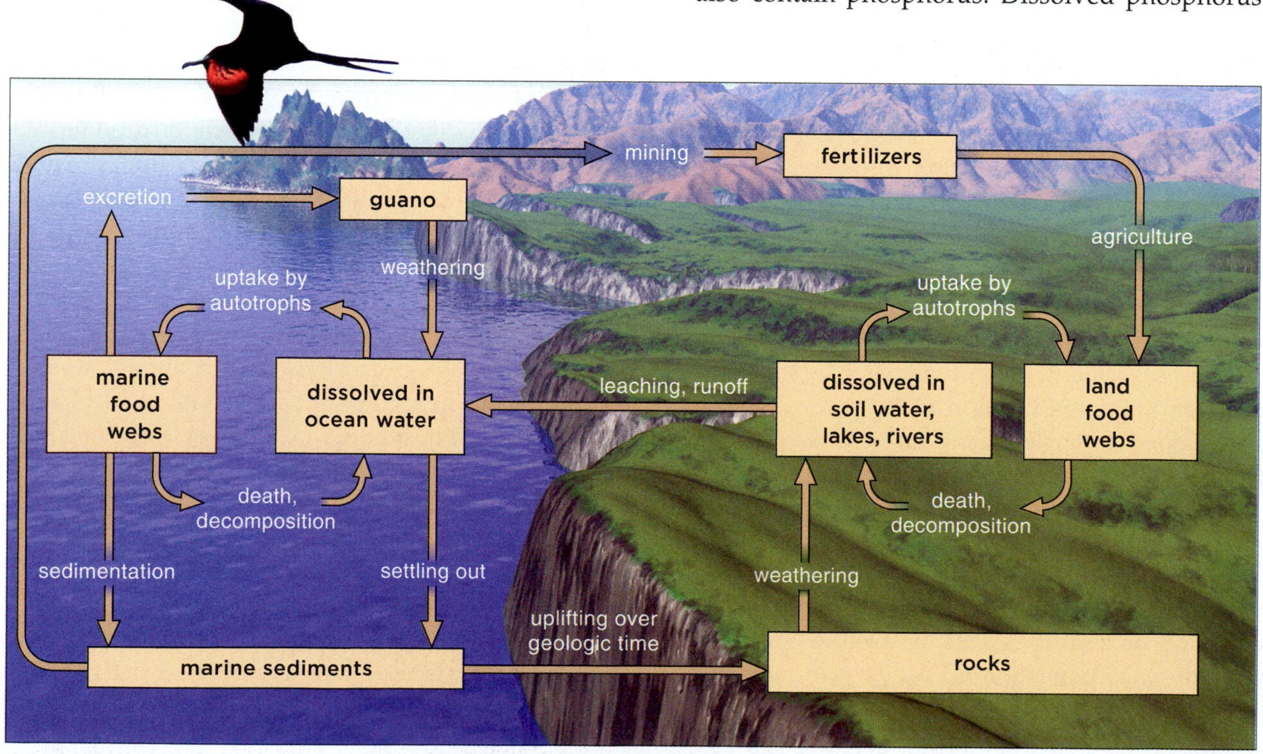

Figure 47.21 Animated Phosphorus cycle. In this sedimentary cycle, phosphorus moves mainly in the form of phosphate ions (PO_4^{3-}) to the ocean. It moves through phytoplankton of marine food webs, then to fishes that eat plankton. Seabirds eat the fishes, and their droppings (guano) accumulate on islands. Humans collect and use guano as a phosphate-rich fertilizer.

gets into aquatic ecosystems can promote destructive algal blooms. Like the plants, algae require nitrogen, phosphorus, and other ions to keep growing. In many freshwater ecosystems, nitrogen-fixing bacteria keep the nitrogen levels high, so phosphorus becomes the limiting factor. When phosphate-rich pollutants pour in, algal populations soar and then crash. As aerobic decomposers break down remains of dead algae, the water becomes depleted of the oxygen that fishes and other organisms require.

Eutrophication refers to nutrient enrichment of any ecosystem that is otherwise low in nutrients. It can occur naturally, but human activities often accelerate it, as the experiment shown in Figure 47.22 demonstrated. Eutrophication of a lake is difficult to reverse. It can take years for excess nutrients that encourage algal growth to be depleted.

Take-Home Message

What is the phosphorus cycle?

- The phosphorus cycle is a sedimentary cycle that moves this element from its main reservoir (Earth's crust), through soils and sediments, aquatic habitats, and bodies of living organisms.

Figure 47.22 A eutrophication experiment. Researchers put a plastic curtain across a channel between two basins of a natural lake. They added nitrogen, carbon, and phosphorus to the water on one side of the curtain (here, the *lower* part of the lake) and added nitrogen and carbon to the water on the other side. Within months, the basin with phosphorous was eutrophic, with a dense algal bloom (*green*) covering its surface.

Summary

Section 47.1 An **ecosystem** consists of an array of organisms along with nonliving components of their environment. There is a one-way flow of energy into and out of an ecosystem, and a cycling of materials among resident species. All ecosystems have inputs and outputs of energy and nutrients.

Sunlight supplies energy to most ecosystems. **Primary producers** convert sunlight energy into chemical bond energy. They also take up the nutrients that they, and all consumers, require. Herbivores, carnivores, omnivores, **decomposers**, and **detritivores** are **consumers**.

Energy moves from organisms at one **trophic level** to organisms at another. Organisms are at the same trophic level if they are an equal number of steps away from the energy input into the ecosystem. A **food chain** shows one path of energy and nutrient flow among organisms. It depicts who eats whom.

- *Use the animation on CengageNOW to learn about energy flow and nutrient cycling.*

Section 47.2 Food chains interconnect as **food webs**. The efficiency of energy transfers is always low, so most ecosystems have no more than four or five trophic levels. In a **grazing food chain**, most energy captured by producers flows to herbivores. In **detrital food chains**, most energy flows from producers directly to detritivores and decomposers. Both types of food chains interconnect in nearly all ecosystems.

- *Use the animation on CengageNOW to explore a food web.*

Section 47.3 A system's **primary production** is the rate at which producers capture and store energy in their tissues. It varies with climate, seasonal changes, nutrient availability, and other factors.

Energy pyramids and **biomass pyramids** depict how energy and organic compounds are distributed among the organisms of an ecosystem. All energy pyramids are largest at their base. If producers get eaten as fast as they reproduce, the biomass of consumers can exceed that of producers, so the biomass pyramid is upside down.

- *Use the animation on CengageNOW to see how energy flows through one ecosystem.*

Section 47.4 With **biological magnification**, a chemical substance is passed from organisms at each trophic level to those above and becomes increasingly concentrated in body tissues.

Section 47.5 In a **biogeochemical cycle**, water or some nutrient moves from an environmental reservoir, through organisms, then back to the environment.

Section 47.6 In the **water cycle**, evaporation, condensation, and precipitation move water from its main reservoir —oceans—into the atmosphere, onto land, then back to oceans. **Runoff** is water that flows over ground into streams. A **watershed** is an area where all precipitation drains into a specific waterway. Water in **aquifers** and in the soil is **groundwater**. Use of irrigation can cause

IMPACTS, ISSUES REVISITED Bye-Bye, Blue Bayou

In 2006, China overtook the United States as the country that emits the most carbon dioxide. Still, an average American life-style causes about 20 tons of carbon emissions per year. That's more than four times the emissions of an average person in China. It's also more than twice that of people in western Europe. Automotive emissions are one factor; fuel efficiency standards in both China and Europe are more stringent than they are in the United States.

How would you vote?
Should the United States increase fuel efficiency standards for cars and trucks to lower carbon dioxide output? See CengageNow for details, then vote online.

---

salinization—salt buildup—in soil. **Desalinization** is an energy-intensive method of obtaining fresh water from salt water.

- *Use the animation on CengageNOW to learn about the water cycle.*

Section 47.7 The **carbon cycle** moves carbon from reservoirs in rocks and seawater, through its gaseous forms (methane and CO_2) in the air, and through ecosystems. Deforestation and the burning of wood and fossil fuels are adding more carbon dioxide to the atmosphere than the oceans can absorb.

- *Use the animation on CengageNOW to observe the flow of carbon through its global cycle.*

Section 47.8 The **greenhouse effect** refers to the ability of certain gases to trap heat in the lower atmosphere. It warms Earth's surface. Human activities are putting larger than normal amounts of greenhouse gases, including carbon dioxide, into the atmosphere. The rise in these gases correlates with a rise in global temperatures (**global warming**) and other climate changes.

- *Use the animation on CengageNOW to explore the greenhouse effect and global warming.*

Section 47.9 The **nitrogen cycle** is an atmospheric cycle. Air is the main reservoir for N_2, a gaseous form of nitrogen that plants cannot use. In **nitrogen fixation**, certain bacteria take up N_2 and form ammonia. **Ammonification** releases ammonia from organic remains. **Nitrification** involves conversion of ammonium to nitrite and then nitrate, which plants are able to take up. Some nitrogen is lost to the atmosphere by **denitrification** carried out by bacteria. Human activities add nitrogen to ecosystems; for example, through fossil fuel burning (which releases nitrogen oxides) and application of fertilizers. The added nitrogen can disrupt ecosystem processes.

- *Use the animation on CengageNOW to learn how nitrogen is cycled in an ecosystem.*

Section 47.10 The **phosphorus cycle** is a sedimentary cycle; Earth's crust is the largest reservoir and there is no major gaseous form. Phosphorus is often the factor that limits population growth of plant and algal producers. Excessive inputs of phosphorus to an aquatic ecosystem can accelerate **eutrophication**.

- *Use the animation on CengageNOW to learn how phosphorus is cycled in an ecosystem.*

Self-Quiz

Answers in Appendix III

1. In most ecosystems, the primary producers use energy from ______ to build organic compounds.
   a. sunlight
   b. heat
   c. breakdown of wastes and remains
   d. breakdown of inorganic substances in the habitat

2. Organisms at the lowest trophic level in a tallgrass prairie are all ______ .
   a. at the first step away from the original energy input
   b. autotrophs
   c. heterotrophs
   d. both a and b
   e. both a and c

3. Decomposers are commonly ______ .
   a. fungi b. plants c. bacteria d. a and c

4. All organisms at the first trophic level ______ .
   a. capture energy from a nonliving source
   b. obtain carbon from a nonliving source
   c. would be at the bottom of an energy pyramid
   d. all of the above

5. Primary productivity on land is affected by ______ .
   a. nutrient availability
   b. amount of sunlight
   c. temperature
   d. all of the above

6. If biological magnification occurs, the ______ will have the highest levels of toxins in their systems.
   a. producers
   b. herbivores
   c. primary carnivores
   d. top carnivores

7. Most of Earth's fresh water is ______ .
   a. in lakes and streams
   b. in aquifers and soil
   c. frozen as ice
   d. in bodies of organisms

8. Earth's largest carbon reservoir is ______ .
   a. the atmosphere
   b. sediments and rocks
   c. seawater
   d. living organisms

9. Carbon is released into the atmosphere by ______ .
   a. photosynthesis
   b. aerobic respiration
   c. burning fossil fuels
   d. b and c

10. Greenhouse gases ______ .
    a. slow the escape of heat energy from Earth into space
    b. are produced by natural and human activities
    c. are at higher levels than they were 100 years ago
    d. all of the above

11. The ______ cycle is a sedimentary cycle.
    a. water
    b. carbon
    c. nitrogen
    d. phosphorus

12. Earth's largest phosphorus reservoir is ______ .
    a. the atmosphere
    b. guano
    c. sediments and rocks
    d. living organisms

Data Analysis Exercise

To assess the impact of human activity on the carbon dioxide level in Earth's atmosphere, it helps to take a long view. One useful data set comes from deep core samples of Antarctic ice. The oldest ice core that has been fully analyzed dates back a bit more than 400,000 years. Air bubbles trapped in the ice provide information about the gas content in Earth's atmosphere at the time the ice formed. Combining ice core data with more recent direct measurements of atmospheric carbon dioxide—as in Figure 47.23—can help scientists put current changes in the atmospheric carbon dioxide into historical perspective.

1. What was the highest carbon dioxide level between 400,000 B.C. and 0 A.D.?
2. During this period, how many times did carbon dioxide reach a level comparable to that measured in 1980?
3. The industrial revolution occurred around 1800. What was the trend in carbon dioxide level in the 800 years prior to this event? What about in the 175 years after it?
4. Was the rise in the carbon dioxide level between 1800 and 1975 larger or smaller than the rise between 1980 and 2007?

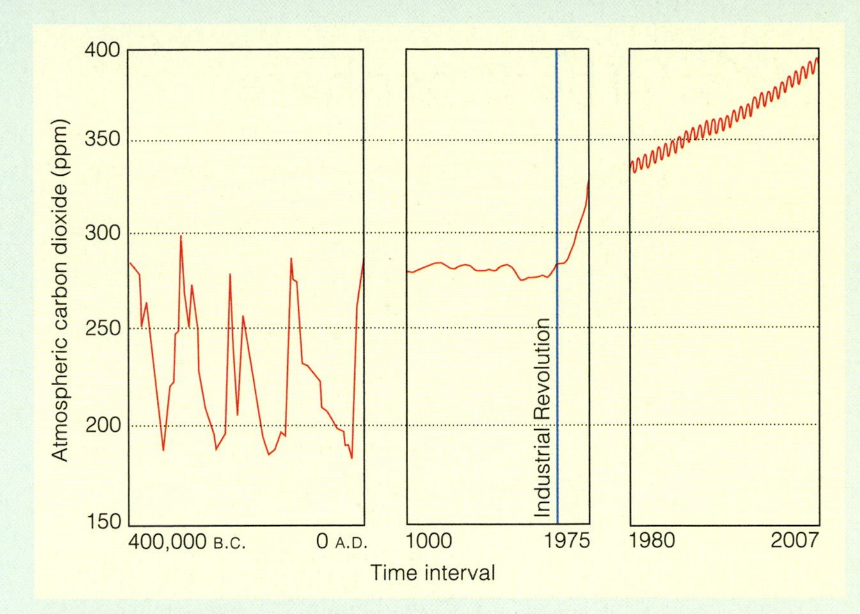

Figure 47.23 Changes in atmospheric carbon dioxide levels (in parts per million). Direct measurements began in 1980. Earlier data are based on ice cores.

13. Plant growth requires ______ uptake from the soil.
    a. nitrogen
    b. carbon
    c. phosphorus
    d. both a and c
    e. all of the above
14. Nitrogen fixation converts ______ to ______ .
    a. nitrogen gas; ammonia
    b. nitrates; nitrites
    c. ammonia; nitrogen gas
    d. ammonia; nitrates
    e. nitrogen gas; nitrogen oxides
15. Match each term with its most suitable description.

___producers	a. steps from energy source
___herbivores	b. feed on small bits of organic matter
___decomposers	c. degrade organic wastes and remains to inorganic forms
___detritivores	d. capture sunlight energy
___trophic level	e. feed on plants
___biological magnification	f. toxins accumulate

■ *Visit CengageNOW for additional questions.*

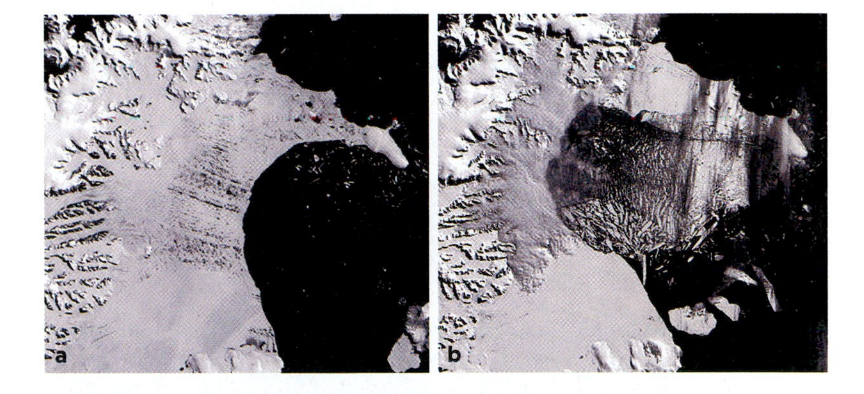

Figure 47.24 Antarctica's Larsen B ice shelf in (**a**) January and (**b**) March 2002. About 720 billion tons of ice broke from the shelf, forming thousands of icebergs. Some of the icebergs project 25 meters (82 feet) above the surface of the ocean. About 90 percent of an iceberg's volume is hidden underwater.

Critical Thinking

1. Marguerite has a vegetable garden in Maine. Eduardo has one in Florida. What are some of the variables that influence primary production in each place?
2. Where does your water come from? A well, a reservoir? Beyond that, what area is included within your watershed and what are the current flows like? Visit the Science in Your Watershed site at water.usgs.gov/wsc and research these questions.
3. Look around you and name all of the objects, natural or manufactured, that might be contributing to amplification of the greenhouse effect.
4. Polar ice shelves are vast, thickened sheets of ice that float on seawater. In March 2002, 3,200 square kilometers (1,250 square miles) of Antarctica's largest ice shelf broke free from the continent and shattered into thousands of icebergs (Figure 47.24). Scientists knew the ice shelf was shrinking and breaking up, but this event was the single largest loss ever observed at one time. Why should this concern people who live in more temperate climates?
5. Nitrogen-fixing bacteria live throughout the ocean, from its sunlit upper waters to 200 meters (650 feet) beneath its surface. Recall that nitrogen is a limiting factor in many habitats. What effect would an increase in populations of marine nitrogen-fixers have on primary productivity in the waters? What effect would that change have on carbon uptake in those waters?

48 The Biosphere

IMPACTS, ISSUES Surfers, Seals, and the Sea

Professional surfer Ken Bradshaw has ridden a lot of waves, but one in particular stands out. In January of 1998, he found himself off the coast of Hawaii riding the biggest wave he had ever seen (Figure 48.1). It towered more than 12 meters (39 feet) high and gave him the ride of a lifetime.

That wave was one manifestation of a climate event that happens about every three to seven years. During such an event, Pacific waters along the west coast of South America and westward become warmer than normal. This change in water temperature leads to shifts in marine currents and wind patterns, and causes wave-generating winter storms.

The rise in water temperature also disrupts currents that normally carry nutrients from the deep ocean toward western coasts of the Americas. The resulting nutrient shortage slows the growth of marine primary producers, causing cascading effects throughout marine food webs. One effect, which most often begins around Christmas, is a shortage of fish in waters near the coast of Peru. Peruvian fisherman noted this pattern and named the periodic climate effect El Niño, meaning "the baby boy," in reference to the birth of Jesus.

The decline in fish populations during an El Niño can have devastating effects on marine mammals that normally feed on those fish. During the 1997–1998 El Niño, about half of the sea lions on the Galápagos Islands starved to death. California's population of northern fur seals also suffered a sharp decline.

The temperature change in Pacific waters during the 1997–1998 El Niño was the largest on record, and it affected climates around the world. Giant waves, including the one that Bradshaw rode, battered eastern Pacific coasts. Heavy rains caused massive flooding and landslides in California and Peru. At the same time, less rain than normal fell in Australia and Indonesia, leading to crop failures and wildfires.

As you will learn in this chapter, the circulation pattern of water in Earth's oceans is just one of the physical factors that affect the distribution of species through the biosphere. We define the biosphere as all the places where we find life on Earth. It includes the hydrosphere (the ocean, ice caps, and other bodies of water, liquid and frozen), the lithosphere (Earth's rocks, soils, and sediments), and the lower portions of the atmosphere (gases and particles that envelop Earth).

See the video! Figure 48.1 A powerful El Niño caused this enormous wave in the Pacific. It also affected fish populations, causing sea lion pups (photo at *left*) and seals to starve.

Key Concepts

Air circulation patterns

Air circulation patterns start with regional differences in energy inputs from the sun, Earth's rotation and orbit, and the distribution of land and seas. These factors give rise to the great weather systems and regional climates. **Sections 48.1, 48.2**

Ocean circulation patterns

Interactions among ocean currents, air circulation patterns, and landforms produce regional climates, which affect where different organisms can live. **Section 48.3**

Land provinces

Biogeographic realms are vast regions characterized by species that evolved nowhere else. They are divided into biomes characterized mainly by the dominant vegetation. Sunlight intensity, moisture, soil, and evolutionary history vary among biomes. **Sections 48.4–48.11**

Water provinces

Water provinces cover more than 71 percent of Earth's surface. All freshwater and marine ecosystems have gradients in light availability, temperature, and dissolved gases that vary daily and seasonally. The variations influence primary productivity. **Sections 48.12–48.16**

Applying the concepts

Understanding interactions among the atmosphere, ocean, and land can lead to discoveries about specific events—in one case, recurring cholera epidemics—that impact human life. **Section 48.17**

Links to Earlier Concepts

- With this chapter, you reach the highest level of organization in nature (Section 1.1).
- You will learn more about soils (29.1), distribution of primary productivity (47.3), carbon-fixing pathways (7.7), and the effects of deforestation (Chapter 23 introduction).
- Our discussions of aquatic provinces will draw on your knowledge of properties of water (2.5), acid rain (2.6, 47.9), the water cycle (47.6), and eutrophication (47.10).
  You will learn more about coral reefs (25.5) and life at hydrothermal vents (20.2).
- You will be reminded of the effects of fossil fuel use (23.5), including global warming (47.8). You will learn about threats to the ozone layer (20.3).
- The chapter ends with an example of a scientific approach to problem solving (1.6, 1.7).

How would you vote? We cannot stop an El Niño from happening, but we might be able to minimize its severity. Would you support the use of taxpayer dollars to fund research into the causes and effects of El Niño? See CengageNOW for details, then vote online.

48.1 Global Air Circulation Patterns

- How much solar energy reaches Earth's surface varies from place to place and with the season.
- Link to Fossil fuels 23.5

Air Circulation and Regional Climates

Climate refers to average weather conditions, such as cloud cover, temperature, humidity, and wind speed, over time. Regional climates differ because the factors that influence winds and ocean currents—intensity of sunlight, the distribution of land masses and seas, and elevation—vary from place to place.

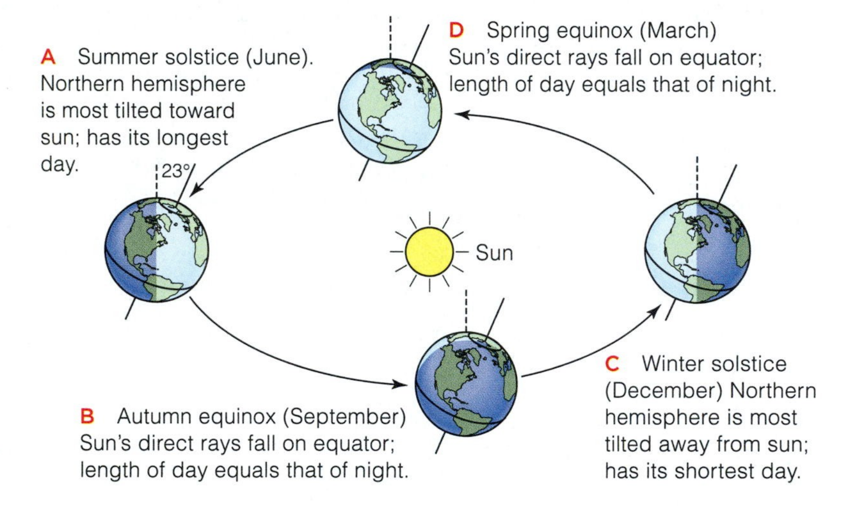

Figure 48.2 **Animated** Earth's tilt and yearly rotation around the sun cause seasonal effects. The 23° tilt of Earth's axis causes the Northern Hemisphere to receive more intense sunlight and have longer days in summer than in winter.

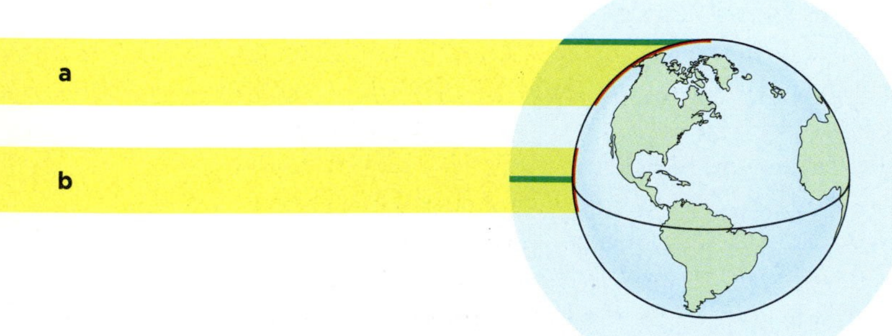

Figure 48.3 Variation in intensity of solar radiation with latitude. For simplicity, we depict two equal parcels of incoming radiation on an equinox, a day when incoming rays are perpendicular to Earth's axis.

Rays that fall on high latitudes (**a**) pass through more atmosphere (*blue*) than those that fall near the equator (**b**). Compare the length of the *green* lines. Atmosphere is not to scale.

Also, energy in the rays that fall at the high latitude is spread over a greater area than energy that falls on the equator. Compare the length of the *red* lines.

Each year, Earth rotates around the sun in an elliptical path (Figure 48.2). Seasonal changes arise because Earth's axis is not perpendicular to the plane of this ellipse, but rather is tilted about 23 degrees. In June, when the Northern Hemisphere is angled toward the sun, it receives more intense sunlight and has longer days than the Southern Hemisphere (Figure 48.2*a*). In December, the opposite occurs (Figure 48.2*c*). Twice a year—on spring and autumn equinoxes—Earth's axis is perpendicular to incoming sunlight. On these days, every place on Earth receives 12 hours of daylight and 12 hours of darkness (Figure 48.2*b*,*d*).

On any particular day, equatorial regions get more sunlight energy than higher latitudes for two reasons (Figure 48.3). First, fine particles of dust, water vapor, and greenhouse gases absorb some solar radiation or reflect it back into space. Because sunlight traveling to high latitudes passes through more atmosphere to reach Earth's surface than light traveling to the equator, less energy reaches the ground. Second, energy in any incoming parcel of sunlight is spread out over a smaller surface area at the equator than at the higher latitudes. As a result of these factors, Earth's surface warms more at the equator than at the poles.

This regional difference in surface warming is the start of global air circulation patterns (Figure 48.4). Warm air can hold more moisture than cooler air and is less dense, so it rises. Near the equator, air warms, picks up moisture from the oceans, and rises (Figure 48.4*a*). Air cools when it rises to higher altitudes and flows north and south, releasing moisture as rain that supports lush tropical rain forests. Deserts often form at latitudes of about 30°, where the drier and cooler air descends (Figure 48.4*b*). Farther north and south, the air picks up moisture again. It rises, and then releases moisture at latitudes of about 60° (Figure 48.4*c*). In the polar regions cold air that holds little moisture descends (Figure 48.4*d*). Precipitation is sparse, and polar deserts form.

Prevailing winds do not blow directly north and south because Earth's rotation and curvature influence the air circulation pattern. Air masses are not attached to Earth's surface, so as an air mass moves north or south this surface rotates beneath it, rotating faster at the equator than the poles. As a result, when viewed from Earth's surface, air masses that move north or south will seem to be deflected east or west, with the deflection greatest at high latitude (Figure 48.4*e*,*f*).

Regional winds occur where the presence of land masses cause differences in air pressure near Earth's surface. Because land absorbs and releases heat faster than water does, air rises and falls faster over land

Initial Pattern of Air Circulation

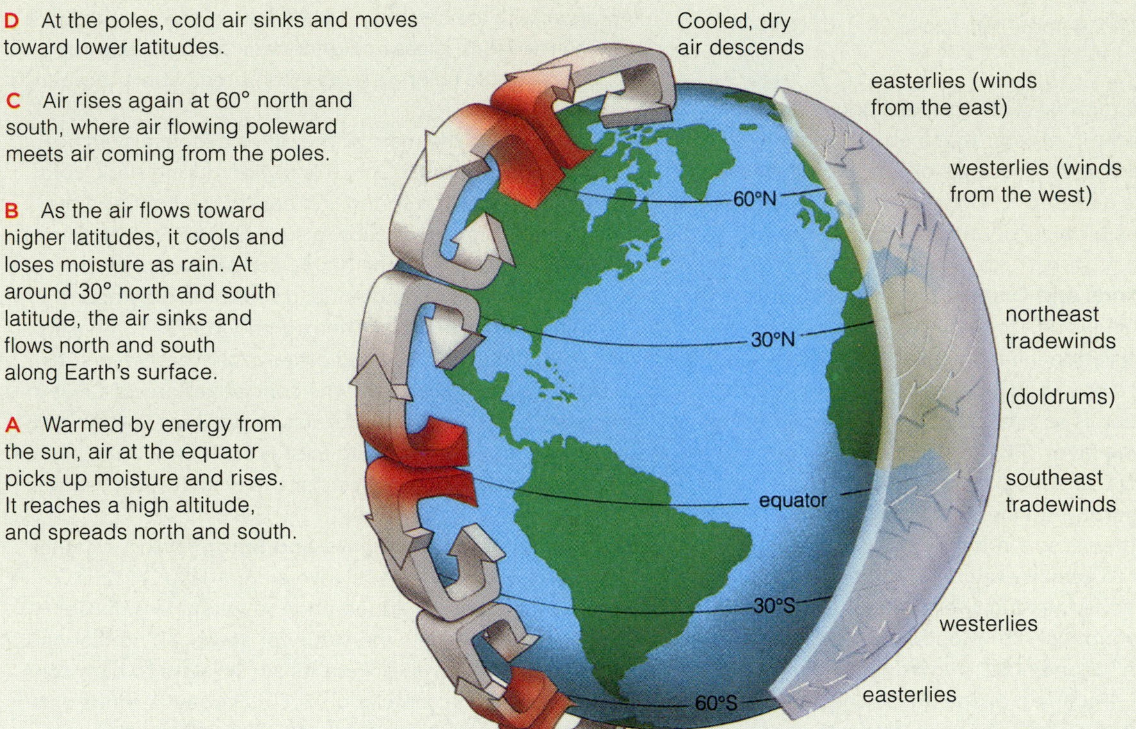

Prevailing Wind Patterns

E Major winds near Earth's surface do not blow directly north and south because of Earth's rotation. Winds deflect to the right of their original direction in the Northern Hemisphere and to the left in the Southern Hemisphere.

F For example, air moving from 30° south toward the equator is deflected to the left (west), as the southeast trade winds. The winds are named by the direction from which they blow.

Figure 48.4 Animated Global air circulation patterns and their effects on climate.

Figure It Out: What is the direction of prevailing winds in the central United States? *Answer: Winds blow from west to east.*

than it does over the ocean. Air pressure is lowest where air rises and greatest where air sinks.

Harnessing the Sun and Wind

The need for energy to support human activities continues to increase. Fossil fuels, including gasoline and coal, are nonrenewable energy sources (Section 23.5). Solar and wind energy are renewable. The amount of solar energy that Earth receives per year is about 10 times the energy of all fossil fuel reserves combined.

Solar energy can be harnessed directly to heat air or water that can then be pumped through buildings to heat them. Solar energy can also be captured by photovoltaic cells and used to generate electricity. The electricity can be used directly, stored in a battery, or used to form oxygen and hydrogen gases from water. Proponents of solar–hydrogen energy argue that it could end smog, oil spills, and acid rain without any of the risks of nuclear power. Hydrogen gas can fuel cars and heat buildings. However, hydrogen is a small molecule that leaks easily from pipelines or containers. How increased leakage of hydrogen into the air would affect the environment is unknown.

We use solar energy indirectly by harnessing winds. Wind energy is only practical where winds blow faster than 8 meters per second (18 miles per hour). Winds seldom blow constantly, but wind energy can charge batteries to supply power even on still days. Energy from winds of North and South Dakota alone could meet 80 percent of the United States' energy needs.

Wind farms have drawbacks. Turbine blades can be noisy and can kill birds and bats. Large facilities may alter local weather patterns. Also, some people see wind farms as a form of "visual pollution" that ruins otherwise scenic views and lowers property values.

Take-Home Message

What causes global air circulation patterns and differences in climate?

- Longitudinal differences in the amount of solar radiation reaching Earth produce global air circulation patterns.
- Earth's shape and rotation also affect air circulation patterns.

48.2 Something in the Air

- Particles and gases act as air pollutants that endanger human health and disrupt ecosystems.
- Links to Acid rain 2.6 and 47.9, Ozone 20.3, CFCs 47.8

A **pollutant** is a natural or synthetic substance released into soil, air, or water in greater than natural amounts; it disrupts normal processes because organisms evolved in its absence, or are adapted to lower levels of it. Today, air pollution threatens biodiversity and human health.

Swirling Polar Winds and Ozone Thinning High in Earth's atmosphere, molecules of ozone (O_3) absorb most of the ultraviolet (UV) radiation in incoming sunlight. Between 17 and 27 kilometers above sea level (10.5 and 17 miles), the ozone concentration is so great that scientists refer to this region as the **ozone layer** (Figure 48.5*a*).

In the mid-1970s, scientists started to notice that the ozone layer was getting thinner. Its thickness had always varied a bit with the season, but now there was steady decline from year to year. By the mid-1980s, the spring ozone thinning over Antarctica was so pronounced that people were calling it an "ozone hole" (Figure 48.5*b*).

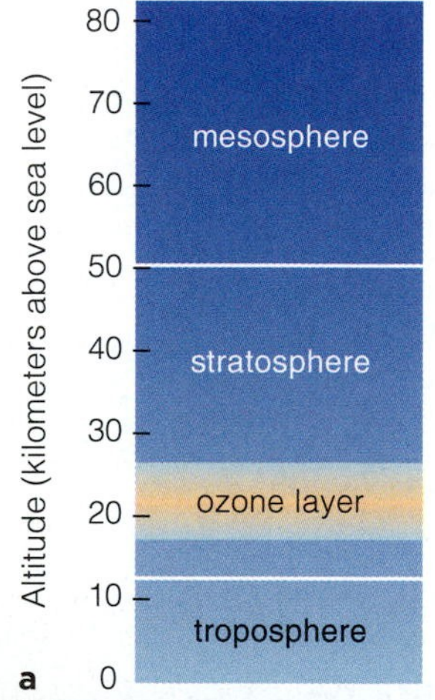

Declining ozone quickly became an international concern. With a thinner ozone layer, people would be exposed to more UV radiation and get more skin cancers (Section 14.5). Higher UV levels also harm wildlife, which do not have the option of rubbing on more sunscreen. Higher UV levels might even harm plants and other producers, slowing rates of photosynthesis and release of oxygen into the atmosphere.

Chlorofluorocarbons, or CFCs, are the main ozone destroyers. These odorless gases were once widely used as propellants in aerosol cans, as coolants, and in solvents and plastic foam. CFCs interact with ice crystals and UV light in the stratosphere. These reactions release chlorine radicals that degrade ozone. A single chlorine radical can break apart thousands of ozone molecules.

Ozone thins the most at the poles because swirling winds concentrate CFCs in this region during dark, cold polar winters. In the spring, increasing daylight and the presence of ice clouds allow a surge in the formation of chlorine radicals from the highly concentrated CFCs.

In response to the potential threat posed by ozone thinning, developed countries agreed in 1992 to phase out the production of CFCs and other ozone destroyers. As a result of that agreement, the concentrations of CFCs in the atmosphere are now starting to decline (Section 47.8). However, they are expected to stay high enough to significantly affect the ozone layer for the next twenty years.

No Wind, Lots of Pollutants, and Smog Often, weather conditions cause a thermal inversion: A layer of cool, dense air becomes trapped under a warm, less dense layer. Trapped air sets the stage for **smog**, an atmospheric condition in which air pollutants accumulate to high concentration. The accumulation occurs because winds cannot disperse pollutants trapped under a thermal inversion layer (Figure 48.6). Thermal inversions have contributed to some of the highest recorded air pollution levels.

Industrial smog forms as a gray haze over cities that burn a lot of coal and other fossil fuels during cold, wet winters. Photochemical smog forms above big cities in warm climate zones. Photochemical smog is most dense over cities in natural topographic basins, such as Los Angeles and Mexico City. Exhaust fumes from vehicles contain nitric oxide, a pollutant that combines with oxygen and forms nitrogen dioxide. Exhaust fumes also contain

Figure 48.5 Animated (**a**) The atmospheric layers. Ozone concentrated in the stratosphere helps shield life from UV radiation. (**b**) Seasonal ozone thinning above Antarctica in 2001. *Dark blue* represents the low ozone concentration, at the ozone hole's center.

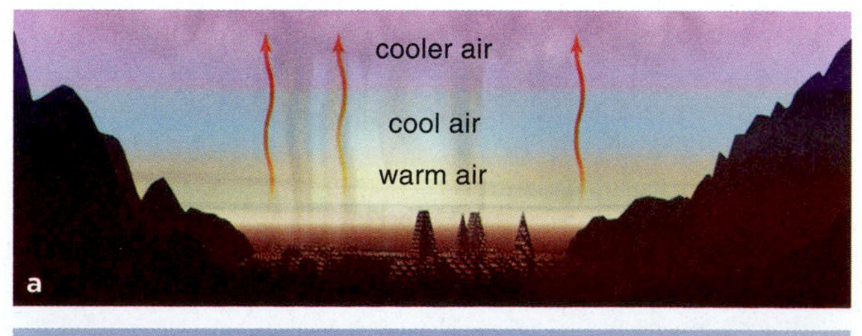

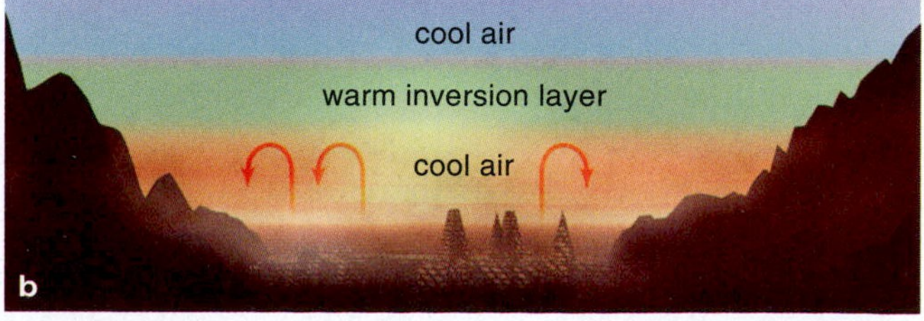

Figure 48.6 (**a**) Normal air circulation in smog-forming regions. (**b**) Air pollutants trapped under a thermal inversion layer.

hydrocarbons that react with nitrogen dioxide to form ozone and other photochemical oxidants. A high ozone level in the lower atmosphere harms plants and animals.

Winds and Acid Rain Coal-burning power plants, smelters, and factories emit sulfur dioxides. Vehicles, power plants that burn gas and oil, and nitrogen-rich fertilizers emit nitrogen oxides. In dry weather, airborne oxides coat dust particles and fall as dry acid deposition. In moist air, they form nitric acid vapor, sulfuric acid droplets, and sulfate and nitrate salts. Winds typically disperse these pollutants far from their source. They fall to Earth in rain and snow. We call this a wet acid deposition, or **acid rain**.

The pH of typical rainwater is about 5 (Section 2.6). Acid rain can be 10 to 100 times more acidic—as potent as lemon juice! It corrodes metals, marble, rubber, plastics, nylon stockings, and other materials. It alters soil pH and can kill trees (Section 47.9) and other organisms.

Rain in much of eastern North America is thirty to forty times more acidic than it was even a few decades ago (Figure 48.7*a*). The heightened acidity has caused fish populations to vanish from more than 200 lakes in the Adirondack Mountains of New York (Figure 48.7*b*). It also is contributing to the decline of forests.

Windborne Particles and Health Pollen, fungal spores, and other natural particles are carried aloft by winds, along with pollutant particles of many sizes. Inhaling small particles can irritate nasal passages, the throat, and lungs. It triggers asthma attacks and can increase their severity. The smallest particles are most likely to reach the lungs, where they can interfere with respiratory function.

Exhaust from vehicles is a major source of particulate pollution. Diesel-fueled engines are the worst offenders because they emit more of the smallest, most dangerous particles than their gasoline-fueled counterparts.

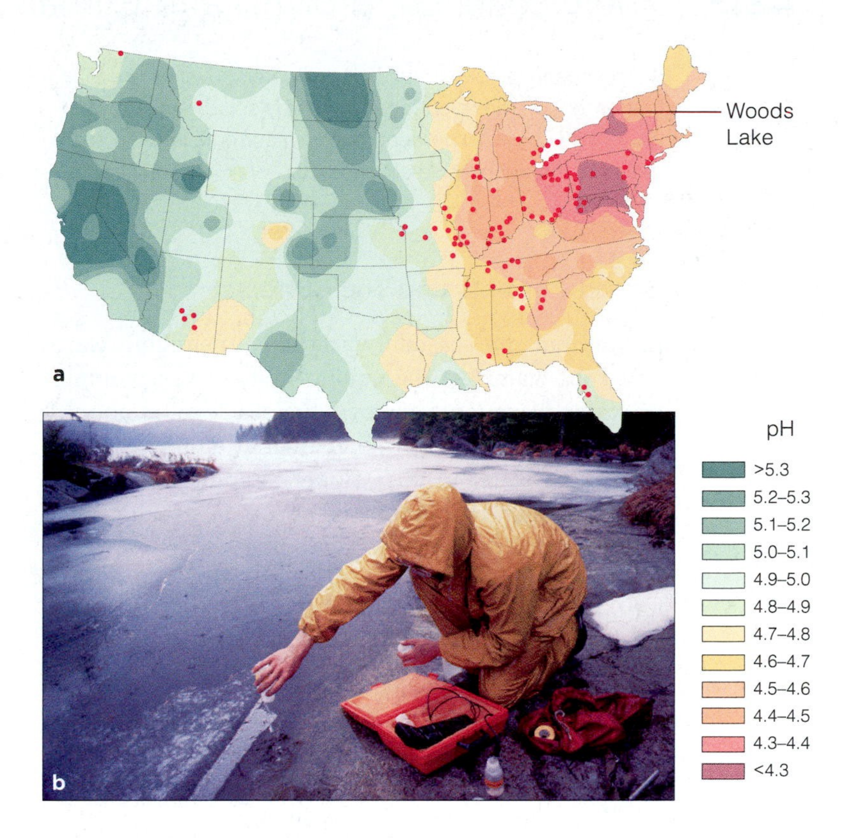

Figure 48.7 Animated (**a**) Average precipitation acidities in the United States in 1998. (**b**) Biologist measuring the pH of New York's Woods Lake. In 1979, the lake water's pH was 4.8. Since then, experimental addition of calcite to soil around the lake has successfully raised the pH of the water to more than 6.

Regardless of their source, air pollutants travel on the winds across continents and the open ocean. As Figure 48.8 shows, airborne pollutants do not stop at national borders. We all share the same air.

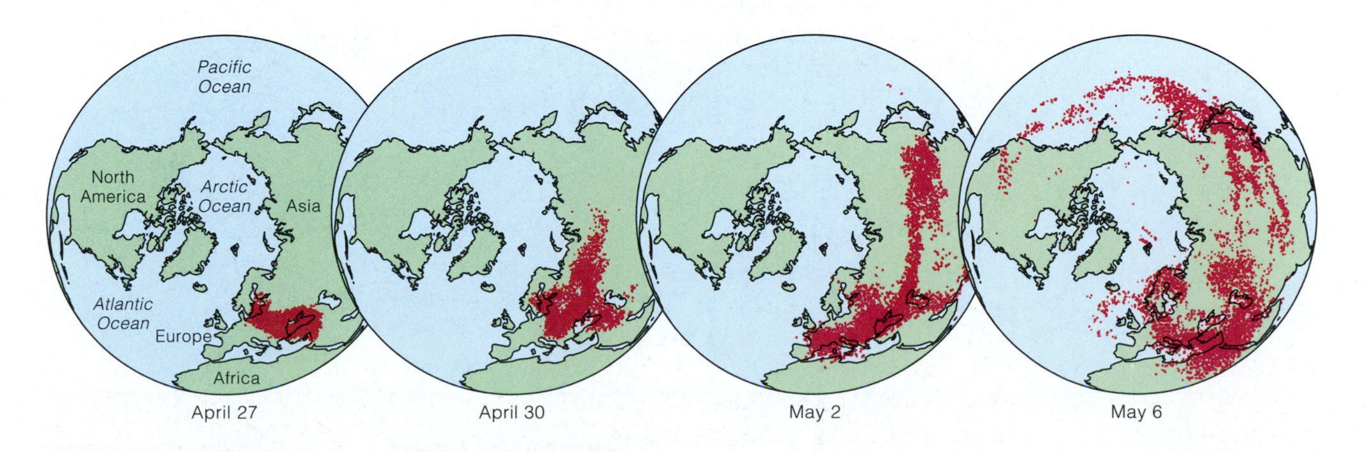

Figure 48.8 Global distribution of radioactive fallout released during the 1986 meltdown of the Chernobyl nuclear power plant in Ukraine. The meltdown allowed radioactive particles to enter the air, then winds dispersed them around the world. The incidence of thyroid cancers in Ukraine and neighboring Belarus continues to rise, a legacy of childhood exposure to high radiation levels.

48.3 The Ocean, Landforms, and Climates

■ The ocean, a continuous body of water, covers more than 71 percent of Earth's surface. Driven by solar heat and wind friction, its upper 10 percent moves in currents that distribute nutrients through marine ecosystems.

Ocean Currents and Their Effects

Latitudinal and seasonal variations in sunlight warm and cool water. At the equator, where vast volumes of water warm and expand, the sea level is about 8 centimeters (3 inches) higher than at either pole. The volume of water in this "slope" is enough to get sea surface water moving in response to gravity, most often toward the poles. The moving water warms air above it. At midlatitudes, oceans transfer 10 million billion calories of heat energy per second to the air!

Enormous volumes of water flow as ocean currents. The force of major winds, Earth's rotation, and topography determine the directional movement of these currents. Surface currents circulate clockwise in the Northern Hemisphere and counterclockwise in the Southern Hemisphere (Figure 48.9).

Swift, deep, and narrow currents of nutrient-poor water flow away from the equator along the east coast of continents. Along the east coast of North America, warm water flows north, as the Gulf Stream. Slower, shallower, broader currents of cold water parallel the west coast of continents and flow toward the equator.

Ocean currents affect climates. Coasts in the Pacific Northwest are cool and foggy in summer because the cold California current chills the air, so water condenses out as droplets. Boston and Baltimore are muggy in summer because air masses pick up heat and moisture from the warm Gulf Stream, then deliver it to these cities.

Ocean circulation patterns shift over geologic time as land masses move (Section 17.9). Some worry that global warming could also alter these patterns.

Rain Shadows and Monsoons

Mountains, valleys, and other surface features of the land affect climate. Suppose you track a warm air mass after it picks up moisture off California's coast. It moves inland, as wind from the west, and piles up

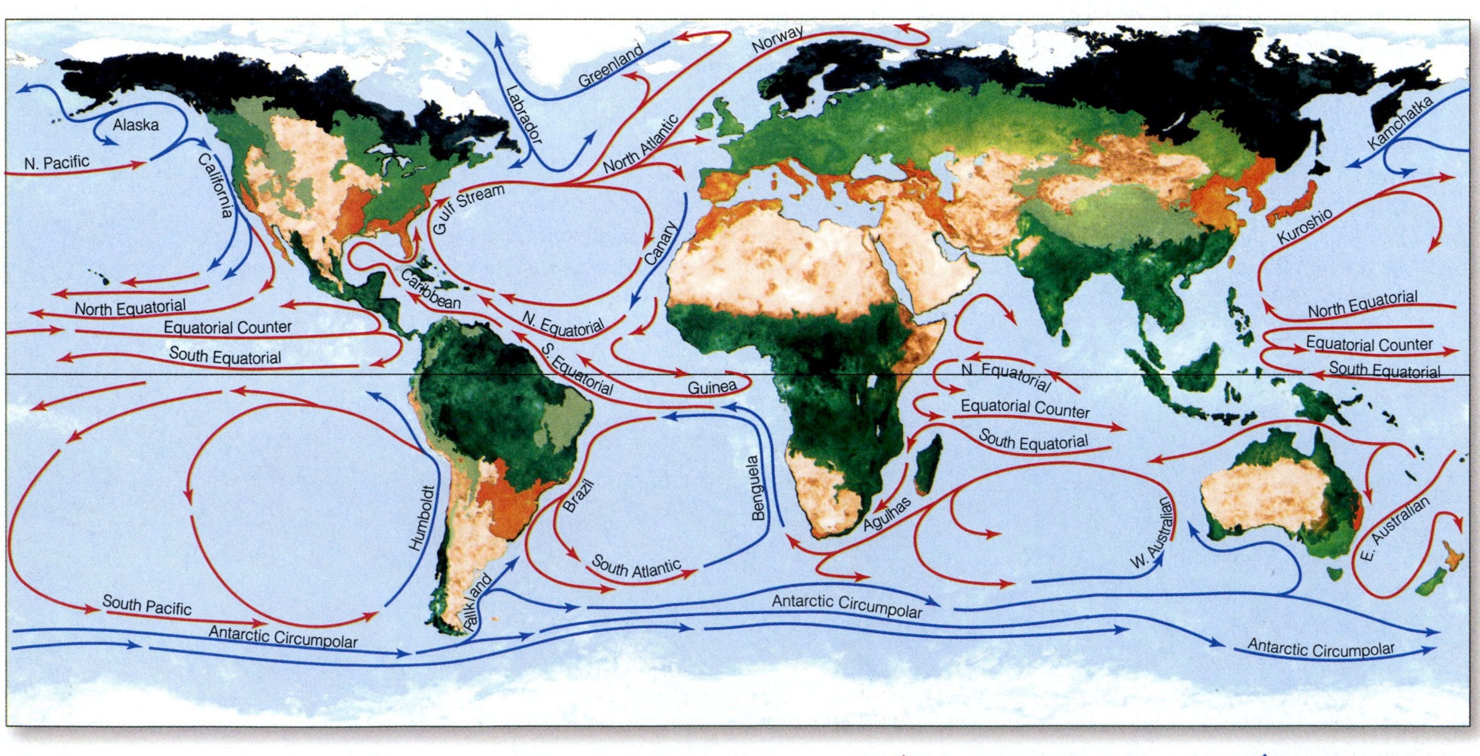

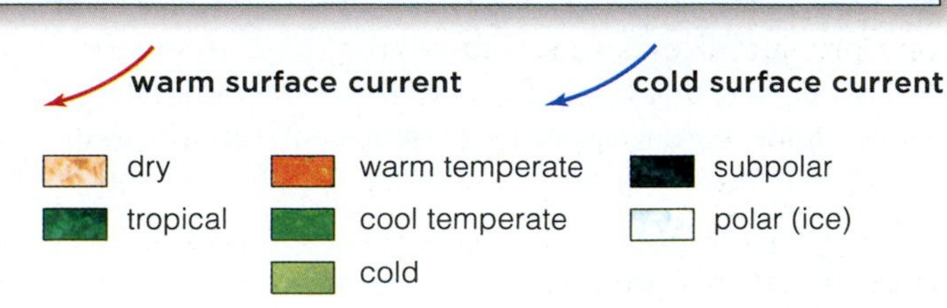

Figure 48.9 Animated Major climate zones correlated with surface currents of the world ocean. Warm surface currents start moving from the equator toward the poles, but prevailing winds, Earth's rotation, gravity, the shape of ocean basins, and landforms influence the direction of flow. Water temperatures, which differ with latitude and depth, contribute to the regional differences in air temperature and rainfall.

Figure 48.10 **Animated** Rain shadow effect. On the side of mountains facing away from prevailing winds, rainfall is light. *Black* numbers signify annual precipitation, in centimeters, averaged on both sides of the Sierra Nevada, a mountain range. *White* numbers signify elevations, in meters.

against the Sierra Nevada. This high mountain range parallels the distant coast. The air cools as it rises in altitude and loses moisture as rain (Figure 48.10). The result is a **rain shadow**—a semiarid or arid region of sparse rainfall on the leeward side of high mountains. "Leeward" is the side facing away from the wind. The Himalayas, Andes, Rockies, and other great mountain ranges cause vast rain shadows.

Differences in the heat capacity of water and land give rise to coastal breezes. In the daytime, water does not warm as fast as the land. Air heated by the warm land rises, and cool offshore air moves in to replace it (Figure 48.11*a*). After sundown, land becomes cooler than the water, so the breezes reverse (Figure 48.11*b*).

Differential heating of water and land also causes **monsoons**, winds that change their direction seasonally. For example, the continental interior of Asia heats up in the summer, so air rises above it. The resulting low pressure draws in moisture from over the warm Indian Ocean to the south, and these north-blowing winds deliver heavy rains. In the winter, the continental interior is cooler than the ocean. As a result, cool, dry winds blowing from the north toward southern coasts, cause a seasonal drought.

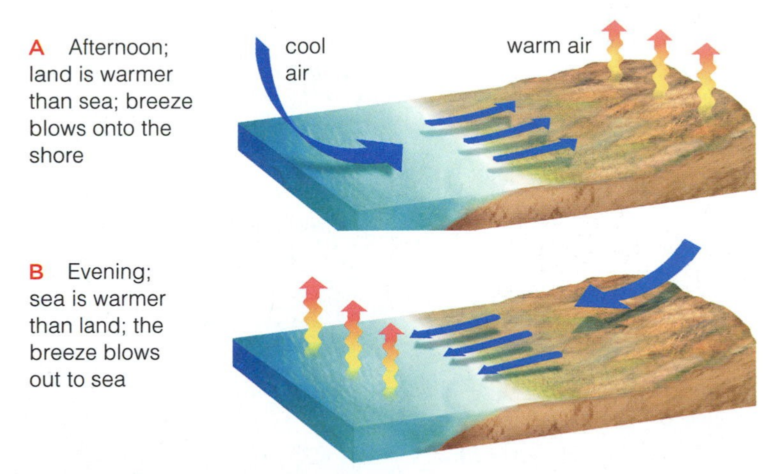

Figure 48.11 **Animated** Coastal breezes.

Take-Home Message

How do ocean currents arise and how do they affect regional climates?

- Surface ocean currents, which are set in motion by latitudinal differences in solar radiation, are affected by winds and by Earth's rotation.
- Collective effects of air masses, oceans, and landforms determine regional temperature and moisture levels.

48.4 Biogeographic Realms and Biomes

- Regions with different physical conditions support different types of organisms.
- Links to Biogeography 17.1, Plate tectonics 17.9

Suppose you live in the coastal hills of California and decide to tour the Mediterranean coast, the southern tip of Africa, and central Chile. In each region, you see highly branched, tough-leafed woody plants that look a lot like the highly branched, tough-leafed chaparral plants back home. Vast geographic and evolutionary distances separate the plants. Why are they alike?

You decide to compare their locations on a global map and discover that American and African desert plants live about the same distance from the equator. Chaparral plants and their distant look-alikes all grow along the western and southern coasts of continents between latitudes 30° and 40°. You have noticed one of many patterns in the global distribution of species.

Early naturalists divided Earth's land masses into six **biogeographic realms**—vast expanses where they could expect to find communities of certain types of plants and animals (Figure 48.12). For example, palm trees and camels live in the Ethiopian realm. In time, the six classic realms became subdivided.

Biomes are finer subdivisions of the land realms, but they are still identifiable on a global scale. Most biomes occur on more than one continent. For instance, dry forest (coded orange in Figure 48.12) covers vast regions of South America, India, and Asia. Similarly, the North American prairie, South American pampa, southern Africa veld, and Eurasian steppe are all types of temperate grasslands (Figure 48.13).

The distribution of biomes is influenced by climate (especially temperature and patterns of rainfall), soil type, and interactions among the array of species that make up their communities. Consumers are adapted to the dominant vegetation. Each species, remember,

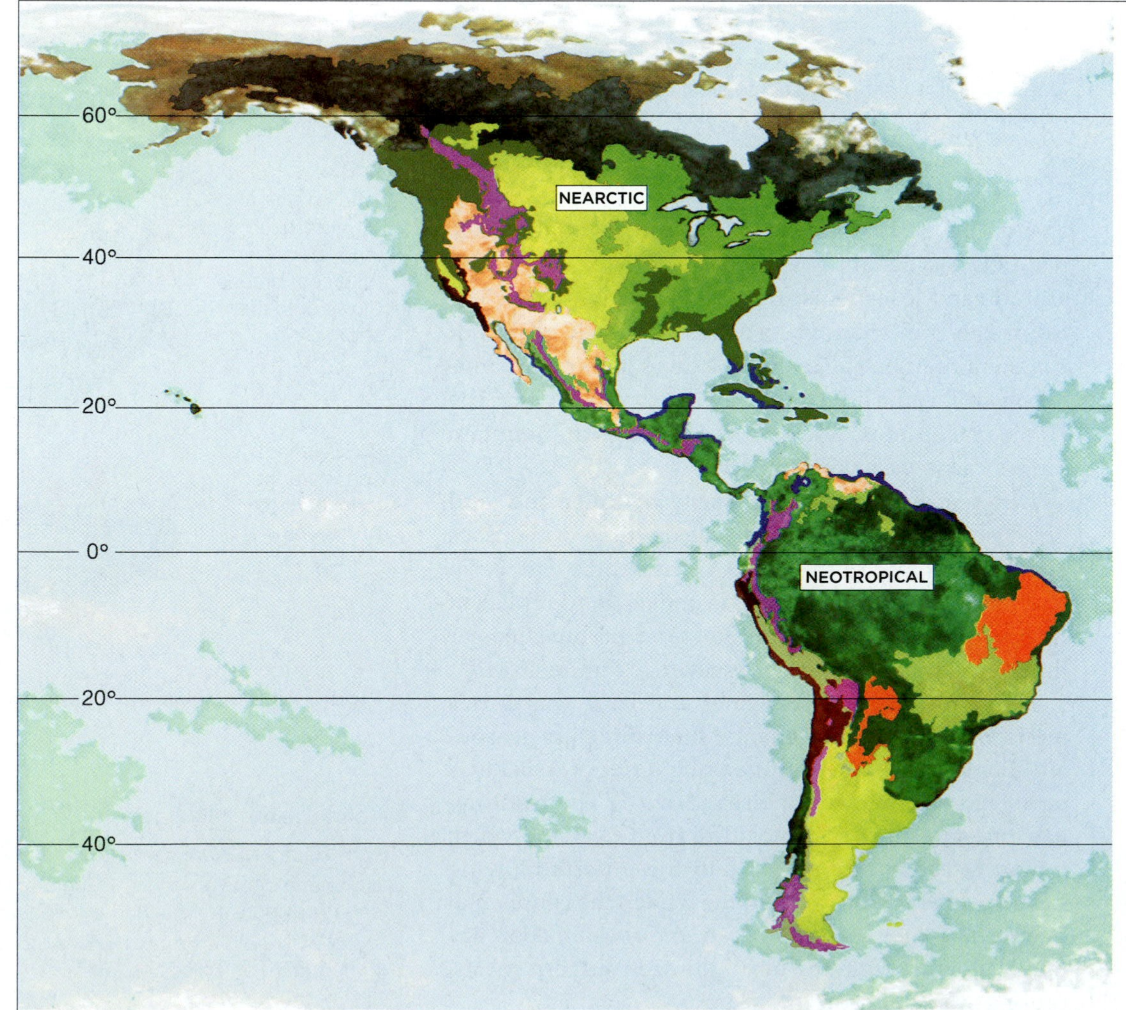

Figure 48.12 Animated Global distribution of major categories of biomes and marine ecoregions.

shows adaptations in its form, function, behavior, and life history pattern.

Distribution of biomes has also been influenced by evolutionary history. For example, species that evolved together on Pangea ended up on different land masses after this supercontinent broke up (Section 17.9).

Similarly, environmental features and evolutionary history helped shape the distribution of species in the seas. Figure 48.12 shows the key marine ecoregions as well as Earth's biomes.

Take-Home Message

What are biomes?

- Biomes are vast expanses of land dominated by distinct kinds of plants that support characteristic communities.
- The global distribution of biomes is a result of topography, climate, and evolutionary history.

Figure 48.13 Two examples of temperate grassland biome. *Top*, Argentine pampa. *Bottom*, Mongolian steppe. See also Figure 48.16.

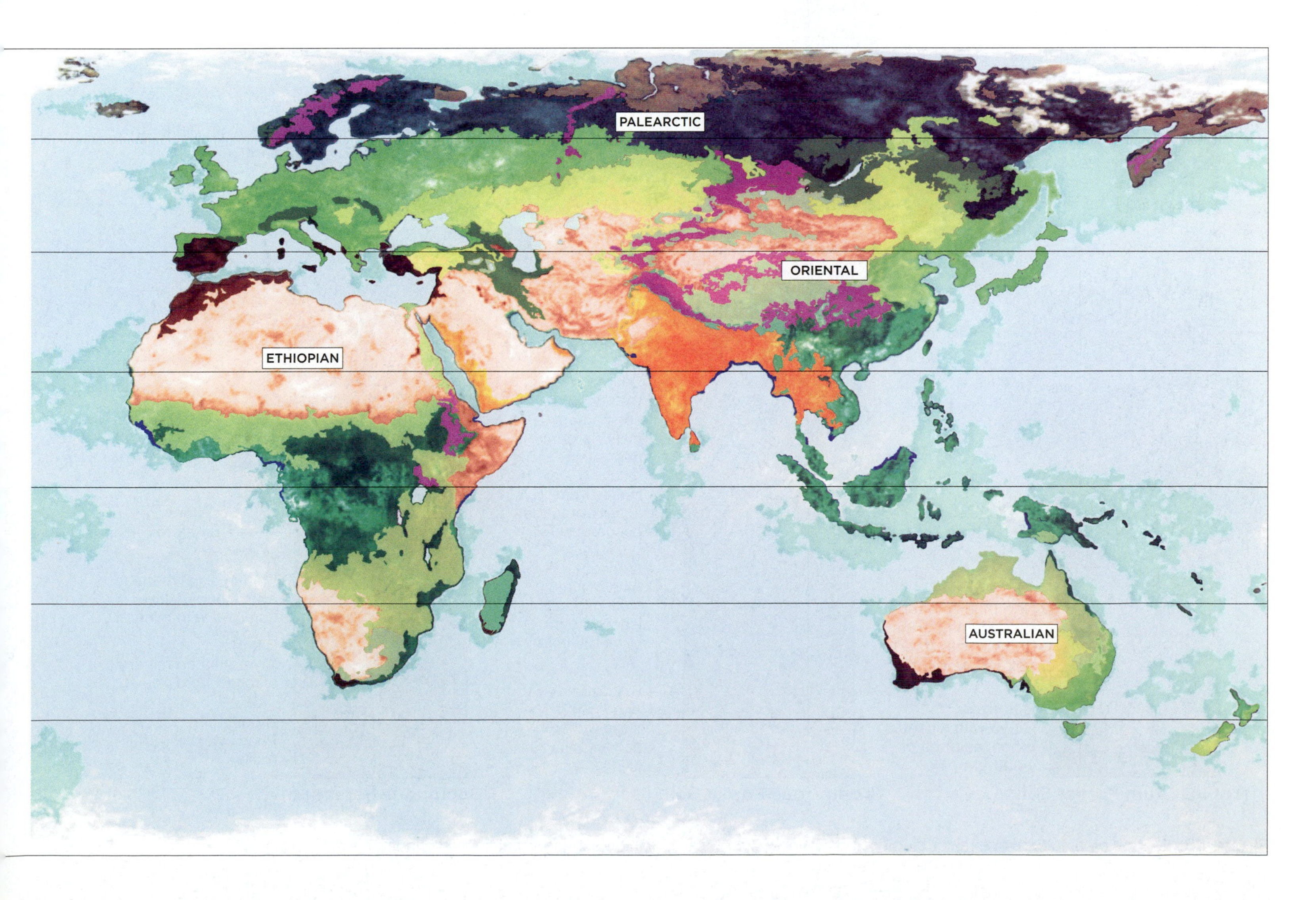

48.5 Soils of Major Biomes

- Plants obtain necessary nutrients from soil. As a result, properties of soil have a big impact on primary production.
- Link to Soil profiles 29.1

Plants obtain water and dissolved mineral ions from soil. Soil consists of mineral particles and decomposing organic matter called humus (Section 29.1). Water and air fill spaces between soil particles.

The properties of soils vary. Clay is richest in minerals, but its fine, close-packed particles drain poorly; there is little air space for roots to take up oxygen. In gravelly or sandy soils, leaching draws off water and mineral ions. Most plants grow best in a soil that is a mixture of different-sized particles and has a moderate amount of humus.

Each biome has a **soil profile**, a layered structure that develops over time (Figures 29.2 and 48.14). The top layers are surface litter (the O horizon) and topsoil (the A horizon). Topsoil is the most important layer for plant growth. Deserts have little topsoil, and their soil is nutrient-poor and high in salts. Grasslands have the deepest topsoil; it can be more than one meter thick. This is why grasslands are favored for conversion to agriculture. In tropical rain forests, decomposition is rapid, so little topsoil can accumulate above the poorly draining lower layers of soil. In coniferous and deciduous temperate forests, decomposition proceeds more slowly, so leaf litter accumulates and upper soil layers tend to be richer.

Take-Home Message

How do soils affect biome characteristics?

- Each biome has a characteristic soil profile, with different amounts of inorganic and organic components.
- Properties of topsoil are the most important for plant growth.

O horizon: Pebbles, little organic matter
A horizon: Shallow, poor soil
B horizon: Evaporation causes salt build-up; leaching removes nutrients
C horizon: Rock fragments from uplands
Desert Soil

A horizon: Alkaline, deep, rich in humus
B horizon: Percolating water enriches layer with calcium carbonates
Grassland Soil

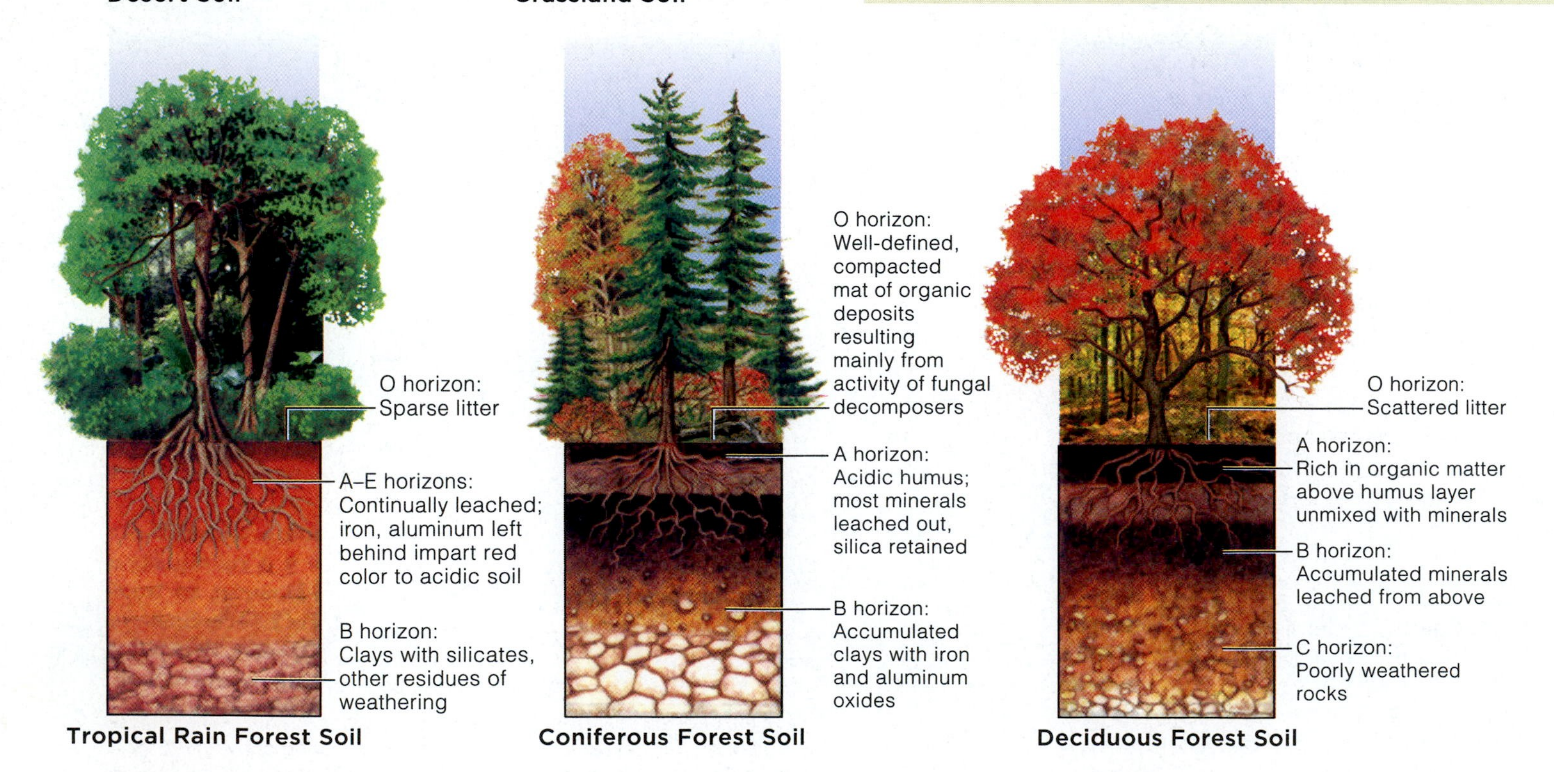

Figure 48.14 Soil profiles for some major biomes. The A horizon, or topsoil, is the most important source of nutrients for plant growth.

48.6 Deserts

- With this section we begin a survey of the major biomes. Our first stop is deserts, which are defined by low rainfall.
- Links to Carbon-fixing pathways 7.7, Atacama Desert 20.6, Desert kangaroo rat 41.3

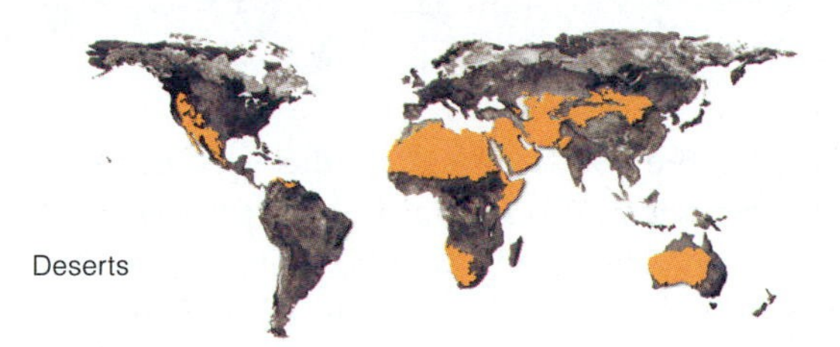

Deserts are regions that receive an average of less than 10 centimeters (4 inches) of rain per year. Most are located at about 30° north and south latitude, where air depleted of moisture sinks. In these regions, the low humidity allows much sunlight to reach the soil surface, so the ground heats fast during the day. Low humidity also causes the ground to cool fast at night. Soils are typically nutrient poor and somewhat salty. Despite these forbidding conditions, some plants and animals survive, especially in areas where moisture is available in more than one season (Figure 48.15).

Many desert plants have adaptations that reduce water loss. Light-colored spines or hairs can help keep humidity around the stomata high and also reflect sunlight. Alternative carbon-fixing pathways also help desert plants conserve water (Section 7.7). Cactuses and agaves are CAM plants and open their stomata only at night. Many annuals that live in deserts are C4 plants. Woody desert shrubs such as mesquite and creosote have extensive, efficient root systems that take up the little water that is available. Mesquite roots have been found as deep as 60 meters beneath the soil surface.

Animals also have adaptations that allow them to conserve water. The desert kangaroo rat discussed in Section 41.3 is a resident of the Sonoran Desert. So are the animals shown in Figure 48.15.

The driest of all deserts may be Chile's cool Atacama Desert, which lies in a rain shadow behind the Andes. Parts of this area are so dry that they were thought to be entirely lifeless. However, scientists recently found bacteria deep in the soil (Section 20.6).

Figure 48.15 Two parts of the same biome—the Sonoran Desert in Arizona. The sun's rays are just as intense in desert lowlands (**a**) as in the uplands (**b**), but differences in water availability, temperature, and soil types influence plant growth. Creosote bush (*Larrea*) dominates lowlands. A greater variety of plants survive in the somewhat wetter and cooler uplands.

Examples of desert animals. (**c**) The Sonoran desert tortoise escapes the heat by burrowing. (**d**) Lesser long-nosed bats spend spring and summer in the Sonoran Desert, where they avoid the daytime heat in caves and abandoned mine shafts. The bats are important pollinators of cactuses and agaves.

Take-Home Message

What features characterize desert biomes?

- A desert gets very little rain and has low humidity. There is plenty of sunlight, but poor soil and lack of water prevent most plants from surviving here.
- Many plants and animals in deserts have adaptations that minimize their need for water.

48.7 Grasslands, Shrublands, and Woodlands

- Where there is more rain than in deserts, grasses take hold. In areas with a bit more rain still, shrubs take hold.
- Links to Soil erosion 29.1, Tallgrass prairie 47.1

Grassland, shrublands, and woodlands

Grasslands

Grasslands form in the interior of continents between deserts and temperate forests (Figure 48.16). Summers are warm, and winters are cold. Annual rainfall of 25 to 100 centimeters (10–40 inches) keeps desert from forming, but is too little water to support forest. Low-growing primary producers tolerate strong winds, sparse and infrequent rain, and intervals of drought. Growth tends to be seasonal. Constant trimming by grazing animals, along with periodic fires, keeps trees and most shrubs from taking hold.

Shortgrass and tallgrass prairie (Figure 48.16*a*,*b*) are North America's main grasslands. Perennial grasses that fix carbon by the water-conserving C4 pathway dominate these biomes. Grass roots extend profusely through the topsoil and help hold it in place, preventing erosion by the constant winds.

During the 1930s, much of the shortgrass prairie of the American Great Plains was plowed under to grow wheat. The strong winds, a prolonged drought, and unsuitable farming practices turned much of the region into what the newspapers of that time called the Dust Bowl. John Steinbeck's historical novel *The Grapes of Wrath*, eloquently describes the human costs of this environmental disaster.

Tallgrass prairie (Section 47.1) once covered 140 million acres, mostly in Kansas. Tall grasses, legumes, and herbaceous plants such as daisies thrived in the continent's interior, which had somewhat richer topsoil and slightly more frequent rainfall than the shortgrass prairie. Nearly all tallgrass prairie has been converted to cropland. The Tallgrass Prairie National Preserve was created in 1996 to protect the little that remains.

Figure 48.16 Three examples of grasslands. (**a**) Tallgrass prairie in eastern Kansas. See also Figure 47.3. (**b**) Bison grazing in shortgrass prairie in South Dakota. (**c**) A herd of wildebeest graze in the African savanna. Figure 48.13 shows additional grasslands.

Figure 48.17 California chaparral. (**a**,**b**) Dominant plants are mostly branching, woody evergreens less than 2 meters (6 feet) tall, with leathery leaves. The leaves often contain oils that deter herbivores and also make the plants highly flammable.

(**c**) A firestorm in the chaparral-covered hills above Malibu. Today, most fires that occur in this biome are caused by humans. (**d**) Toyon (*Heteromeles arbutifolia*), a fire-adapted chaparral shrub, resprouting from its roots after a fire.

Savannas are broad belts of grasslands with a few scattered shrubs and trees. Savannas lie between the tropical forests and hot deserts of Africa, India, and Australia. Temperatures are warm year-round. During the rainy season 90–150 centimeters (35–60 inches) of rain falls. Fires that occur during the dry season help keep forest from replacing the grassland.

Africa's savanna's are famous for their abundant wildlife (Figure 48.16*c*). Herbivores include giraffes, zebras, elephants, a variety of antelopes, and immense herds of wildebeests. Lions and hyenas are carnivores that eat the grazers.

Dry Shrublands and Woodlands

Dry shrublands receive less than 25 to 60 centimeters (10–24 inches) of rain annually. We see them in South Africa, in Mediterranean regions, and in California, where they are known as chaparral. California has about 6 million acres of chaparral (Figure 48.17*a*,*b*).

Rains occur seasonally, and lightning-sparked fires sometimes sweep through shrublands during the dry season. In California, where homes are often built near chaparral, the fires frequently cause property damage (Figure 48.17*c*). Foliage of many chaparral shrubs is highly flammable. However, the plants have adapted to occasional fires. Some grow back from root crowns after a fire (Figure 48.17*d*). Seeds of other chaparral species germinate only after they are exposed to heat or smoke, ensuring that the seeds sprout only when young seedlings face little competition.

Dry woodlands prevail where the annual rainfall is 40 to 100 centimeters (16–40 inches). Drought-tolerant trees are often tall, but do not form a continuous canopy. Examples include the eucalyptus forests of Australia, and the oak forests of California and Oregon.

Take-Home Message

What are grasslands, dry shrublands, and woodlands?

- Grasslands form in the interior of continents. Grasses and other short, nonwoody plants predominate. Activity of grazing animals and occasional fires help prevent trees and shrubs from taking hold.
- Dry shrublands such as California's chaparral also include fire-adapted species—predominantly short, woody shrubs.
- Dry woodlands are dominated by trees that are adapted to withstand seasonal drought.

48.8 More Rain, Broadleaf Forests

- Broadleaf (angiosperm) trees dominate moist forests in both temperate and equatorial regions.
- Links to Pigments 7.1, Plant nutrients 29.1

Semi-Evergreen and Deciduous Broadleaf Forests

Semi-evergreen forests occur in the humid tropics of Southeast Asia and India. These forests include a mix of broadleaf trees that retain leaves year round, and deciduous broadleaf trees. Deciduous trees or shrubs shed leaves once a year, prior to the season when cold or dry conditions would not favor growth. Deciduous trees in a semi-evergreen forest shed their leaves in preparation for the dry season.

Where less than 2.5 centimeters (1 inch) of rain falls in the dry season, **tropical deciduous forests** form. In tropical deciduous forests, most trees shed leaves at the start of the dry season.

Temperate deciduous forests form in parts of eastern North America, western and central Europe, and parts of Asia, including Japan. About 50 to 150 centimeters (about 20–60 inches) of precipitation fall throughout the year. Leaves turn brilliant red, orange, and yellow before dropping in autumn (Figure 48.18 and Section 7.1). Having discarded their leaves, the trees become dormant during the cold winter, when water is locked in snow and ice. In the spring, when conditions again favor growth, deciduous trees flower and new leaves appear. Also during the spring, leaves shed the prior autumn decay and form a rich humus. Rich soil and a somewhat open canopy that lets sunlight through allows many understory plants to flourish.

Tropical Rain Forests

Evergreen broadleaf forests form between latitudes 10° north and south in equatorial Africa, the East Indies, Malaysia, Southeast Asia, South America, and Central America. Yearly rainfall averages 130 to 200 centimeters (50 to 80 inches). Regular rains, combined with an average temperature of 25°C (77°F) and high humidity, support tropical rain forests of the sort shown in the next section. In structure and diversity, these biomes are the most complex. Some trees are 30 meters (100 feet) tall. Many form a closed canopy that stops most sunlight from getting to the forest floor. Vines and epiphytes (plants that grow on another plant, but do not withdraw nutrients from it) grow in the shady canopy.

Decomposition and mineral cycling happen fast in these forests, so litter does not accumulate. Soils are highly weathered, heavily leached, and are very poor nutrient reservoirs.

Figure 48.18 North American temperate deciduous forest. The series above shows changes in a deciduous tree's foliage from winter (*far left*) through spring, summer, and fall.

Take-Home Message

What is a broadleaf forest?

- Conditions in broadleaf forests favor dense stands of trees that form a continuous canopy.
- Deciduous broadleaf trees shed leaves seasonally. Evergreen broadleafs drop them in small numbers throughout the year.

48.9 You and the Tropical Forests

■ The Chapter 23 introduction discussed the deforestation of northern conifer forests. We turn here to factors that currently threaten the once-vast tropical forests.

Southeast Asia, Africa, and Latin America stretch across the tropical latitudes. Developing nations on these continents have the fastest-growing populations and high demands for food, fuel, and lumber. Of necessity, people turn to forests (Figure 48.19). Most of these tropical forests may vanish within our lifetime. That possibility concerns people in highly developed nations, who use most of the world's resources, including forest products.

On purely ethical grounds, the destruction of so much biodiversity is a concern. Tropical rain forests have the greatest variety and numbers of insects, and the world's largest ones. They are homes to the most species of birds and to plants with the largest flowers (*Rafflesia*). Forest canopies and understories support monkeys, tapirs, and jaguars in South America; and apes, leopards, and okapis in Africa. Massive vines twist around tree trunks. Orchids, mosses, lichens, and other organisms grow on branches, absorbing minerals from rains. Communities of microbes, insects, spiders, and amphibians live, breed, and die in small pools of water that collect in furled leaves.

Also, products provided by rain forest species save and enhance human lives. Analysis of compounds in rain forest species can point the way toward new drugs. Quinine, an antimalarial drug, was first derived from an extract of *Cinchona* bark from a tree in the Amazonian rain forest. Two chemotherapy drugs, vincristine and vinblastine, were extracted from the rosy periwinkle (*Catharanthus roseus*), a low-growing plant native to Madagascar's rain forests. Today, these drugs help fight leukemia, lymphoma, breast cancer, and testicular cancer. Many ornamental plants, spices, and foods, including cinnamon, chocolate, and coffee, originated in tropical forests. So did the latex, gums, resins, dyes, waxes, and oils used in tires, shoes, toothpaste, ice cream, shampoo, and condoms.

Conservation biologists decry the loss of forest species and their essential natural services. Yet tropical rain forest loss keeps accelerating. The amount of temperate forest is increasing in North America, Europe, and China, but this rise is overshadowed by the staggering losses of tropical forests elsewhere.

The disappearance of rain forests could influence the atmosphere. The forests take up and store carbon, and they release oxygen. Burning enormous tracts of tropical forest to make way for agriculture releases carbon dioxide, which contributes to global warming (Section 47.8).

Ironically, concern about greenhouse gas release from fossil fuels may encourage rain forest destruction. Areas of rainforest in the Amazon and Indonesia are being cleared to make way for plantations that grow soybeans or palms. Oils from these plants are exported, mainly to Europe, where they are used to produce biodiesel fuel.

Figure 48.19 Tropical rain forests in Southeast Asia and Latin America.

48.10 Coniferous Forests

- Compared to broadleaf trees, conifers are more tolerant of cold and drought, and can withstand poorer soils. Where these conditions occur, coniferous forests prevail.
- Link to Conifers 23.7

Conifers—evergreen gymnosperms with seed-bearing cones—dominate **coniferous forests**. Their leaves are typically needle-shaped, with a thick cuticle. Stomata are sunk below the leaf surface. These adaptations help conifers conserve water during drought or times when soil water is frozen. As a group, conifers tolerate poorer soils and drier habitats than broadleaf trees.

In the Northern Hemisphere, montane coniferous forests extend southward through the great mountain ranges (Figure 48.20*a*). Spruce and fir dominate at the highest elevations, with firs and pines taking over as you move farther down the slopes.

Boreal, or northern, coniferous forests occur in Asia, Europe, and North America in formerly glaciated areas where lakes and streams abound (Figure 48.20*b*). These forests are dominated by pine, fir, and spruce. They are also known as taigas, which means "swamp forests." Most rain falls in summer. Winters are long, cold, and dry. Moose are dominant grazers in this biome.

Conifers also dominate temperate lowlands along the Pacific coast from Alaska into northern California. These coniferous forests hold the world's tallest trees, Sitka spruce to the north and coast redwoods to the south. Large tracts have been logged (Chapter 23).

We find other conifer-dominated ecosystems in the eastern United States. About a quarter of New Jersey is pine barrens, a mixed forest of pitch pines and scrub oaks that grow in sandy, acidic soil. Pine forest covers about one-third of the Southeast. Fast-growing loblolly pines dominate these forests and are a major source of lumber and wood pulp. The pines survive periodic fires that kill most hardwood species. If fires are suppressed, hardwoods will replace the pines.

Take-Home Message

What are coniferous forests?

- Conferous forests consist of hardy evergreen trees able to withstand conditions that most broadleaf trees cannot.

Figure 48.20 (**a**) Montane coniferous forest near Mount Rainier, Washington. (**b**) Taiga in Alberta, Canada.

48.11 Tundra

- Low-growing plants tolerate cold and wind in tundra, which forms at high altitudes and latitudes.
- Link to Global warming 47.8

Arctic tundra forms between the polar ice cap and the belts of boreal forests in the Northern Hemisphere. Most is in northern Russia and Canada. It is Earth's youngest biome; it appeared about 10,000 years ago, when glaciers retreated at the end of the last ice age.

Arctic tundra is blanketed with snow for as long as nine months of the year. Annual snow and rain is usually less than 25 centimeters (10 inches). During a brief summer, plants grow rapidly under the nearly continuous sunlight. Lichens and shallow-rooted, low-growing plants are a base for food webs that include voles, arctic hares, caribou, arctic foxes, wolves, and brown bears (Figure 48.21). Enormous numbers of migratory birds nest here in the summer, when the air is thick with mosquitoes.

Only the surface layer of tundra soil thaws during summer. Below that lies the **permafrost**, a frozen layer 500 meters (1,600 feet) thick in places. Permafrost acts as a barrier that prevents drainage, so the soil above it remains perpetually waterlogged. Cool, anaerobic conditions slow decay, so organic remains can build up. Organic matter in the permafrost makes the arctic tundra one of Earth's greatest stores of carbon.

Figure 48.22 Alpine tundra in Washington's Cascade range.

As global temperatures rise, the amount of frozen soil that melts each summer is increasing. With warmer temperatures, much of the snow and ice that would otherwise reflect sunlight is disappearing. As a result, newly exposed dark soil absorbs heat from the sun's rays, which encourages more melting.

Alpine tundra occurs at high altitudes throughout the world (Figure 48.22). Even in the summer, some patches of snow persist in shaded areas, but there is no permafrost. The alpine soil is well drained, but thin and nutrient-poor. As a result, primary productivity is low. Grasses, heaths, and small-leafed shrubs grow in patches where soil has accumulated to a greater depth. These low-growing plants can withstand strong winds that discourage the growth of trees.

Take-Home Message

What is tundra?

- Arctic tundra prevails at high latitudes, where short, cold summers alternate with long, cold winters. Lichens and short plants grow above a frozen soil layer, the permafrost, which is a reservoir for carbon.
- Alpine tundra, also dominated by short plants, prevails at high altitudes.

Brown bear

Figure 48.21 Arctic tundra in the summer.

48.12 Freshwater Ecosystems

- Freshwater and saltwater provinces cover more of Earth's surface than all land biomes combined. Here we begin our survey of these watery realms.
- Links to Properties of water 2.5, Aquatic respiration 39.2, Food chains 47.2, Water cycle 47.6, Eutrophication 47.10

Lakes

A **lake** is a body of standing fresh water. If it is sufficiently deep, it can be divided into zones that differ in their physical characteristics and species composition (Figure 48.23). Near shore is the littoral zone, from the Latin *litus* for shore. Here, sunlight penetrates all the way to the lake bottom; aquatic plants and algae that attach to the bottom are the primary producers. The lake's open waters include an upper, well-lit limnetic zone, and—if the lake is deep—a dark profundal zone where light does not penetrate. Primary producers in the limnetic zone can include aquatic plants, green algae, diatoms, and cyanobacteria. These organisms serve as food for rotifers, copepods, and other types of zooplankton. In the profundal zone, where there is not enough light for photosynthesis, consumers feed on organic debris that drifts down from above.

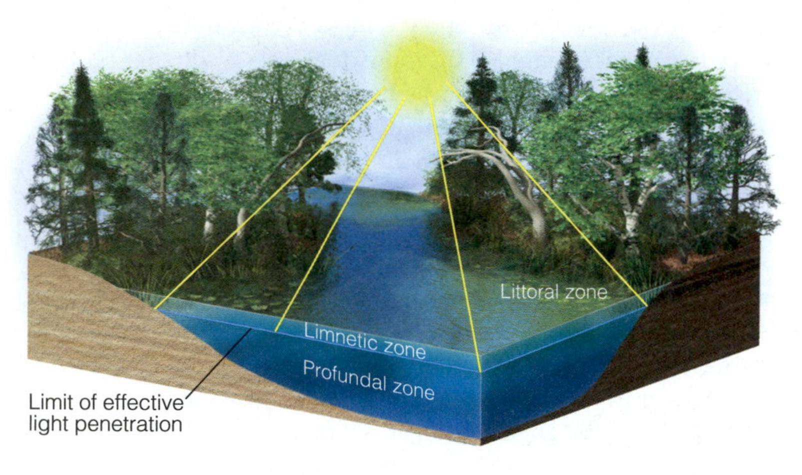

Figure 48.23 Lake zonation. A lake's littoral zone extends around the shore to a depth where rooted aquatic plants stop growing. Its limnetic zone is the open waters where light penetrates and photosynthesis occurs. Below that lies the cool, dark profundal zone, where detrital food chains predominate.

Nutrient Content and Succession Like a habitat on land, a lake undergoes succession; it changes over time (Section 46.8). A newly formed lake is oligotrophic: deep, clear, and nutrient-poor, with low primary productivity (Figure 48.24). Later, sediments accumulate and plants take root. The lake becomes eutrophic.

Eutrophication refers to processes, either natural or artificial, that enrich a body of water with nutrients (Section 47.10).

Seasonal Changes Temperate-zone lakes undergo a seasonal variation in temperature gradients, from the surface to the bottom. During winter, a layer of ice forms at the lake surface. Unlike most substances, water is denser as a liquid than as a solid (ice). As water cools, its density increases, until it reaches 4°C (39°F). Below this temperature, additional cooling decreases

Oligotrophic Lake	Eutrophic Lake
Deep, steeply banked	Shallow with broad littoral
Large deep-water volume relative to surface-water volume	Small deep-water volume relative to surface-water volume
Highly transparent	Limited transparency
Water blue or green	Water green to yellow- or brownish-green
Low nutrient content	High nutrient content
Oxygen abundant through all levels throughout year	Oxygen depleted in deep water during summer
Not much phytoplankton; green algae and diatoms dominant	Abundant, thick masses of phytoplankton; and cyanobacteria dominant
Aerobic decomposers favored in profundal zone	Anaerobic decomposers in profundal zone
Low biomass in profundal zone	High biomass in profundal zone

Figure 48.24 Crater Lake, an oligotrophic lake in a collapsed volcano that filled with snow melt. The chart compares oligotrophic and eutrophic lakes.

water's density—which is why ice floats on water (Section 2.5). In an ice-covered lake, water just under the ice is near its freezing point and at its lowest density. The densest (4°C) water is at the lake bottom (Figure 48.25*a*).

In spring, the air warms and the ice melts. When the temperature of the meltwater rises to 4°C, it sinks. This causes a **spring overturn**, during which oxygen-rich water in surface water moves downward while nutrient-rich water from the lake's depths moves up (Figure 48.25*b*). Winds aid in the overturn.

In the summer, a lake has three layers that differ in their temperature and oxygen content (Figure 48.25*c*). The upper layer is warm and oxygen-rich. It overlies the **thermocline**, a thin layer where temperature falls rapidly. Beneath the thermocline is the coolest water. The thermocline acts as a barrier that keeps the upper and lower layers from combining. During the summer, decomposers use up the oxygen in the lakes deepest waters, and nutrients from the depths cannot escape into surface waters. In autumn, the upper layer cools and sinks, and the thermocline vanishes. During the **fall overturn**, oxygen-rich water moves down while nutrient-rich water moves up (Figure 48.25*d*).

Overturns influence primary productivity. After a spring overturn, longer daylength and an abundance of nutrients support the greatest primary productivity. During the summer, vertical mixing ceases. Nutrients do not move up, and photosynthesis slows. By late summer, nutrient shortages limit growth. Fall overturn brings nutrients to the surface and favors a brief burst of photosynthesis. The burst ends as winter brings shorter days and the amount of sunlight declines.

A Winter. Ice covers the thin layer of slightly warmer water just below it. Densest (4°) water is at bottom. Winds do not affect water under the ice, so there is little circulation.

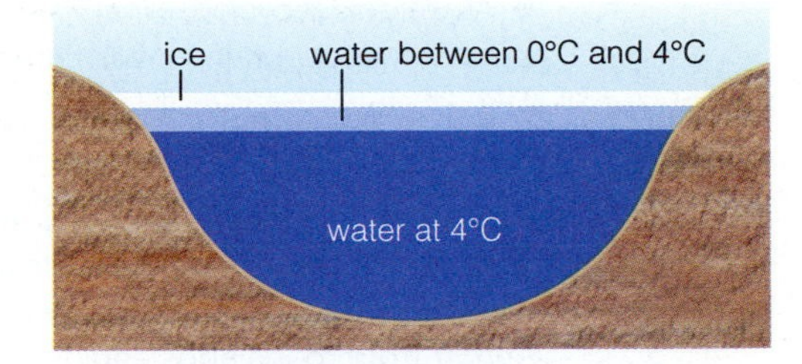

B Spring. Ice thaws. Upper water warms to 4°C and sinks. Winds blowing across water create vertical currents that help overturn water, bringing nutrients up from bottom.

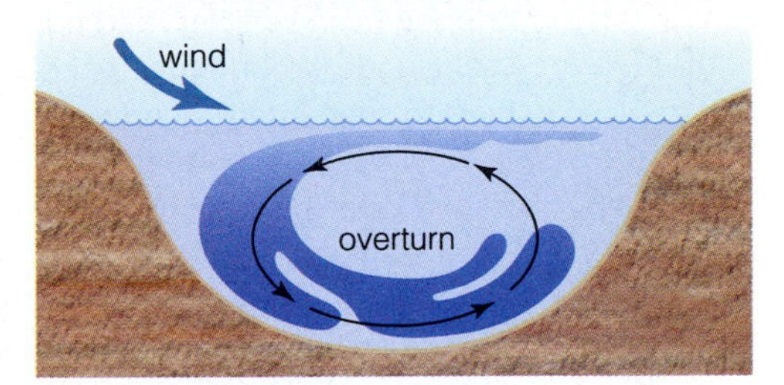

C Summer. Sun warms the upper water, which floats on a thermocline, a layer across which temperature changes abruptly. Upper and lower water do not mix because of this thermal boundary.

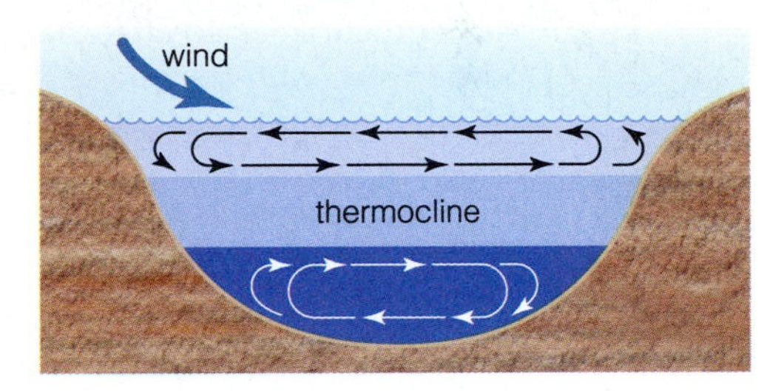

D Fall. Upper water cools and sinks downward, eliminating the thermocline. Vertical currents mix water that was separated during the summer.

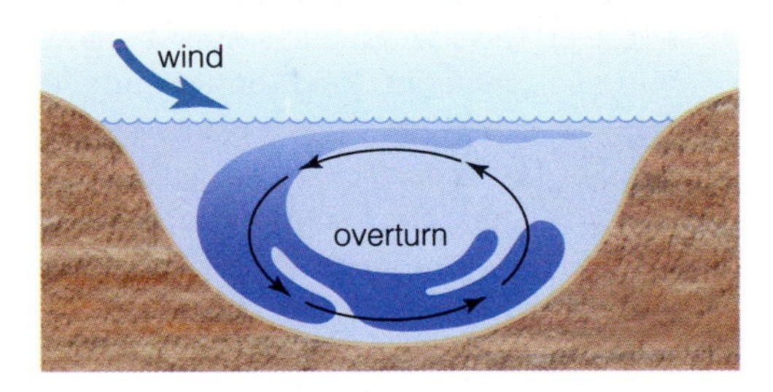

Figure 48.25 Seasonal changes in a temperate zone lake.

Streams and Rivers

Streams are flowing-water ecosystems that begin as freshwater springs or seeps. As they flow downslope, they grow and merge to form rivers. Rainfall, snowmelt, geography, altitude, and shade cast by plants affect flow volume and temperature.

Properties of a stream or river vary along its length. Streambed composition affects solute concentrations, as when limestone rocks dissolve and add calcium. Water that flows rapidly over rocks mixes with air and holds more oxygen than slower-moving, deeper water. Also, cold water holds more oxygen than warm water. As a result, different parts of a stream or river support species with different oxygen needs (Section 39.2).

A stream imports nutrients into many food webs. In forests, trees cast shade and hinder photosynthesis, but dead leaves sustain detrital food chains (Section 47.2). Aquatic species take up and release nutrients as water flows downstream. Nutrients move upstream in the tissues of migratory fish and other animals. The nutrients cycle between aquatic organisms and water as it flows on a one-way course to the sea.

Take-Home Message

What factors affect life in freshwater provinces?

- Lakes have gradients in light, dissolved oxygen, and nutrients.
- Primary productivity varies with a lake's age and—in temperate zones—with the season.
- DIfferent conditions along the length of a stream or river favor different organisms.

48.13 "Fresh" Water?

- All water is recycled. What happens to the water that you send down your drain, or flush down your toilet?
- Link to Groundwater 47.6

Pollutants flow into rivers, lakes, and groundwater from countless sources. Pollutants include sewage, animal wastes, industrial chemicals, fertilizers, and pesticides. Runoff from roads adds engine oil and antifreeze that dripped from vehicles, and rubber residues from tire wear. Leaky underground fuel tanks allow gasoline and other fuels to seep into groundwater.

How do we keep these poisons out of our drinking water? One safeguard is wastewater treatment. There are three stages of treatment. In primary treatment, screens and settling tanks remove large bits of organic material (sludge), which is dried, and burned or dumped in landfills. In secondary treatment, microbes break down any organic matter that remained after the primary treatment. Water is then treated with chlorine or exposed to ultraviolet light to kill disease-causing microorganisms. By now, most organic wastes are gone—but not all nitrogen, phosphorus, toxins, and heavy metals. Tertiary treatment uses chemical filters to remove these contaminants from water but it adds to the cost of treatment. In the United States most water is discharged after secondary treatment.

One variation on standard wastewater treatment is a solar-aquatic system such as the one constructed by biologist John Todd (Figure 48.26). Sewage enters tanks in which aquatic plants grow. Decomposers degrade the wastes and release nutrients that promote plant growth. Heat from sunlight speeds the decomposition. Water next flows through an artificial marsh that filters out algae and organic wastes. Then it flows through other tanks filled with living organisms, including plants that take up metals. After ten days, water flows into a second artificial marsh for final filtering and cleansing. Versions of this system are now used to treat both sewage and industrial wastes.

Figure 48.26 John Todd in the experimental solar-aquatic wastewater treatment facility he designed. Unlike traditional treatments, Todd's system does not require toxic chemicals or emit unpleasant odors. Bacteria, fungi, plants, invertebrates, and fish break down wastes. Solar-aquatic treatment systems are now in use in eight countries around the world.

48.14 Coastal Zones

- Where sea meets shore we find regions of high primary productivity.
- Link to Grazing and detrital food chains 47.2

Wetlands and the Intertidal Zone

Like freshwater ecosystems, estuaries and mangrove wetlands have distinct physical and chemical features, including their depth, water temperature, salinity, and light penetration. An **estuary** is an enclosed coastal region where seawater mixes with nutrient-rich fresh water from rivers and streams (Figure 48.27*a*). Water inflow continually replenishes nutrients, which is one reason estuaries are highly productive.

Primary producers include algae and other types of phytoplankton, and plants that tolerate being submerged at high tide. Detrital food chains are common (Section 47.2). Estuaries are marine nurseries; many larval and juvenile invertebrates and fishes develop in them. Migratory birds use estuaries as rest stops.

Estuaries can be broad and shallow like Chesapeake Bay, Mobile Bay, and San Francisco Bay, or narrow and deep like the fjords of Norway. Many face threats. Fresh water that should refresh them is diverted for human uses. Rivers deliver harmful substances, such as pesticides and fertilizers that entered streams in the runoff from agricultural fields.

In tidal flats at tropical latitudes, we find nutrient-rich mangrove wetlands. "Mangrove" is the common term for certain salt-tolerant woody plants that live in sheltered areas along tropical coasts. The plants have prop roots that extend out from their trunk (Figure 48.27*b*). Specialized cells at the surface of some roots allow gas exchange with air.

Increasing human populations along tropical coasts threaten mangrove wetlands. People have traditionally cut these trees for firewood. A more recent threat is conversion of mangrove wetlands to shrimp farms. The shrimp mainly end up on dinner plates in the United States, Japan, and western Europe. Disappearance of mangrove wetlands threatens the fishes and migratory birds that depend upon them for shelter and food.

Figure 48.27 Wetlands. (**a**) South Carolina salt marsh. Marsh grass (*Spartina*) is the major producer. (**b**) In the Florida Everglades, a mangrove wetland lined with red mangroves (*Rhizophora*).

Figure 48.28 Contrasting coasts. (**a**,**b**) Algae-rich rocky shores where invertebrates abound. (**c**) A sandy shore in Australia shows fewer signs of life. Invertebrates burrow in its sediments.

Rocky and Sandy Coastlines

Rocky and sandy coastlines support ecosystems of the intertidal zone. Biologists divide a shoreline into three vertical zones that differ in physical characteristics and diversity. The upper littoral zone is submerged only at the highest tide of a lunar cycle. It holds the fewest species. The midlittoral zone is submerged during the highest average tide and exposed at the lowest tide. The lower littoral zone, exposed only during the lowest tide of the lunar cycle, has the most diversity.

You can easily see the zonation along a rocky shore (Figure 48.28*a*,*b*). Algae clinging to rocks are primary producers for the prevailing grazing food chains. The primary consumers include a variety of snails.

Zonation is less obvious on sandy shores where detrital food chains start with material washed ashore (Figure 48.28*c*). Some crustaceans eat detritus in the upper littoral zone. Nearer to the water, other invertebrates feed as they burrow through the sand.

Take-Home Message

What kinds of ecosystems occur along coastlines?

- We find estuaries where rivers empty into seas. The rivers deliver nutrients that foster high productivity.
- Mangrove wetlands are common along shorelines in tropical latitudes.
- Rocky and sandy shores show zonation, with different zones exposed during different phases of the tidal cycle. Diversity is highest in the zone that is submerged most of the time.

48.15 The Once and Future Reefs

- Coral reefs are highly productive, and greatly threatened.
- Links to Dinoflagellates 22.5, Corals 25.5

Coral reefs are wave-resistant formations that consist primarily of calcium carbonate secreted by generations of coral polyps (Section 25.5). Reef-forming corals live mainly in clear, warm waters between latitudes 25° north and 25° south (Figure 48.29*d*). The mineral-hardened cell walls of red algae such as the one shown at *left* contribute to the structural framework of many reefs. The resulting reef is home to a remarkably diverse array of vertebrate and invertebrate species.

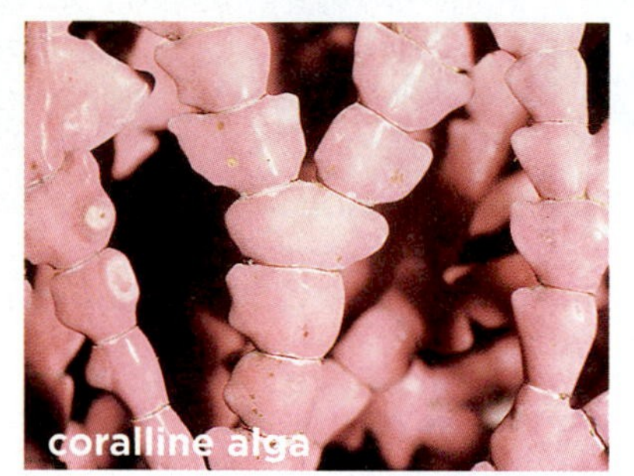

Australia's Great Barrier Reef parallels Queensland for 2,500 kilometers (1,550 miles), and is the largest example of biological architecture. Actually it is a string of reefs, some 150 kilometers (95 miles) across. It supports 500 coral species, 3,000 fish species, 1,000 kinds of mollusks, and 40 kinds of sea snakes. Figure 48.29*e* shows a wealth of warning colors, tentacles, and stealthy behavior—all signs of fierce competition for resources among species jostling for the limited space.

Reefbuilding corals have photosynthetic dinoflagellates (Section 22.5) in their tissues. These protists give a coral its color and provide it with oxygen and sugars. When a coral is stressed, it expels the protists and loses its color, an event called coral bleaching (Figure 48.30). If the coral remains stressed for more than a few months, protists do not return and the coral dies.

Figure 48.30 Coral bleaching on Australia's Great Barrier Reef.

Abnormal, widespread bleaching in the Caribbean and the tropical Pacific began in the 1980s. So did increases in sea surface temperature, which might be a key stress factor. Is the damage one outcome of global warming? If so, as marine biologists Lucy Bunkley-Williams and Ernest Williams suggest, the future looks grim for reefs, which may be destroyed within three decades.

Also, people can directly destroy reefs, as by sewage discharges into nearshore waters of populated islands. Massive oil spills, commercial dredging operations, and mining for coral rock have catastrophic impact. Logging of areas adjacent to reefs increases runoff of nutrients and silt, which can harm reef species.

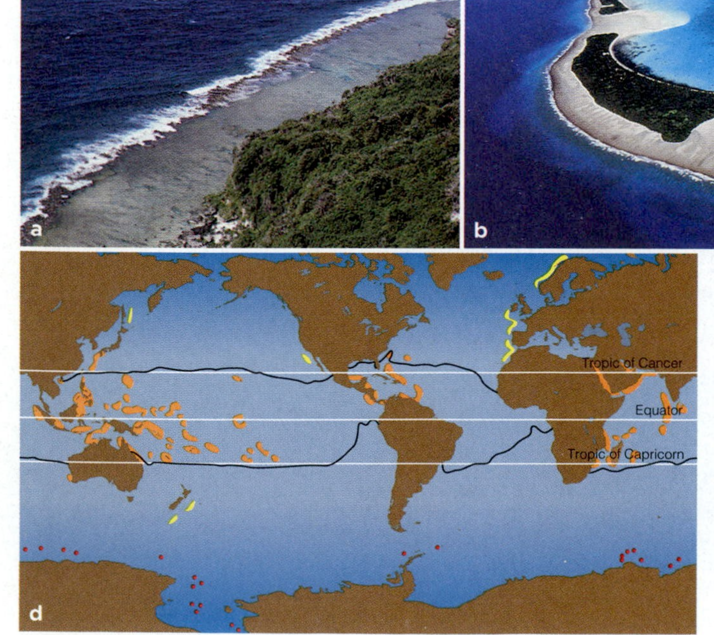

Figure 48.29 Coral reef formations. (**a**) *Fringing* reefs form near land when rainfall and runoff are light, as on the downwind side of young volcanic islands. Many reefs in the Hawaiian Islands and Tahiti are like this.

(**b**) Ring-shaped *atolls* consist of coral reefs and coral debris. They fully or partly enclose a shallow lagoon, often with a channel to open ocean. Biodiversity is not great in the shallow water, which can get too hot for corals. (**c**) *Barrier* reefs parallel the shore of continents and volcanic islands, as in Bora Bora. Behind them are calm lagoons.

(**d**) Distribution map for coral reefs (*orange*) and coral banks (*yellow*). Nearly all reef-building corals live in warm seas, here enclosed in dark lines. Past latitudes 25° north and south, solitary and colonial corals (*red*) form coral banks in temperate seas and in cold seas above continental shelves. (**e**) *Facing page*, a sampling of coral reef biodiversity.

LIONFISH

Fishing nets can break pieces off corals, but some fishermen prefer even more destructive practices such as dropping dynamite in the water. Fish hiding in the coral are blasted out and float to the surface, some dead, but others only stunned. The practice is outlawed nearly everywhere, but enforcement is often lacking.

Capture of fish for the pet trade also has harmful effects. In some places, sodium cyanide is squirted into the water to stun fish, which float to the surface. Most fish that survive being stunned with cyanide are shipped off for sale in pet stores in the United States or Europe.

Invasive species also threaten reefs. In Hawaii, reefs are being overgrown by exotic algae, including several species imported for cultivation during the 1970s.

Reef biodiversity is in danger around the world, from Australia and Southeast Asia to the Hawaiian Islands, the Galápagos Islands, the Gulf of Panama, Florida, and Kenya. For example, the biodiversity on the coral reef off Florida's Key Largo has been reduced by 33 percent since 1970.

PART OF A FIJIAN CORAL REEF

MORAY EEL

NUDIBRANCH

LONGNOSE HAWKFISH AND RED SEA FAN

e

BANDED CORAL SHRIMP

PURPLE TUBE SPONGE

GREEN CORAL POLYP

48.16 The Open Ocean

- Earth's vast oceans are still largely unexplored. We are only beginning to catalogue the diversity they contain.
- Links to Hydrothermal vents 20.2, Chemoautotrophs 21.4

Oceanic Zones and Habitats

Like a lake, an ocean shows gradients in light, nutrient availability, temperature, and oxygen concentration. We refer to the open waters of the ocean as the **pelagic province** (Figure 48.31*a*). These open waters include the neritic zone—the water over continental shelves—together with the more extensive oceanic zone farther offshore. The neritic zone receives nutrients in runoff from land, and is the zone of greatest productivity.

In the ocean's upper, brightly lit waters, photosynthetic microorganisms are the primary producers, and grazing food chains predominate. Depending on the region, some light may penetrate as far as 1,000 meters beneath the sea surface. Below that, organisms live in darkness, and organic material that drifts down from above is the basis of detrital food chains. At the top of food webs, carnivores range from the familiar sharks and squids to giant colonial cnidarians and the bizarre deep-sea angler fishes (Figure 48.32*a,b*). In what may be the greatest circadian migrations, many species rise thousands of feet at night to feed in upper waters, then move down in the morning.

The **benthic province** is the ocean bottom—its rocks and sediments. Benthic biodiversity is greatest on the margins of continents, or the continental shelves. The benthic province also includes some largely unexplored concentrations of biodiversity on seamounts and at hydrothermal vents.

Seamounts are undersea mountains that stand 1,000 meters or more tall, but are still below the sea surface (Figure 48.31*b*). They attract large numbers of fishes and are home to many marine invertebrates (Figure 48.32*c*). Like islands, seamounts often are home to species that evolved there and are found nowhere else.

The abundance of life at seamounts makes them attractive to commercial fishing vessels. Fish and other organisms are often harvested by trawling, a fishing technique in which a large net is dragged along the bottom, capturing everything in its path. The process is ecologically devastating; trawled areas are stripped bare of life, and silt stirred up by the giant, weighted nets suffocates filter-feeders in adjacent areas.

Superheated water that contains dissolved minerals spews out from the ocean floor at **hydrothermal vents**. When this heated, mineral-rich water mixes with cold seawater, the minerals settle out as extensive deposits. Chemoautotrophic prokaryotes (Section 21.4) can get energy from these deposits. The prokaryotes serve as primary producers for food webs that include diverse invertebrates, such as tube worms and brittle stars (Figure 48.32*d–f*). As explained in Section 20.2, one hypothesis holds that life originated on the sea floor in such heated, nutrient-rich places.

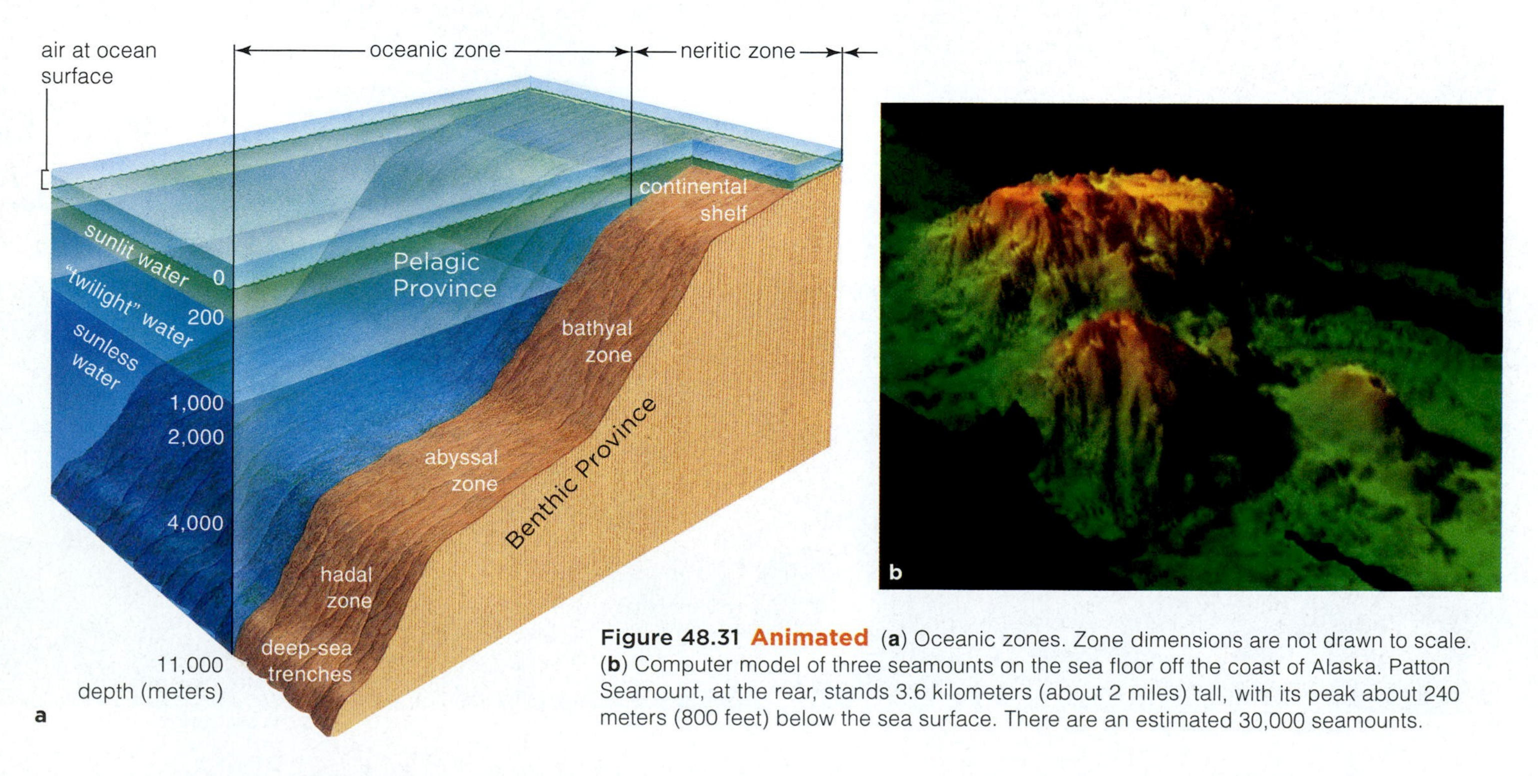

Figure 48.31 Animated (**a**) Oceanic zones. Zone dimensions are not drawn to scale. (**b**) Computer model of three seamounts on the sea floor off the coast of Alaska. Patton Seamount, at the rear, stands 3.6 kilometers (about 2 miles) tall, with its peak about 240 meters (800 feet) below the sea surface. There are an estimated 30,000 seamounts.

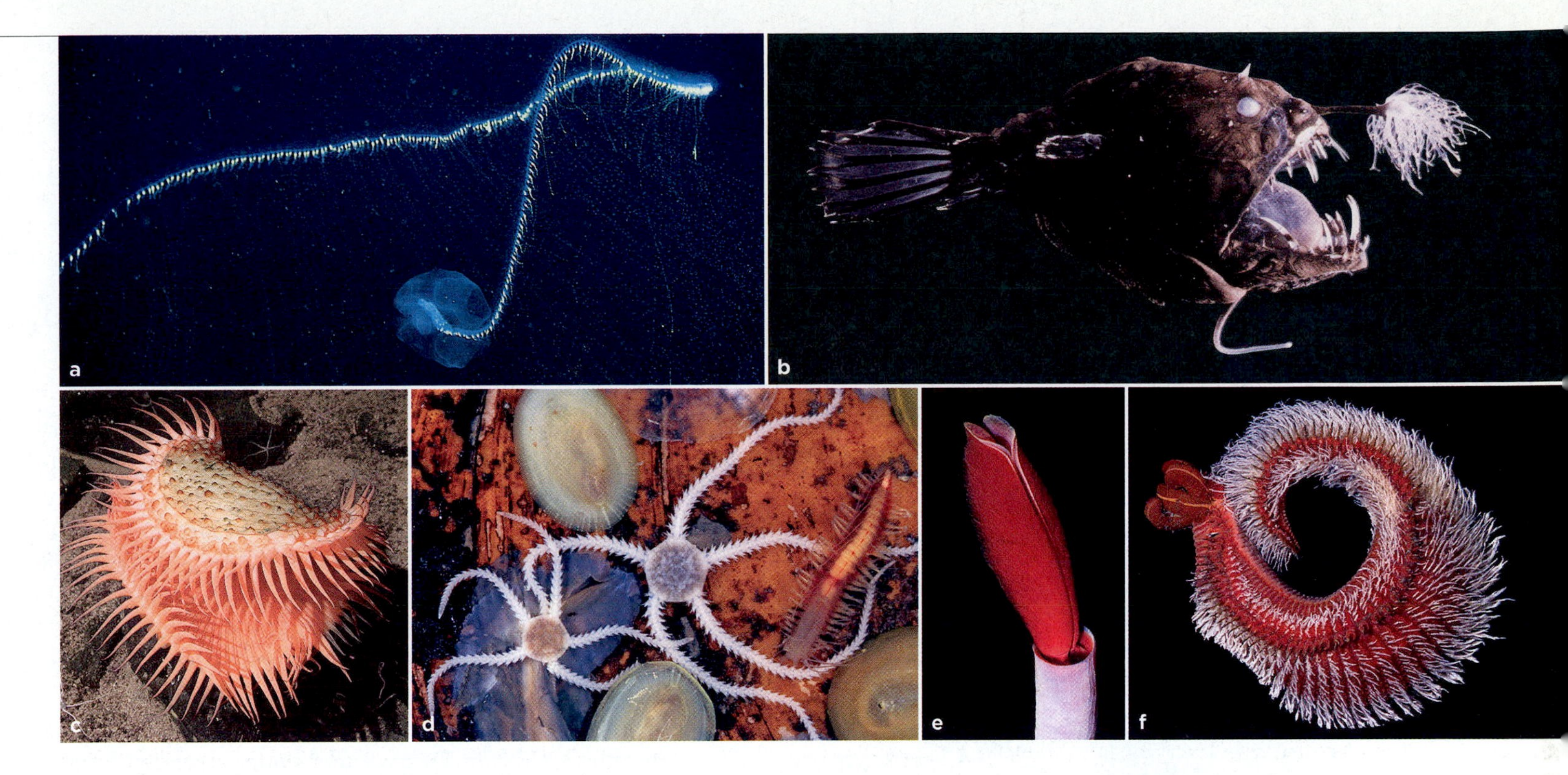

Figure 48.32 What lies beneath: a vast, largely unexplored world of marine life. (**a**) A siphonophore, *Praya dubia*, one of the colonial relatives of corals and jellyfish can be 40 meters (130 feet) long. (**b**) Deep-sea angler fish with bioluminescent lures.

(**c**) A flytrap anemone, from Davidson Seamount, just off the California coast.

Residents of hydrothermal vent communities: (**d**) brittle stars, limpets, and a polychaete worm; (**e**) tube worm, a polychaete; (**f**) Pompei-worm, another polychaete.

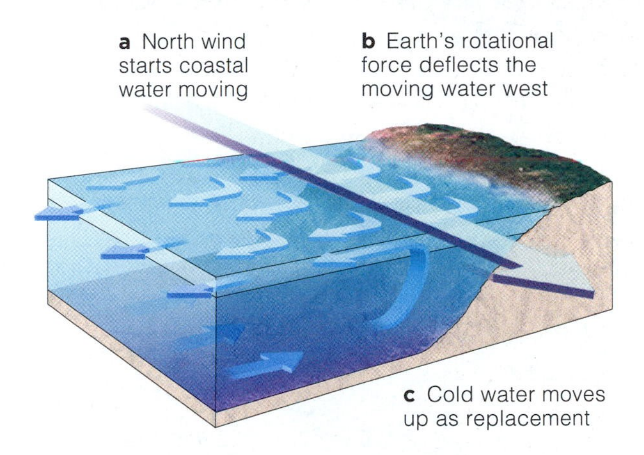

Figure 48.33 Coastal upwelling in the Northern Hemisphere.

Upwelling—A Nutrient Delivery System

Deep, cold ocean waters are rich in nutrients. By the process of **upwelling**, this nutrient-laden water moves upward along the coasts of continents. Winds set the coastal waters in motion. For example, in the Northern Hemisphere, prevailing winds that blow from north to south parallel to the west coasts of continents start the surface waters moving (Figure 48.33). Upwelling occurs as Earth's rotation deflects the masses of slow-moving water away from a coast and cold, deep water moves up vertically in its place.

In the Southern Hemisphere, winds from the south tug surface water away from the coast. Cold, deeper water of the Humboldt Current moves in to replace it. Nutrients in this water sustain phytoplankton that are the basis for a rich fishery.

Every three to seven years, surface waters of the western equatorial Pacific Ocean warm up, causing a change in the direction of the wind. This warming most often happens around Christmas, so fishermen in Peru named this event **El Niño**, as discussed in the chapter introduction. The name became part of a more inclusive term: the El Niño Southern Oscillation, or ENSO. The next section takes a closer look at some of the consequences of this recurring event.

Take-Home Message

What factors affect life in ocean provinces?

- Oceans have gradients in light, dissolved oxygen, and nutrients. Nearshore and well-lit zones are the most productive and species-rich.
- On the sea floor, pockets of diversity occur on seamounts and around hydrothermal vents.
- Upwelling brings nutrient-rich water from deep regions of the sea to surface waters along coasts.

48.17 Climate, Copepods, and Cholera

- Events in the atmosphere and oceans, and on land, interconnect in ways that profoundly affect the world of life.
- Link to Copepods 25.14

An El Niño Southern Oscillation, or ENSO, is defined by changes in sea surface temperatures and in the air circulation patterns. "Southern oscillation" refers to a seesawing of the atmospheric pressure in the western equatorial Pacific—Earth's greatest reservoir of warm water and warm air. It is the source of heavy rainfall, which releases enough heat energy to drive global air circulation patterns.

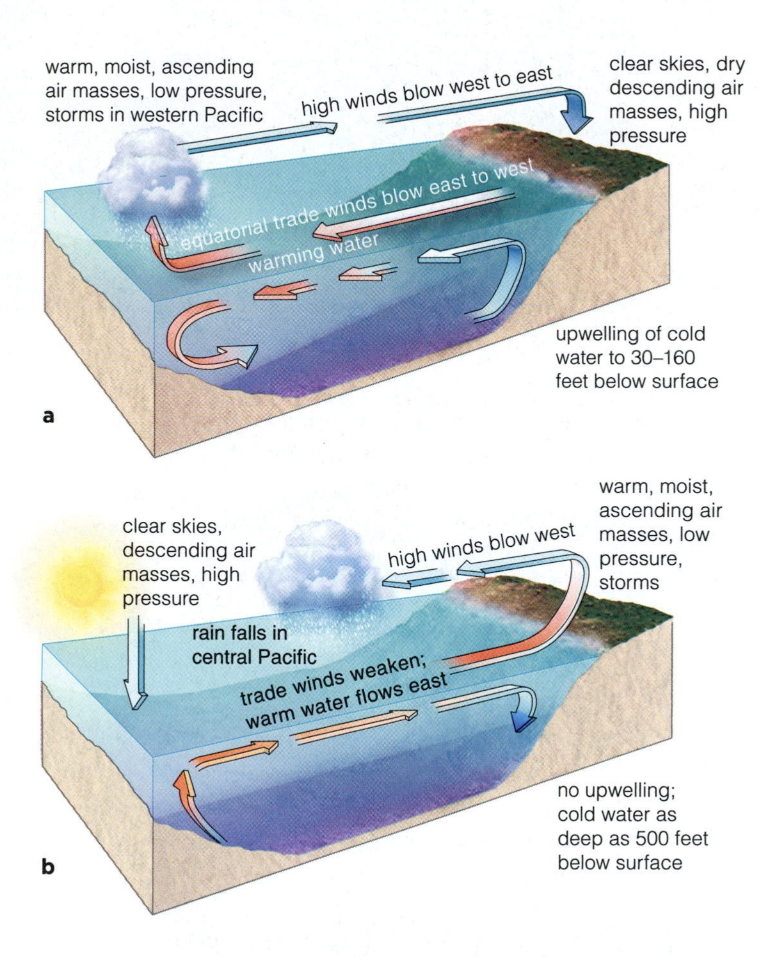

Figure 48.34 Animated (**a**) Westward flow of cold, surface water between ENSOs. (**b**) Eastward dislocation of warm water during El Niño.

Between ENSOs, the warm waters and heavy rains move westward (Figure 48.34*a*). During an ENSO, the prevailing surface winds over the western equatorial Pacific pick up speed and "drag" surface waters east (Figure 48.34*b*). As they do, the westward transport of water slows down. Sea surface temperatures rise, evaporation accelerates, and air pressure falls. These changes affect weather worldwide.

El Niño episodes persist for 6 to 18 months. Often they are followed by a **La Niña** episode in which the Pacific waters become cooler than usual. Other years, waters are neither warmer nor colder than average.

As noted in the chapter introduction, 1997 ushered in the most powerful El Niño event of the century. The average sea surface temperatures in the eastern Pacific rose by 5°C (9°F). This warmer water extended 9,660 kilometers (6,000 miles) west from the coast of Peru.

The 1997–1998 El Niño/La Niña roller-coaster had extraordinary effects on the primary productivity in the equatorial Pacific. With the massive eastward flow of nutrient-poor warm water, photoautotrophs were almost undetectable in satellite photos that measure primary productivity (Figure 48.35*a*).

During the La Niña rebound, cooler, nutrient-rich water welled up to the sea surface and was displaced westward all along the equator. As satellite images revealed, upwelling had sustained an algal bloom that stretched across the equatorial Pacific (Figure 48.35*b*).

During the 1997–1998 El Niño event, 30,000 cases of cholera were reported in Peru alone, compared with only 60 cases from January to August in 1997. People knew that water contaminated by *Vibrio cholerae* causes epidemics of cholera (Figure 48.36*b*). The disease agent triggers severe diarrhea. Bacteria-contaminated feces enter the water supply and individuals who use the tainted water become infected.

Figure 48.35 Satellite data on primary productivity in the equatorial Pacific Ocean. The concentration of chlorophyll in the water was used as the measure.

(**a**) During the 1997–1998 El Niño episode, a massive amount of nutrient-poor water moved to the east, and so photosynthetic activity was negligible.

(**b**) During a subsequent La Niña episode, massive upwelling and westward displacement of nutrient-rich water led to a vast algal bloom that stretched all the way to the coast of Peru.

a Near-absence of phytoplankton in the equatorial Pacific during an El Niño.

b Huge algal bloom in the equatorial Pacific in the La Niña rebound event.

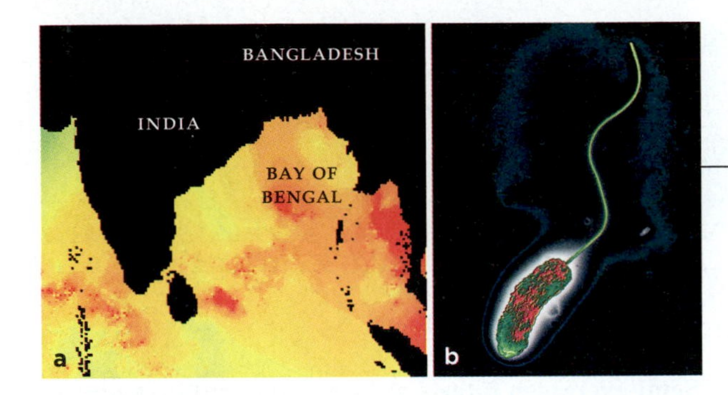

What people did not know was where *V. cholerae* remained between cholera outbreaks. It could not be found in humans or in water supplies. Even so, the cholera would often break out simultaneously in far apart places—usually coastal cities where the urban poor draw water from rivers near the sea.

Marine biologist Rita Colwell had been thinking about the fact that humans are not the host between outbreaks. Was there an environmental reservoir for the pathogen? Maybe, but nobody had detected it in water samples subjected to standard culturing.

Then Colwell had a flash of insight: What if no one could find the pathogen because it changes its form and enters a dormant stage between outbreaks?

During one cholera outbreak in Louisiana, Colwell realized that she could use an antibody-based test to detect a protein unique to *V. cholerae*'s surface. Later, tests in Bangladesh revealed bacteria in fifty-one of fifty-two samples of water. Standard culture methods had missed it in all but seven samples.

V. cholerae survives in rivers, estuaries, and seas. As Colwell knew, plankton also thrive in these aquatic environments. She decided to restrict her search for the unknown host to warm waters near Bangladesh, where outbreaks of cholera occur seasonally (Figure 48.36). It was here that Colwell discovered a dormant *V. cholerae* stage inside copepods, a type of tiny marine crustacean (Section 25.14). Copepods eat phytoplankton, so the abundance of copepods—and of *V. cholerae* cells inside them—increases and decreases with the abundance of phytoplankton.

Colwell suspected that water temperature changes in the Bay of Bengal were tied to cholera outbreaks, so she looked at medical reports for the 1990–1991 and 1997–1998 El Niño episodes. She found that the number of reported cholera cases rose four to six weeks after an El Niño event began. El Niño brings warmer water with more nutrients to the Bay of Bengal, encouraging the growth of phytoplankton. This added food then increases the number of cholera-carrying copepods.

Figure 48.36 (**a**) Satellite data on rising sea surface temperatures in the Bay of Bengal. *Red* shows warmest summer temperatures. (**b**) *Vibrio cholerae*, the agent of cholera. Copepods host a dormant stage of this bacterium that waits out adverse environmental conditions that do not favor its growth and reproduction. (**c**) A typical Bangladeshi waterway from which water samples were drawn for analysis. (**d**) In Bangladesh, Rita Colwell comparing samples of unfiltered and filtered drinking water.

Today, Colwell and Anwarul Huq, a Bangladeshi scientist, are investigating salinity and other factors that may relate to outbreaks. Their goal is to design a model for predicting where cholera will occur next. They advised women in Bangladesh to use sari cloth as a filter to remove *V. cholerae* cells from the water (Figure 48.36*d*). The copepod hosts are too big to pass through the thin cloths, which can be rinsed in clean water, sun-dried, and used again. This inexpensive, simple method has cut cholera outbreaks by half.

Take-Home Message

What occurs during an El Niño event?

- During an El Niño event, changes in ocean temperatures and winds alter currents, affecting weather, marine food webs, and human health.

IMPACTS, ISSUES REVISITED Surfers, Seals, and the Sea

It is becoming increasingly clear that "normal" weather depends on what time frame you consider. Cycles of heating and cooling in the Pacific Ocean alter conditions over the course of 3 to 7 years. Evidence of longer-term cycles is also emerging. Some cycles seem to have a period as long as 50 to 70 years. These findings suggest that long-term plans that are based on current weather and climate conditions may be short-sighted.

How would you vote?
Is supporting studies of El Niño and the other long term climate cycles a good use of government funds? See CengageNOW for details, then vote online.

Summary

Sections 48.1, 48.2 Global air circulation patterns affect **climate** and the distribution of communities. The patterns are set into motion by latitudinal variations in incoming solar radiation. Air circulation patterns are influenced by Earth's daily rotation and annual path around the sun, the distribution of landforms and seas, and elevations of landforms. Solar energy, and the winds that it causes, are renewable, clean sources of energy.

Human put **pollutants** into the atmosphere. Use of CFCs depletes the **ozone layer** in the upper atmosphere and allows more UV radiation to reach Earth's surface.

Smog, a form of air pollution, occurs when fossil fuels are burned in warm, still air above cities. Coal-burning power plants also are big contributors to **acid rain**, which alters habitats and kills many organisms.

- *Use the animation on CengageNOW to learn how Earth's tilt affects seasons, how sunlight drives air circulation, how CFCs destroy ozone, and how acid rain forms.*

Section 48.3 Latitudinal and seasonal variations in sunlight warm sea surface water and start currents. The currents distribute heat energy worldwide and influence the weather patterns. Ocean currents, air currents, and landforms interact in shaping global temperature zones, as when the presence of coastal mountains causes a **rain shadow** or **monsoon** rains fall seasonally.

- *Use the animation on CengageNOW to learn about ocean currents, rain shadows, and coastal breezes.*

Sections 48.4, 48.5 **Biogeographic realms** are vast areas with communities of plants and animals found nowhere else. **Biomes** are somewhat smaller regions with a particular type of dominant vegetation. Regional variations in climate, elevation, **soil profiles**, and evolutionary history affect the distribution of biomes.

- *Use the animation on CengageNOW to see the distribution of biomes and compare some of their soil profiles.*

Sections 48.6–48.11 **Deserts** form around latitudes 30° north and south. Vast **grasslands** form in the interior of midlatitude continents. Slightly moister southern or western coastal regions support **dry woodlands** and **dry shrublands**.

From the equator to latitudes 10° north and south, high rainfall, high humidity, and mild temperatures can support **evergreen broadleaf forests**.

Semi-evergreen forests and **tropical deciduous forests** form between latitudes 10° and 25°, depending on how much of the annual rainfall occurs in a prolonged dry season. **Temperate deciduous forests** form at higher latitudes. Where a cold, dry season alternates with a cold, rainy season, **coniferous forests** dominate. Conifers also are favored in temperate areas with poor soil.

Low-growing, hardy plants of the **arctic tundra** occur at high latitudes, where there is a layer of **permafrost**. At high altitudes, similar plants grow as **alpine tundra**.

Sections 48.12–48.15 Most **lakes**, streams, and other aquatic ecosystems have gradients in the penetration of sunlight, water temperature, and in dissolved gases and nutrients. These characteristics vary over time and affect primary productivity.

In temperate-zone lakes, a **spring overturn** and a **fall overturn** cause vertical mixing of waters and trigger a burst of productivity. In summer, a **thermocline** prevents upper and lower waters from mixing.

Coastal zones support diverse ecosystems. Among these, the coastal wetlands, **estuaries**, and **coral reefs** are especially productive.

Sections 48.16, 48.17 Life persists throughout the ocean. Diversity is highest in sunlit waters at the top of the **pelagic province**, or ocean waters. In the **benthic province**—the seafloor—diversity is high near deep-sea **hydrothermal vents** and on **seamounts**.

Upwelling is an upward movement of deep, cool, often nutrient-rich ocean water, typically along the coasts of continents. An **El Niño** event is a warming of eastern Pacific waters that triggers changes in rainfall and other weather patterns around the world. A **La Niña** is a cooling of these same waters; it too influences global weather patterns.

- *Use the interaction on CengageNOW to learn about oceanic zones and to observe how an El Niño event affects ocean currents and upwelling.*

Self-Quiz

Answers in Appendix III

1. Solar radiation drives the distribution of weather systems and so influences ________ .
   a. temperature zones
   b. rainfall distribution
   c. seasonal variations
   d. all of the above

Data Analysis Exercise

To try to predict the effect of El Niño or La Niña events in the near future, the National Oceanographic and Atmospheric Administration collects information about sea surface temperature (SST) and atmospheric conditions. They compare monthly temperature averages in the eastern equatorial Pacific Ocean to historical data and calculate the difference (the degree of anomaly) to determine if El Niño conditions, La Niña conditions, or neutral conditions are developing. El Niño is a rise in the average SST above 0.5°C. A decline of the same amount is La Niña. Figure 48.37 shows data for nearly 39 years.

1. When did the greatest positive temperature deviation occur during this time period?

2. What type of event, if any, occurred during the winter of 1982–1983? What about the winter of 2001–2002?

3. During a La Niña event, less rain than normal falls in the American West and Southwest. In the time interval shown, what was the longest interval without a La Niña event?

4. What type of conditions were in effect in the fall of 2007 when California suffered severe wildfires?

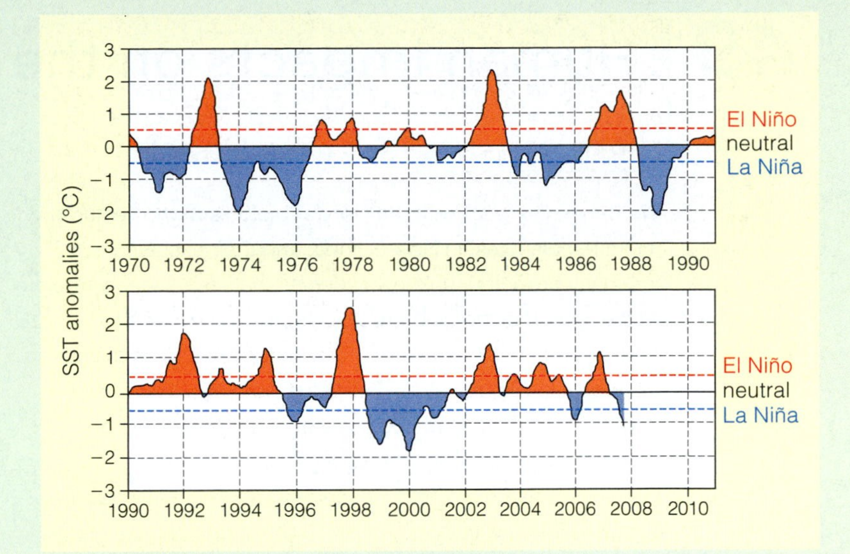

Figure 48.37 Sea surface temperature anomalies (differences from the historical mean) in the eastern equatorial Pacific Ocean. A rise above the dotted *red* line is an El Niño event, a decline below the *blue* line is La Niña.

2. ______ shields living organisms against the sun's UV wavelengths.
a. A thermal inversion
b. Acid precipitation
c. The ozone layer
d. The greenhouse effect

3. Regional variations in the global patterns of rainfall and temperature depend on ______.
a. global air circulation
b. ocean currents
c. topography
d. all of the above

4. A rain shadow is a reduction in rainfall ______.
a. on the inland side of a coastal mountain range
b. during an El Niño event
c. that occurs seasonally in the tropics

5. Air masses rise ______.
a. at the equator
b. at the poles
c. as air cools
d. all of the above

6. Biomes are ______.
a. water provinces
b. water and land zones
c. land regions
d. partly characterized by dominant plants
e. both c and d

7. Biome distribution depends on ______.
a. climate
b. elevation
c. soils
d. all of the above

8. Grasslands most often predominate ______.
a. near the equator
b. at high altitudes
c. in interior of continents
d. b and c

9. Permafrost underlies ______, and is a vast store of carbon.
a. arctic tundra
b. alpine tundra
c. coniferous forests
d. all of the above

10. During ______, deeper, often nutrient-rich water moves to the surface of a body of water.
a. spring overturns
b. fall overturns
c. upwellings
d. all of the above

11. Chemoautotrophic prokaryotes are the primary producers for food webs ______.
a. in grasslands
b. in deserts
c. on coral reefs
d. at hydrothermal vents

12. Match the terms with the most suitable description.

___tundra	a. equatorial broadleaf forest
___chaparral	b. partly enclosed by land where fresh water and seawater mix
___desert	c. type of grassland with trees
___savanna	d. has low-growing plants at high latitudes or elevations
___estuary	e. at latitudes 30° north and south
___boreal forest	f. mineral-rich, superheated water supports communities
___tropical rain forest	g. conifers dominate
___hydrothermal vents	h. dry shrubland

- *Visit CengageNOW for additional questions.*

Critical Thinking

1. London, England, is at the same latitude as Calgary in Canada's province of Alberta. However, the mean January temperature in London is 5.5°C (42°F), whereas in Calgary it is minus 10°C (14°F). Compare the locations of these two cities and suggest a reason for this temperature difference.

2. Increased industrialization in China has environmentalists worried about air quality elsewhere. Are air pollutants from Beijing more likely to end up in eastern Europe or the western United States? Why?

3. The use of off-road recreational vehicles may double in the next twenty years. Enthusiasts would like increased access to government-owned deserts. Some argue that it's the perfect place for off-roaders because "There's nothing there." Explain whether you agree, and why.

49 Human Impacts on the Biosphere

IMPACTS, ISSUES A Long Reach

We began this book with the story of biologists who ventured into a remote forest in New Guinea, and their excitement at the many previously unknown species that they encountered. At the far end of the globe, a U.S. submarine surfaced in Arctic waters and discovered polar bears hunting on the ice-covered sea (Figure 49.1). The bears were about 270 miles from the North Pole and 500 miles from the nearest land.

Even such seemingly remote regions are no longer beyond the reach of human explorers—and human influence. You already know that increasing levels of greenhouse gases are raising the temperature of Earth's atmosphere and seas. In the Arctic, the warming is causing sea ice to thin and to break up earlier in the spring. This raises the risk that polar bears hunting far from land will become stranded and unable to return to solid ground before the ice thaws.

Polar bears are top predators and their tissues contain a surprisingly high amount of mercury and organic pesticides. The pollutants entered the water and air far away, in more temperate regions. Winds and ocean currents deliver them to polar realms. Contaminants also travel north in the tissues of migratory animals such as seabirds that spend their winters in temperate regions and nest in the Arctic.

In places less remote than the Arctic, effects of human populations have a more direct effect. As we cover more and more of the world with our dwellings, factories, and farms, less appropriate habitat remains for other species. We also put species at risk by competing with them for resources, overharvesting them, and introducing nonnative competitors.

It would be presumptuous to think that we alone have had a profound impact on the world of life. As long ago as the Proterozoic, photosynthetic cells were irrevocably changing the course of evolution by enriching the atmosphere with oxygen. Over life's existence, the evolutionary success of some groups assured the decline of others. What is new is the increasing pace of change and the capacity of our own species to recognize and affect its role in this increase.

A century ago, Earth's physical and biological resources seemed inexhaustible. Now we know that many practices put into place when humans were largely ignorant of how natural systems operate take a heavy toll on the biosphere. The rate of species extinctions is on the rise and many types of biomes are threatened. These changes, the methods scientists use to document them, and the ways that we can address them, are the focus of this chapter.

Figure 49.1 Three polar bears investigate an American submarine that surfaced in ice-covered Arctic waters.

Key Concepts

The newly endangered species

Human activities have accelerated the rate of extinctions. Habitat loss, degradation, and fragmentation lead to extinctions, as do species introductions and overharvesting. **Sections 49.1, 49.2**

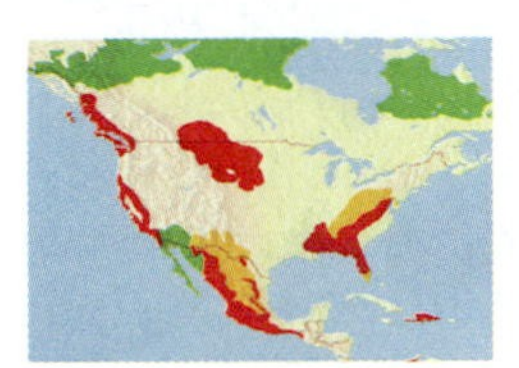

Assessing biodiversity

Our knowledge of species is biased toward large land animals. Conservation biologists assess the state of ecosystems and their biodiversity, with the goal of preserving as much of it as possible. **Sections 49.3, 49.4**

Harmful practices

Building homes, using energy, purchasing products, raising crops, and discarding trash all have harmful environmental effects that endanger species and ecosystems. **Sections 49.5–49.7**

Sustainable solutions

All nations have biological wealth that can benefit human populations. Recognizing the value of biodiversity and putting it to use in sustainable ways is good for Earth and all of its species. **Section 49.8**

Links to Earlier Concepts

- You already know about extinction (Section 18.12) and how mass extinctions were used to create the geologic time scale (17.8). Here we take a look at how human population growth and resource use (45.7, 45.9), including use of fossil fuel (23.5) are accelerating extinctions.
- You will learn how human activities can cause inbreeding by disrupting gene flow (18.8, 18.9). You will also be reminded of the effects of aquifer depletion (47.6), acid rain (47.9), soil erosion (29.1), and greenhouse gas emissions (47.8). You will see how transpiration (29.3) affects local rainfall patterns.
- We will look at the story of lichens and pollution (18.4) from another perspective, and see another example of the effects of pathogenic oomycotes (22.8).

How would you vote? The Arctic contains reserves of gas, oil, and minerals. The United States has a claim to some Arctic territory. Should it push for protection of the Arctic rather than exploitation of these resources? See CengageNOW for details, then vote online.

49.1 The Extinction Crisis

- Extinction is a natural process, but we are accelerating it.
- Links to Geologic time scale 17.8, Extinction 18.12

Era	Period	Major extinction under way
CENOZOIC	QUATERNARY	With high population growth rates and cultural practices (e.g., agriculture, deforestation), humans become major agents of extinction.
	— 1.8 mya —	
	TERTIARY	
	— 65.5 —	*Major extinction event*
MESOZOIC	CRETACEOUS	Slow recovery after Permian extinction, then adaptive radiations of some marine groups and plants and animals on land. Asteroid impact at K–T boundary, 85% of all species disappear from land and seas.
	— 145.5 —	
	JURASSIC	
	— 199.6 —	
	TRIASSIC	
	— 251 —	*Major extinction event*
PALEOZOIC	PERMIAN	Pangea forms; land area exceeds ocean surface area for first time. Asteroid impact? Major glaciation, colossal lava outpourings, 90%–95% of all species lost.
	— 299 —	
	CARBONIFEROUS	
	— 359 —	*Major extinction event*
	DEVONIAN	More than 70% of marine groups lost. Reef builders, trilobites, jawless fishes, and placoderms severely affected. Meteorite impact, sea level decline, global cooling?
	— 416 —	
	SILURIAN	
	— 443 —	*Major extinction event*
	ORDOVICIAN	Second most devastating extinction in seas; nearly 100 families of marine invertebrates lost.
	— 488 —	
	CAMBRIAN	
	— 542 —	*Major extinction event*
	(Precambrian)	Massive glaciation; 79% of all species lost, including most marine microorganisms.

a

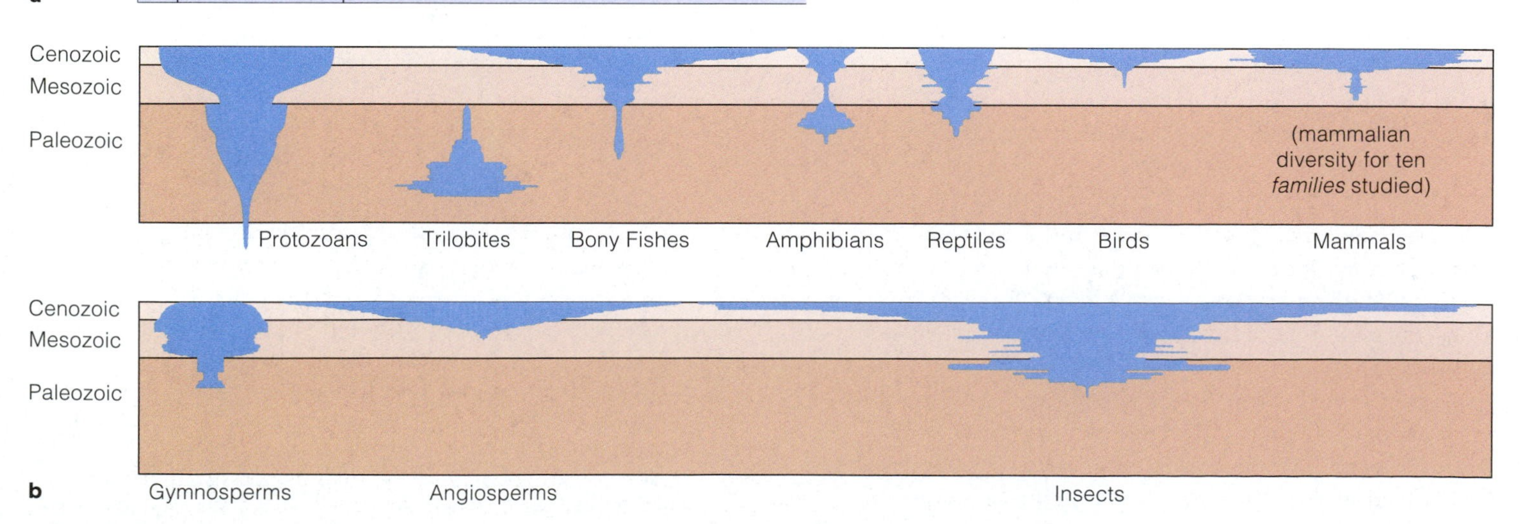

Figure 49.2 Animated (**a**) Dates of the five greatest mass extinctions and recoveries in the past. Compare Figure 17.14. (**b**) Species diversity over time for a sampling of taxa. The width of each *blue* shape represents the number of species in that lineage. Notice the variation among lineages.

Mass Extinctions and Slow Recoveries

Extinction, like speciation, is a natural process (Section 18.12). Species arise and become extinct on an ongoing basis. Based on several lines of evidence, scientists estimate that 99 percent of all species that have ever lived are now extinct.

The rate of extinction picks up dramatically during a mass extinction, when many kinds of organisms in many different habitats all become extinct in a relatively short period. Five great mass extinctions mark the boundaries for the geologic time periods (Section 17.8). With each mass extinction event, biodiversity plummeted both on land and in the oceans. Afterwards, the surviving species underwent adaptive radiations. Each time, biodiversity recovered extremely slowly. It took at least 10 million years for diversity to return to the level that preceded the extinction event. Figure 49.2*a* reviews the major extinctions and recoveries.

This pattern of extinctions is a composite of what happened to the major taxa. However, lineages differ in their time of origin, their tendency to branch and give rise to new species, and how long they endure. If we consider the number of species as the measure of success for any lineage, not all lineages are equally successful. Figure 49.2*b*, illustrates how the number of species changed over time in some major lineages. Expansion of one lineage sometimes occurred at the same time as contraction of another, as when a decline in gymnosperms accompanied the adaptive radiation of angiosperms.

Figure 49.3 Drawing of a dodo (*Raphus cucullatus*). Extinct since the late 1600s, it was larger than a turkey and flightless.

Figure 49.4 Living or extinct? Colorized photo of an ivory-billed woodpecker (*Campephilus principalis*). It is, or was, North America's largest woodpecker and a native of the Southeastern states.

The Sixth Great Mass Extinction

We are currently in the midst of a mass extinction. The current extinction rate is estimated to be 100 to 1,000 times above the typical background rate, putting it on a par with the five major extinction events. Unlike those early events, this one cannot be blamed on some natural catastrophe such as an asteroid impact. Rather, this mass extinction is the outcome of the success of a single species—humans—and their effect on Earth.

The ongoing extinction event may have begun as early as 60,000 years ago. The estimated arrival time of humans in Australia and North America correlates with a rise in the extinction rate for large mammals. Climate change certainly played a role in the declines, but hunting may have been a contributing factor.

It is easier to pin the blame for recent extinctions on humans. The World Conservation Union has compiled a list of more than 800 documented extinctions that occurred since 1500. As one example, the dodo (Figure 49.3) was a big, flightless bird that lived on the island of Mauritius in the Indian Ocean. Dodos were plentiful in 1600, when Dutch sailors first arrived on the island, but 80 or so years later the birds were extinct. Some were eaten by the sailors. However, destruction of nests and habitat by rats, cats, and pigs that accompanied the humans probably had a greater effect.

Extinctions of animals tend to garner more press than those of plants. Disappearances of large land animals, especially birds and mammals, are usually well documented. We know less about the losses of small animals, especially invertebrates. Historically, the losses of microorganisms, protists, and fungi have been almost entirely undocumented.

It can be difficult to determine whether a species is entirely extinct. As its numbers dwindle, sightings will become rare, but a few individuals may survive in isolated pockets of habitat. Consider, for example, the ivory-billed woodpecker, a spectacular bird that is native to swamp forests of the American Southeast (Figure 49.4). Lumbering of these forests caused the species' decline, and it was believed to have become extinct in the 1940s. A possible sighting in Arkansas in 2004 led to an extensive hunt for evidence of the bird's survival. By the end of 2007, this search had produced some blurry photos, snippets of video, and a few recordings of what may or may not be ivory-bill calls and knocks. Definitive proof that the bird still lives remains elusive.

If the ivory-billed woodpecker is still around, it is an **endangered species**, a species that has population levels so low that it faces extinction in all or part of its range. A **threatened species** is one that is likely to become endangered in the near future. Nearly all species that are currently endangered or threatened owe their precarious position to human influences, as detailed in the next section.

Take-Home Message

How are humans affecting the pattern of extinctions?

- Humans are causing a rise in the rate of extinction.
- Previous mass extinctions occurred as a result of global catastrophes. Species diversity takes millions of years to recover after a mass extinction.
- Many species are currently endangered or threatened as a result of human activity, in what is being called the sixth great mass extinction.

49.2 Current Threats to Species

- Expansion of human populations and the accompanying industrialization threatens countless species.
- Links to Inbreeding and gene flow 18.8 and 18.9, Aquifer depletion 47.6, Greenhouse gases 47.8

Habitat Loss, Fragmentation, and Degradation

Each species requires a particular type of habitat, and any loss, degradation, or fragmentation of that habitat reduces population numbers. An **endemic species**, one that is confined to the limited area in which it evolved, is more likely to go extinct than a species with a more widespread distribution.

Species with highly specific resource requirements are particularly vulnerable to habitat alterations. For example, giant pandas (Figure 49.5) are endemic to China's bamboo forests and feed mainly on this plant. As China's human population soared, bamboo was cut for building materials and to make room for farms. As the bamboo forests disappeared, so did pandas. Their numbers, which may have once been as high as 100,000, fell to 1,000 or so animals in the wild.

In addition to habitat loss, pandas are affected by habitat fragmentation; suitable panda habitat is now limited to widely separated patches atop mountains. Because of this fragmentation, pandas facing adverse conditions in one area cannot move to new site. The fragmentation also hampers the dispersal of young females. This decreases gene flow, effectively dividing the population into smaller subunits. The small group size encourages inbreeding and reduces the genetic diversity of the species as a whole. Current efforts to save giant pandas involve protecting existing habitat, creating corridors of suitable habitat to connect now isolated preserves, and captive breeding programs.

In the United States, habitat loss affects nearly all of the more than 700 species of threatened or endangered flowering plants. For example, the conversion of prairies and meadows to farms and housing developments has put both the eastern and western species of prairie fringed orchids (*Platanthera*) on the federal threatened species list (Figure 49.6*a*).

Humans also degrade habitat in less direct ways. For example, Edwards Aquifer in Texas consists of water-filled, underground, limestone formations that supply drinking water to the city of San Antonio. Excessive withdrawals of water from this aquifer, along with pollution of the water that recharges it, endanger the species that live in the aquifer, such as the Texas blind salamander (Figure 49.6*b*). Biologists find this species of interest because it demonstrates the evolutionary effects of many generations of life in total darkness.

Figure 49.5 Giant panda (*Ailuropoda melanoleuca*), one of the best-known endangered species. The panda's diet consists almost entirely of bamboo. Destruction and fragmentation of China's bamboo forests threaten its survival.

Acid rain, pesticide residues, fertilizer runoff, and emissions of greenhouse gases also degrade habitats and contribute to the decline of species. The chapter introduction explained how melting of polar ice may harm polar bears. Recognition of this threat may lead to the listing of this species as endangered.

Overharvesting and Poaching

When European settlers first arrived in North America, they found 3 billion to 5 billion passenger pigeons. In the 1800s, commercial hunting caused a steep decline in the bird's numbers. The last time anyone saw a wild passenger pigeon was 1900—and he shot it. The last captive bird died in 1914.

We are still overharvesting species. The crash of the Atlantic codfish population, described in Section 45.6, is one recent example. As another, the white abalone was the first marine invertebrate listed as threatened in the United States. Commercial harvesting of white abalones accelerated during the 1970s. By 1990, only 1 percent of the population remained.

Biologist Boris Worm estimates that populations of about 29 percent of commercially harvested marine fish and invertebrates have already collapsed—annual catch for these species is now less than 10 percent of the recorded maximum. If current trends continue, all populations of marine species that we now harvest for commercial sale could collapse by 2050.

Poaching, the illegal harvest of species, is another threat, especially in less-developed countries. People who have few other sources of protein will kill and eat

Figure 49.6 Two North American species under threat. (**a**) Habitat destruction threatens the eastern fringed prairie orchid (*Platanthera leucophaea*). (**b**) Aquifer depletion and pollution endanger Texas blind salamanders, *Typhlomolge rathbuni*. Generations of life in a dark aquifer, where there is no selection against mutations that impair eye development, have reduced this species' eyes to tiny black spots.

local animals, regardless of the animals' endangered status. Endangered species are also collected or killed for profit. It is a sad commentary on human nature that the rarer a species becomes, the higher the price it fetches on the black market. Globalization means that species can be sold to high bidders anywhere in the world. For example, rhino horn from endangered animals in Africa ends up as a traditional medicine in Asia and as knife handles in Yemen.

Species Introductions

Exotic predators (Section 46.9) are another threat. For example, rats that expanded their range by stowing away on ships now endanger many island species. Rats eat bird eggs and nestlings. They also devour other small animals such as snails. Humans also unintentionally dispersed the brown tree snake, which is native to Samoa. The arrival of this snake in Guam resulted in the extinction of most birds endemic to the island, and endangers the three that remain.

Exotic species often outcompete native ones. In the American Southeast, introduced vines such as kudzu (Figure 46.22*b*) and Japanese honeysuckle overgrow and threaten native low-growing plants. In California's mountain streams, competition from European brown trout and eastern brook trout—both introduced for sport fishing—endangers native golden trout.

Exotic pathogens also cause species declines. For example, avian malaria was unknown in Hawaii until it was carried to the islands by introduced birds and dispersed by introduced mosquitoes. Avian malaria is contributing to the extinction of native honeycreepers (birds described in the Chapter 19 introduction).

Interacting Effects

A species most often becomes endangered because of a number of concurrent threats. Often, the decline or loss of one species endangers another. For example, running buffalo clover (*Trifolium stoloniferum*) and the buffalo, or bison, that grazed on it were once common in the Midwest. The plants thrived in the open woodlands that the buffalo favored. Here, the soil was enriched by the hefty herbivore's droppings and periodically disturbed by its hooves. Buffalo helped to disperse the clover's seeds, which survive passage through the animal's gut. When buffalo were hunted to near extinction, clover populations declined. Now listed as an endangered species, the clover is further threatened by conversion of habitat for human use, competition from introduced plants, and attacks by introduced insects and pathogens.

Take-Home Message

How do human activities endanger existing species?

- Species decline when humans destroy or fragment natural habitat by converting it to human use, or degrade it through pollution or withdrawal of an essential resource.
- Humans also directly cause declines by overharvesting species and by poaching.
- Global travel and trade can introduce exotic species that harm native ones.
- Most endangered species are affected by multiple threats.

FOCUS ON RESEARCH

49.3 The Unknown Losses

■ We have only begun to evaluate the threats to many groups of species, especially the microbial ones.

Endangered species listings have historically focused on vertebrates. Biologists have just begun to evaluate the threats to invertebrates and plants. Our impact on protists and fungi is essentially unknown, and the World Conservation Union's IUCN Red List of Threatened Species (Table 49.1) does not even address prokaryotes.

In a 2006 article, microbiologist Tom Curtis made a plea for increased research on microbial ecology and microbial diversity. He argued that we have barely begun to comprehend the vast number of microbial species and to understand their importance.

Curtis concluded by writing, "I make no apologies for putting microorganisms on a pedestal above all other living things. For if the last blue whale choked to death on the last panda, it would be disastrous but not the end of the world. But if we accidentally poisoned the last two species of ammonia-oxidizers, that would be another matter. It could be happening now and we wouldn't even know . . . " The ammonia-oxidizing bacteria are essential because they make nitrogen available to plants.

Table 49.1 Global List of Threatened Species (2007)*

	Described Species	Evaluated for Threats	Found to Be Threatened
Vertebrates			
Mammals	5,416	4,863	1,094
Birds	9,956	9,956	1,217
Reptiles	8,240	1,385	422
Amphibians	6,199	5,915	1,808
Fishes	30,000	3,119	1,201
Invertebrates			
Insects	959,000	1,255	623
Mollusks	81,000	2,212	978
Crustaceans	40,000	553	460
Corals	2,175	13	5
Others	130,200	83	42
Land Plants			
Mosses	15,000	92	79
Ferns and allies	13,025	211	139
Gymnosperms	980	909	321
Angiosperms	258,650	10,771	7,899
Protists			
Green algae	3,715	2	0
Red algae	5,956	58	9
Brown algae	2,849	15	6
Fungi			
Lichens	10,000	2	2
Mushrooms	16,000	1	1

* IUCN–WCU Red List, available online at www.iucnredlist.org

49.4 Assessing Biodiversity

■ Conservation biologists are busy surveying and seeking ways to protect the world's existing biodiversity.

■ Links to Lichens and pollution 18.4, Oomycotes 22.8

Conservation Biology

Biologists recognize three levels of **biodiversity**: genetic diversity, species diversity, and ecosystem diversity. The rate of decline in biodiversity is accelerating at all three levels. Conservation biology addresses these declines. The goals of this relatively new field of biology are (1) to survey the range of biodiversity, (2) to investigate the evolutionary and ecological origins of biodiversity, and (3) to find ways to maintain and use biodiversity in ways that benefit human populations. The objective is to conserve as much biodiversity as possible by using it in sustainable ways.

Monitoring Indicator Species

Habitat damage and loss can affect different species in different ways. An **indicator species** is a species that alerts biologists to habitat degradation and impending loss of diversity when its populations decline. As one example, biologists can assess the health of a stream by monitoring certain fish and invertebrates. A decline in a trout population can be an early sign of problems in a freshwater habitat because trout do not tolerate pollutants or low oxygen levels.

Lichens function as indicators of habitat quality on land. Because lichens absorb mineral ions from dust in the air, they are harmed by air pollution. The lichens absorb toxic metals such as mercury and lead, and cannot get rid of them. Section 18.4 described how, with the onset of the industrial revolution, decline of lichens in England's forests selected for a particular coloration pattern among forest moths.

Identifying Regions at Risk

With so many species at risk, conservation biologists are working to identify the **hot spots**, habitats that are rich in endemic species and under great threat. The idea is that once identified, hot spots can take priority in worldwide conservation efforts.

The identification of a hot spot involves making an inventory of the organisms in some limited area, such as an isolated valley. Sampling quadrats and capture-mark–recapture studies identify species present in the area, and allow an estimation of their population size (Section 45.2). The Chapter 1 introduction highlights one exploratory survey in New Guinea.

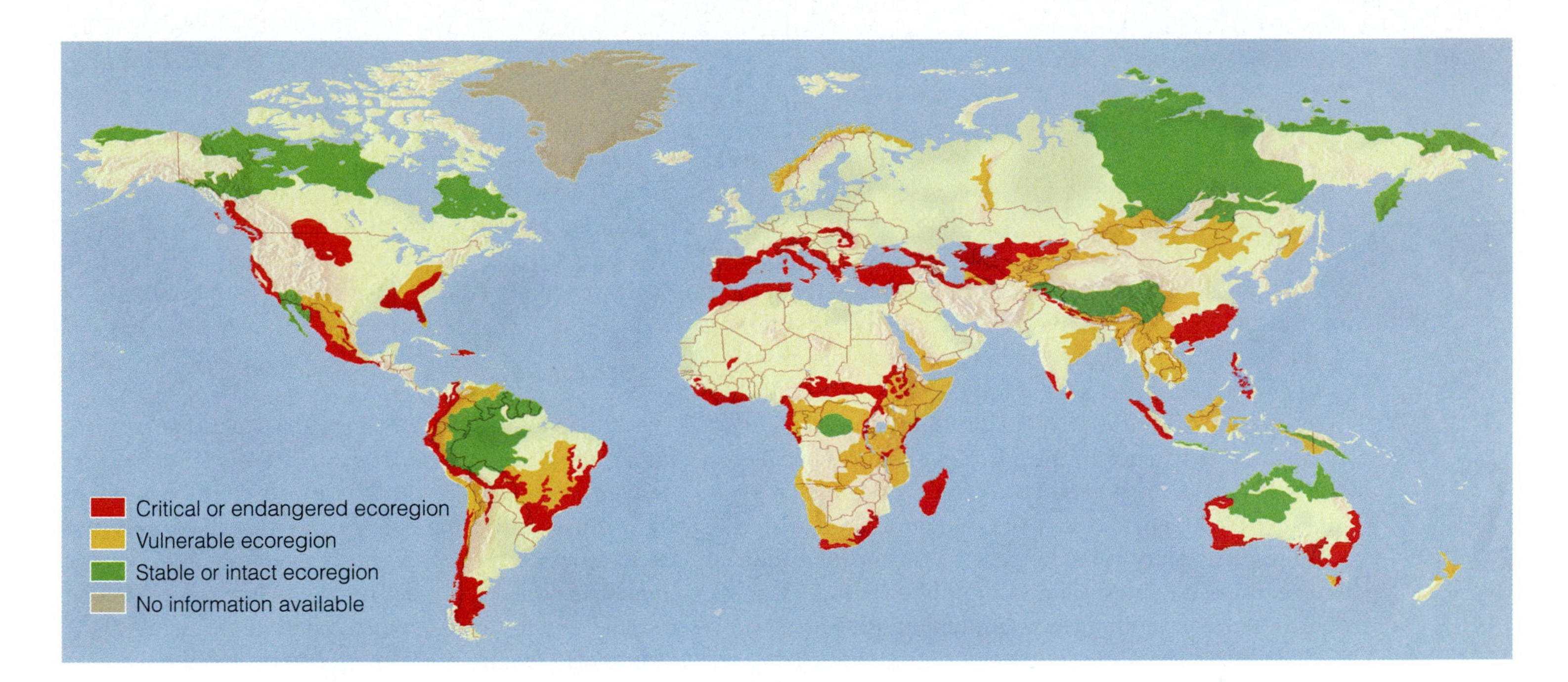

Figure 49.7 The location and current conservation status of the land ecoregions deemed most important by the World Wildlife Fund.

On a broader scale, conservation biologists define **ecoregions**, which are land or aquatic regions characterized by climate, geography, and the species found within them. The most widely used ecoregion system was developed by conservation scientists of the World Wildlife Fund. These scientists defined 867 distinctive land ecoregions. Figure 49.7 shows the locations and conservation status of ecoregions that are considered the top priority for conservation efforts.

The goal of prioritizing ecoregions is to save representative examples of all of Earth's existing biomes. By focusing on hotspots and critical ecoregions, rather than on individual endangered species, scientists hope to maintain ecosystem processes that naturally sustain biological diversity.

Table 49.2 lists the critical/endangered ecoregions located partially or entirely in the United States. Each has a large number of endemic species and is under threat. As one example, the Klamath-Siskiyou forest in southwestern Oregon and northwestern California is home to many rare conifers. Logging is the main threat to this region. However, a newly introduced conifer pathogen, *Phytophthora lateralis*, is also a concern. It is a relative of the oomycote protist that causes sudden oak death (Section 22.8). Two endangered birds, the northern spotted owl and the marbled murlet, nest in old-growth portions of the forest. Endangered coho salmon breed in streams that run through the forest.

Table 49.2 Critical or Endangered Ecoregions in the U.S.

Ecoregion	Area (sq. km)	Major Threats
Northern prairie	700,000	Conversion to pasture or farms; oil and gas development
Klamath-Siskiyou coniferous forest	50,300	Logging, exotic root disease spread by road building
Pacific temperate rain forest	295,000	Logging
Sierra Madre pine-oak forests	289,000	Overgrazing, logging, overuse for recreation
California chaparral and woodlands	121,000	Establishment of exotic species, overgrazing, fire suppression
Nevada coniferous forest	53,000	Logging, urban expansion
Southeastern coniferous and broadleaf forests	585,000	Logging, suppression of fire, urban expansion

Take-Home Message

How do conservation biologists help protect biodiversity?

- Conservation biologists assess Earth's species richness and create systems for prioritizing conservation efforts.
- Hot spots are areas that include many endemic species and face a high degree of threat. Ecoregions are larger areas characterized by physical factors and species composition.

49.5 Effects of Development and Consumption

- As human populations soar, their need for energy and other resources puts pressure on native species.
- Links to Acid rain 47.9, Resource use 45.9, Fossil fuel 23.5

Effects of Urban and Suburban Development

When homes, factories, and shopping centers replace undisturbed habitat, biodiversity declines. Worldwide, people continue to migrate from rural areas into cities at an ever accelerating pace (Figure 49.8).

In the United States, expansion of urban and suburban areas is a factor in the declines of many species. The threatened Florida sandhill crane is being pressured by the expanding city of Orlando. In Nevada, a small population of relict leopard frogs, once thought to be extinct, is hanging on in rapidly growing Clark County. Another endangered amphibian, the Houston toad, now survives only between the growing cities of Austin and Houston. In northern California, new housing developments near San Francisco may harm the endangered Mission Blue butterfly.

Proximity to human development affects different species in different ways. Exotic plants introduced to beautify suburban yards can release seeds that become established in the wild and outcompete natives such as California's chapparal species. Dogs and cats that are allowed to roam can kill wild animals or change their behavior in ways that interfere with breeding. Roads interrupt and restrict land animal movement and so hamper gene flow. Lighting at night also has negative impacts. For example, light from cities along tropical beaches can disorient endangered sea turtle hatchlings as they attempt to make their way to the sea. Night-flying migratory birds that use light to navigate tend to collide with tall, well-lit buildings.

Effects of Resource Consumption

The life-style of people in industrial nations requires large quantities of resources, and the extraction and delivery of these resources affects biodiversity. In the United States, the size of the average family has declined since the 1950s, but the size of the average home has doubled. Larger homes require more lumber to build and to furnish, which encourages logging. Big homes also require more energy to heat and to cool.

Most of the energy used in developed countries is supplied by fossil fuels—petroleum, natural gas, and coal (Figure 49.9). You already know that use of these nonrenewable fuels contributes to global warming and acid rain. In addition, extraction and transportation of these fuels have negative impacts. Oil harms many species when it leaks from pipelines or from ships. Strip mining for coal degrades the immediate area and it often lowers the water quality of nearby streams.

Figure 49.8 Cities displace wild species and require huge amounts of resources. In 2008, for the first time, a majority of the human population was city dwellers.

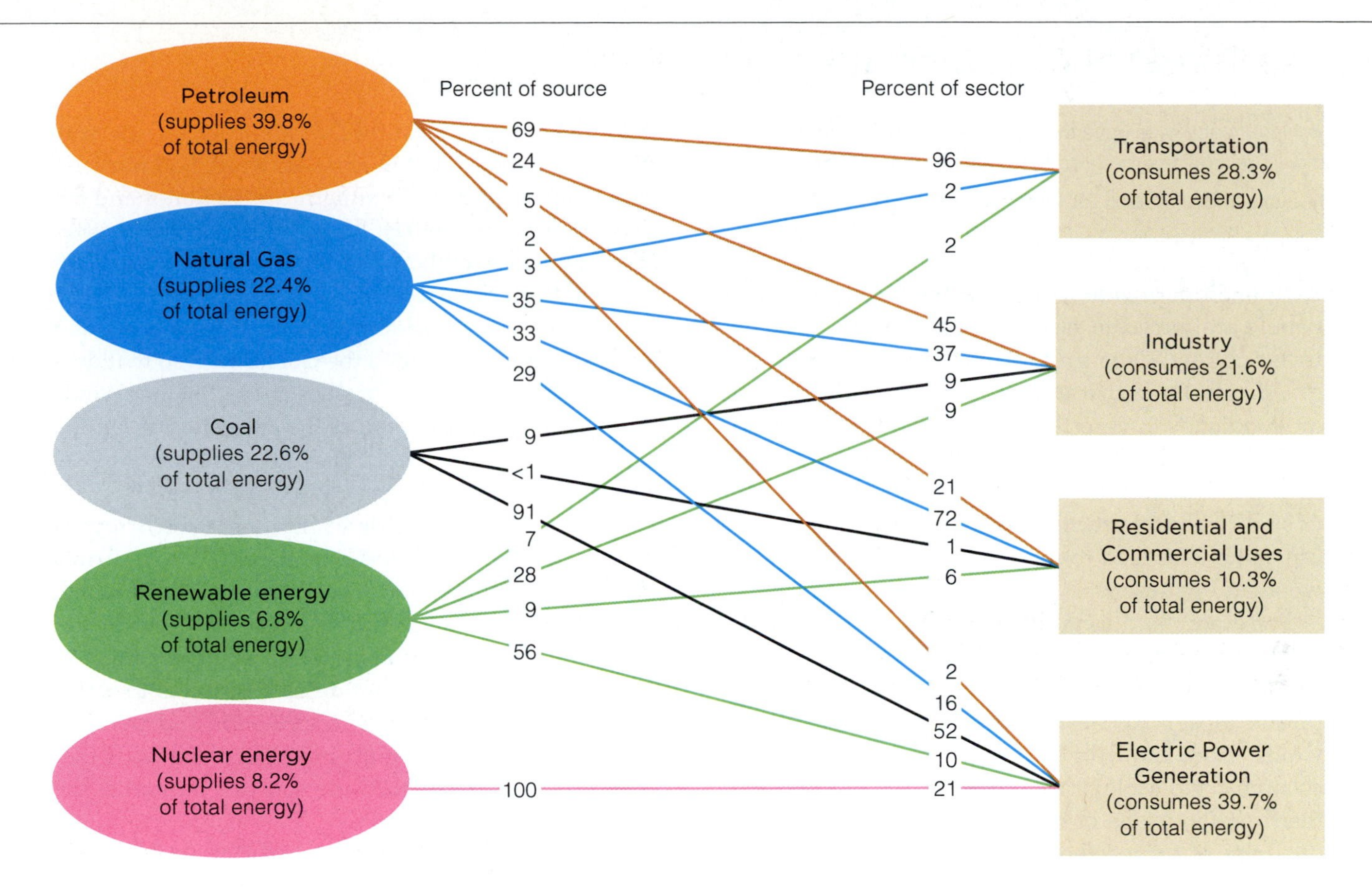

Figure 49.9 United States energy consumption in 2006 by source and sector of use. For example, 69 percent of the petroleum used in the United States goes to transportation, and petroleum supplies 96 percent of the transportation sector's energy needs.

Figure It Out: What percent of coal used in the United States goes to generate electricity? What percentage the country's electric power is derived from coal?

Answer: 91%, 52%

For example, the runoff from coal mines in Tennessee is poisoning endangered freshwater mussels that live in water downstream from the mines.

Even renewable energy sources can cause problems. For example, dams in rivers of the Pacific Northwest generate renewable hydroelectric power. However, the dams also prevent endangered salmon from returning to streams above the dam to breed. As these salmon populations have declined, so have endangered killer whales (orcas) that feed on adult salmon in the ocean. Such effects are not exactly what people have in mind when they think about "green energy."

The reality is that all commercially produced energy has some kind of negative environmental impact. The best way to minimize that impact is to use less energy.

Extracting and delivering materials that sustain the economies of developed countries also has environmental costs. Petroleum is used not only as fuel, but also as raw material for plastics, a topic we will return to later in this chapter. Surface mines extract essential minerals, such as the copper we use in computers and other electronic products, and the manganese used to manufacture steel for building, cars, and appliances. Surface mining strips an area of vegetation and soil, creating an ecological dead zone. It puts dust into the air, creates mountains of rocky waste, and sometimes contaminates nearby waterways.

Where do raw materials in manufactured products you buy come from? With globalization, it's hard to know. Mines in developing countries operate under regulations that are usually less strict than those in the United States or less stringently enforced, so their environmental impact is even greater.

Take-Home Message

How do development and resource use affect biodiversity?

- Expansion of cities and suburbs has a negative impact on biodiversity. Developed areas displace wild species and also harm them indirectly, as by introducing competing plants or causing light pollution.
- Processes that extract or capture energy, both renewable and nonrenewable, can destroy or degrade habitat.
- Obtaining the raw materials used in consumer products frequently involves degradation of the environment, which can reduce biodiversity.

49.6 The Threat of Desertification

- Human activities have the potential not only to harm individual species, but also to transform entire biomes.
- Links to Soil erosion 29.1, Transpiration 29.3

As human populations increase, greater numbers of people are forced to farm in areas that are ill-suited to agriculture. Others allow livestock to overgraze in grasslands. **Desertification**, a conversion of grassland or woodlands to desertlike conditions, is one result.

Deserts naturally expand and contract over time as climate conditions vary (Section 26.15). However, poor agricultural practices that encourage soil erosion can sometimes lead to rapid shifts from grassland or woodland to desert.

For example, during the mid-1930s, large portions of prairie on the southern Great Plains were plowed under to plant crops. This plowing exposed the deep prairie topsoil to the force of the region's constant winds. Coupled with a drought, the result was an economic and ecological disaster. Winds carried more than a billion tons of topsoil aloft as sky-darkening dust clouds turned the region into what came to be known as the Dust Bowl (Figure 49.10). Tons of displaced soil fell to earth as far away as New York City and Washington, D.C.

Desertification now threatens vast areas. In Africa, the Sahara Desert is expanding south into the Sahel region. Overgrazing in this region strips grasslands of their vegetation and allows winds to erode the soil. Winds carry the soil aloft and westward. Soil particles land as far away as the southern United States and the Caribbean (Figure 49.11). As described in the Chapter 24 introduction, pathogens traveling on dust particles threaten Caribbean corals.

In China's northwestern regions, overplowing and overgrazing has expanded the Gobi Desert so that dust clouds periodically darken skies above Beijing. Winds carry some of this dust across the Pacific to the United States. In an effort to hold back the desert, China has now planted billions of trees as a "Green Wall."

Drought encourages desertification, which results in more drought in a positive feedback cycle. Plants cannot thrive in a region where the topsoil has blown away. With less transpiration (Section 29.3), less water enters the atmosphere, so local rainfall decreases.

The best way to prevent desertification is to avoid farming in areas subject to high winds and periodic drought. If these areas must be utilized, methods that do not repeatedly disturb the soil can minimize the risk of desertification.

Take-Home Message

What is desertification?

- Desertification turns productive grassland or woodland into a desertlike region in which little grows.

Figure 49.10 A giant dust cloud about to descend on a farm in Kansas during the 1930s. A large portion of the southern Great Plains was then known as the Dust Bowl. Drought and poor agricultural practices allowed winds to strip tons of topsoil from the ground and carry it aloft.

Figure 49.11 Modern-day dust cloud blowing from Africa's Sahara Desert out into the Atlantic ocean. The Sahara's area is increasing as a result of a long-term drought, overgrazing, and stripping of woodlands for firewood.

49.7 The Trouble With Trash

- Recycling saves limited resources, and it also keeps dangerous trash out of habitats where it can do harm.
- Link to Groundwater and aquifers 47.6

Six billion people use and discard a lot of stuff. Where does all the waste go? Historically, unwanted material was simply buried in the ground or dumped out at sea. Trash was out of sight, and also out of mind. We now know that chemicals in buried trash can contaminate groundwater and aquifers. Waste dumped into the sea harms marine life. For example, seabirds eat floating bits of plastic and feed them to their chicks, with deadly results (Figure 49.12).

In 2006, the United States generated 251 million tons of garbage, which averages out to 2.1 kilograms (4.6 pounds) per person per day. By weight, about a third of that material was recycled, but there is plenty of room for improvement. Two-thirds of plastic soft drink bottles and three-quarters of glass bottles were not recycled. Nonrecycled trash now gets burned in high-temperature incinerators or placed in engineered landfills lined with material that minimizes the risk of groundwater contamination. No solid municipal waste can legally be dumped at sea.

Nevertheless, plastic and other garbage constantly enters our coastal waters. Foam cups and containers from fast-food outlets, plastic shopping bags, plastic water bottles, and other material discarded as litter ends up in storm drains. From there it is carried to streams and rivers, which can convey it to the sea. A seawater sample taken near the mouth of the San Gabriel River in southern California had 128 times as much plastic as plankton by weight.

Once in the ocean, trash can persist for a surprisingly long time. Components of a disposable diaper will last for more than 100 years, as will fishing line. A plastic bag will be around for more than 50 years, and a cigarette filter for more than 10.

To reduce the impact of plastic trash, choose more durable objects over disposable ones, and avoid buying plastic when other, less environmentally harmful alternatives exist. If you use plastic, be sure to recycle or dispose of it properly.

Take-Home Message

What are the ecological effects of trash?

- Trash, especially plastics, often ends up in the oceans where it harms marine life.
- You can minimize your environmental impact by avoiding disposable goods and by recycling.

Figure 49.12 (**a**) A recently deceased Laysan albatross chick, dissected to reveal the contents of its gut. (**b**) Scientists found more than 300 pieces of plastic inside the bird. One of the pieces had punctured its gut wall, resulting in its death. The chick was fed the plastic by its parents, who gathered the material from the ocean surface, mistaking it for food.

49.8 Maintaining Biodiversity and Human Populations

- Managing biodiversity can sustain biological wealth, while also providing economic opportunity.
- Link to Human population growth 45.7

Bioeconomic Considerations

Every nation enjoys three forms of wealth—material, cultural, and biological wealth. Its biological wealth—biodiversity—can be a source of food, medicine, and other products. However, protecting biological wealth is often a tricky proposition. Even in developed countries, people often oppose environmental protections because they fear such measures will have adverse economic consequences. However, taking care of the environment can make good economic sense. People can both preserve and profit from their biological wealth. We now turn to some success stories.

Sustainable Use of Biological Wealth

Using Genetic Diversity One particularly observant Mexican college student discovered *Zea diploperennis,* a wild maize long believed to be extinct. It had disappeared from most of its range, but a relic population clung to life in a 900-acre region of mountain terrain near Jalisco. Unlike domesticated corn, *Z. diploperennis* is perennial and resistant to most viruses. Gene transfers from this wild species into crop plants may boost production of corn for hungry people in Mexico and elsewhere. In recognition of the potential value of *Z. diploperennis* species, the Mexican government has set aside its mountainous habitat as a biological reserve, the first ever created to protect a wild relative of an important crop plant.

Discovering Useful Chemicals Many species make chemical compounds that could serve as medicines or other commercial products. Most developing countries do not have laboratories to test species for potential products, but large pharmaceutical companies in other countries do. The National Institute of Biodiversity of Costa Rica collects and identifies species that look promising, and sends away extracts of these species for chemical analyses. If a product of one of the species is marketed, Costa Rica will share in the profits, which are earmarked for conservation programs.

Ecotourism Setting aside species-rich preserves and encouraging tourists to visit them can have biological and economic benefits. For example, during the 1970s, George Powell was studying birds in the Monteverde Cloud Forest in Costa Rica. This forest was rapidly being cleared and Powell got the idea to buy part of it as a nature sanctuary. His efforts inspired individuals and conservation groups to donate funds, and much of the forest is now protected as a private nature reserve.

Figure 49.13 Strip logging. The practice may protect biodiversity as it permits logging on tropical slopes. A narrow corridor paralleling the land's contours is cleared. A roadbed is made at the top to haul away logs. After a few years, saplings grow in the cleared corridor. Another corridor is cleared above the roadbed. Nutrients leached from exposed soil trickle into the first corridor. There they are taken up by saplings, which benefit from all the nutrient input by growing faster. Later, a third corridor is cut above the second one—and so on in a profitable cycle of logging, which the habitat sustains over time.

The reserve's plants and animals include more than 100 mammalian species, 400 bird species, and 120 species of amphibians and reptiles. The reserve is one of the few habitats left for the jaguar, ocelot, puma, and their relatives.

More than 50,000 tourists now visit the Monteverde Cloud Forest Reserve each year. Ecotourism centered on this reserve provides employment to local people and has other beneficial effects. For example, a nonprofit school set up inside the reserve helps educate the area's children.

Sustainable Logging A tropical forest yields wood for local needs and for export to developed countries. However, severe erosion often follows the logging of forested slopes. Gary Hartshorn devised a method to minimize erosion. As explained in Figure 49.13, strip logging allows cycles of logging. It yields sustainable economic benefits for local loggers, while minimizing effects of erosion and maximizing forest biodiversity.

Responsible Ranching Developed countries are also implementing conservation practices that sustain biological wealth. For example, **riparian zones** are narrow corridors of vegetation along a stream or river. They are of great ecological importance. Plants in a riparian zone act as a line of defense against flood damage by sponging up water during spring runoffs and summer storms. Shade cast by a canopy of taller shrubs and trees in a riparian zone helps conserve water during droughts. A riparian zone provides wildlife with food, shelter, and shade, particularly in arid and semiarid regions. In the western United States, 67 to 75 percent of the endemic species spend all or part of their life cycle in riparian zones. Among them are 136 kinds of songbirds, some of which will nest only in the plants of a riparian zone.

In ranch country, cattle tend to congregate near rivers and streams, which are often their only source of water. There, the animals trample and feed until grasses and herbaceous shrubs are gone. It takes only a few head of cattle to destroy a riparian zone. All but 10 percent of the riparian vegetation of Arizona and New Mexico is already gone, mainly into the stomachs of grazing cattle.

To preserve the biodiversity in riparian zones, cattle on some ranches are now kept away from riverbanks and provided with an alternate water source. Figure 49.14 shows how excluding cattle from the riparian zone can make a difference. Once cattle are excluded, native vegetation regrows fast and the biodiversity of the habitat is restored.

Figure 49.14 Riparian zone restoration. The photos show Arizona's San Pedro River, before restoration (*above*) and after (*below*).

These examples show how we can put our knowledge of biological principles to use. The health of our planet depends on our ability to recognize that the principles of energy flow and of resource limitation, which govern the survival of all systems of life, do not change. It is our biological and cultural imperative to come to terms with these principles, and to ask ourselves this: What will be our long-term effect on the world of life?

Take-Home Message

How do we meet human needs and sustain biodiveristy?

- Any nation has biological wealth, which people will tend to protect if they recognize its value.
- Sustainable practices allow people to benefit economically from biological resources without destroying them.

IMPACTS, ISSUES REVISITED A Long Reach

Global warming is thawing ice in the Arctic and opening access to this continent, which was previously protected from development by the lack of shipping lanes. Eight countries, including the United States, Canada, and Russia, control parts of the Arctic and have rights to its oil, gas, and mineral deposits. Conservationists worry that exploitation of these resources will put more pressure on arctic species already vulnerable to extinction.

How would you vote?

Should conservation of the Arctic receive priority over exploitation of its resources? See CengageNOW for details, then vote online.

Summary

Section 49.1 The current rate of species loss is high enough to suggest that an extinction crisis is under way. After other mass extinctions, it has taken millions of years for biodiversity to recover to its previous level.

Human-caused extinctions may have begun when humans first entered the Americas and Australia. Many more-recent extinctions definitely resulted from human activity. **Endangered species** currently face a high risk of extinction. **Threatened species** are likely to become endangered in the future.

- *Use the animation on CengageNOW to view extinction patterns for different taxa.*

Section 49.2 **Endemic species**, which evolved in one place and are present only in that habitat, are highly vulnerable to extinction. Species with highly specialized resource needs are also especially vulnerable.

Humans cause habitat loss, degradation of habitat, and habitat fragmentation, all of which can endanger a species. Humans also directly reduce populations and endanger species by overharvesting. Species introductions also cause declines of native species. In most cases, a species becomes endangered because of multiple factors. Sometimes, a decline in one species as a result of human activity leads to decline of another species.

Sections 49.3, 49.4 Our knowledge of existing species is limited and biased toward vertebrates. We know little about the abundance and diversity of microbial species that carry out essential ecosystem processes.

We recognize three levels of **biodiversity**: genetic diversity, species diversity, and ecosystem diversity. All are threatened. The field of conservation biology surveys the range of biodiversity, investigates its origins, and identifies ways to maintain and use it in ways that benefit human populations.

An **indicator species** is one that is especially sensitive to environmental change and can be monitored to determine the health of an ecosystem.

Given that resources are limited, biologists attempt to identify **hot spots**, regions rich in endemic species and under a high level of threat. Biologists also identify **ecoregions**, larger regions characterized by their physical characteristics as well as the species in them.

The biologists prioritize ecoregions, with the goal of identifying those whose conservation will ensure that a representative sample of all of Earth's current biomes remains intact. The United States holds part or all of several ecoregions that are considered critical or endangered by international conservation organizations.

Section 49.5 Growth of cities and suburbs displaces wild species. Proximity to humans can also put stress on some species, as by the effects of introduced competitors or predators, or by ill effects of nighttime lighting.

People in developed countries contribute to species extinctions by their pattern of resource consumption. Their use of large amounts of fossil fuels contributes to global warming and also has other adverse effects on the environment. Renewable energy sources can also degrade habitat. Extracting fuel and mineral resources requires mining and other processes that pollute and otherwise make habitat unsuitable for native organisms.

Section 49.6 **Desertification** is the conversion of grassland or woodland to desertlike conditions. In the 1930s, a drought and poor agricultural practices caused desertification of a portion of the Great Plains, which became known as the Dust Bowl.

Desertification is currently a problem in China and in Africa. Some effects of desertification are felt far from the problem site because winds can pick up soil and carry it for long distances.

Section 49.7 Production of large amounts of trash is another threat to biodiversity. Plastic that enters oceans is particularly harmful and persistent.

Section 49.8 All nations have biological wealth; they have unique species that are of value to humans. People tend to preserve biological wealth when they recognize and benefit economically from its existence.

Preserving areas and using them for ecotourism can benefit local people while protecting endangered species.

Resources can also be harvested sustainably, as by strip logging of tropical forests on mountain slopes. This method minimizes erosion and ensures that there is always tree cover.

Developed nations also benefit by using their biological wealth in a sustainable fashion. For example, cattle ranching can have adverse effects on riparian zones, which are areas of high species diversity on river banks. Responsible ranchers exclude cattle from riparian zones to sustain biodiversity.

Data Analysis Exercise

Winds carry chemical contaminants produced and released at temperate latitudes to the Arctic, where the chemicals enter food webs. By the process of biological magnification (Section 47.4), top carnivores in arctic food webs—such as polar bears and people—end up with high doses of these chemicals. For example, indigenous arctic people who eat a lot of local wildlife tend to have unusually high levels of polychlorinated biphenyls, or PCBs, in their bodies. The Arctic Monitoring and Assessment Programme studies the effects of these chemicals on health and reproduction. Figure 49.15 shows the effect of PCBs on the sex ratio at birth in indigenous populations in the Russian Arctic.

1. Which sex is was most common in offspring of women with less than one microgram per millileter of PCB in serum?

2. At what PCB concentrations were women more likely to have daughters?

3. In some Greenland villages, nearly all recent newborns are female. Would you expect PCB levels in those villages to be above or under 4 micrograms per milliliter?

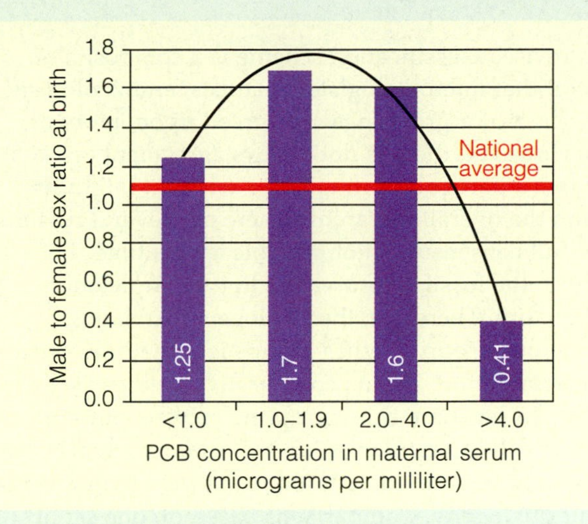

Figure 49.15 Effect of maternal PCB concentration on sex ratio of newborns in indigenous populations in the Russian Arctic. The red line indicates the average sex ratio for births in Russia—1.06 males per female.

Self-Quiz

Answers in Appendix III

1. True or false? Most species that evolved have already become extinct.

2. Dodos were driven to extinction ______ .
   a. when humans arrived in North America
   b. by overharvesting and introduced species
   c. as a result of global warming
   d. both a and b

3. An ______ species has population levels so low it is at great risk of extinction in the near future.
   a. endemic c. indicator
   b. endangered d. exotic

4. Gene flow among populations is hampered by ______ .
   a. habitat fragmentation c. poaching
   b. species introductions d. all of the above

5. ______ , native to the United States, have now been driven to extinction.
   a. Dodos c. Pandas
   b. Passenger pigeons d. Bison (buffalo)

6. Which of the following has the most representatives among the known endangered species?
   a. bacteria c. vertebrates
   b. fungi d. invertebrates

7. An ______ species can be monitored to gauge the health of its environment.
   a. endemic c. indicator
   b. endangered d. exotic

8. A(n) ______ is an area that conservation biologists consider a high priority for preservation.
   a. hot spot c. biome
   b. ecoregion d. biogeographic province

9. True or false? Artificial lighting harms some species.

10. Dams built to provide renewable hydroelectric power have caused declines in populations of ______ .
    a. salmon c. sea turtles
    b. killer whales d. both a and b

11. In the United States most plastic enters oceans by way of ______ .
    a. littering c. offshore drilling
    b. careless boaters d. municipal dumping

12. Match the organisms with their descriptions.

___riparian zone	a. evolved and found in one area
___hot spot	b. cause of some dust storms
___endemic species	c. locals benefit from visitors
___indicator species	d. many species, under threat
___desertification	e. less erosion, sustains forest
___ecotourism	f. species-rich area near river
___strip logging	g. highly sensitive to changes
___conservation biology	h. assesses and seeks ways to preserve biodiversity

■ *Visit CengageNOW for additional questions.*

Critical Thinking

1. Many biologists think that global climate change resulting from greenhouse gas emissions is the single greatest threat to existing biodiversity. List some negative effects that climate change could have on native species in your area.

2. In one seaside community in New Jersey, the U.S. Fish and Wildlife Service suggested removing feral cats (domestic cats that live in the wild) in order to protect some endangered wild birds (plovers) that nested on the town's beaches. Many residents were angered by the proposal, arguing that the cats have just as much right to exist as the birds. Do you agree?

Appendix I. Classification System

This revised classification scheme is a composite of several that microbiologists, botanists, and zoologists use. The major groupings are agreed upon, more or less. However, there is not always agreement on what to name a particular grouping or where it might fit within the overall hierarchy. There are several reasons why full consensus is not possible at this time.

First, the fossil record varies in its completeness and quality. Therefore, the phylogenetic relationship of one group to other groups is sometimes open to interpretation. Today, comparative studies at the molecular level are firming up the picture, but the work is still under way. Also, molecular comparisons do not always provide definitive answers to questions about phylogeny. Comparisons based on one set of genes may conflict with those comparing a different part of the genome. Or comparisons with one member of a group may conflict with comparisons based on other group members.

Second, ever since the time of Linnaeus, systems of classification have been based on the perceived morphological similarities and differences among organisms. Although some original interpretations are now open to question, we are so used to thinking about organisms in certain ways that reclassification often proceeds slowly.

A few examples: Traditionally, birds and reptiles were grouped in separate classes (Reptilia and Aves); yet there are compelling arguments for grouping the lizards and snakes in one group and the crocodilians, dinosaurs, and birds in another. Many biologists still favor a six-kingdom system of classification (archaea, bacteria, protists, plants, fungi, and animals). Others advocate a switch to the more recently proposed three-domain system (archaea, bacteria, and eukarya).

Third, researchers in microbiology, mycology, botany, zoology, and other fields of inquiry inherited a wealth of literature, based on classification systems that have been developed over time in each field of inquiry. Many are reluctant to give up established terminology that offers access to the past.

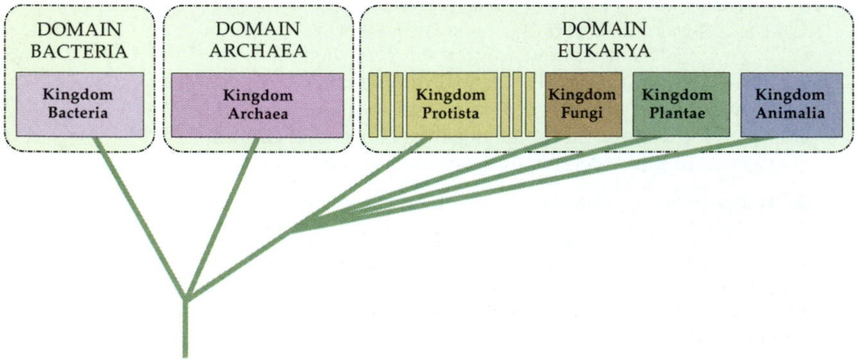

DOMAIN BACTERIA
Kingdom Bacteria
DOMAIN ARCHAEA
Kingdom Archaea
DOMAIN EUKARYA
Kingdom Protista
Kingdom Fungi
Kingdom Plantae
Kingdom Animalia

For example, botanists and microbiologists often use *division*, and zoologists *phylum*, for taxa that are equivalent in hierarchies of classification.

Why bother with classification frameworks if we know they only imperfectly reflect the evolutionary history of life? We do so for the same reasons that a writer might break up a history of civilization into several volumes, each with a number of chapters. Both are efforts to impart structure to an enormous body of knowledge and to facilitate retrieval of information from it. More importantly, to the extent that modern classification schemes accurately reflect evolutionary relationships, they provide the basis for comparative biological studies, which link all fields of biology.

Bear in mind that we include this appendix for your reference purposes only. Besides being open to revision, it is not meant to be complete. Names shown in "quotes" are polyphyletic or paraphyletic groups that are undergoing revision. For example, "reptiles" comprise at least three and possibly more lineages.

The most recently discovered species, as from the mid-ocean province, are not listed. Many existing and extinct species of the more obscure phyla are also not represented. Our strategy is to focus primarily on the organisms mentioned in the text or familiar to most students. We delve more deeply into flowering plants than into bryophytes, and into chordates than annelids.

PROKARYOTES AND EUKARYOTES COMPARED

As a general frame of reference, note that almost all bacteria and archaea are microscopic in size. Their DNA is concentrated in a nucleoid (a region of cytoplasm), not in a membrane-bound nucleus. All are single cells or simple associations of cells. They reproduce by prokaryotic fission or budding; they transfer genes by bacterial conjugation.

Table A lists representative types of autotrophic and heterotrophic prokaryotes. The authoritative reference, *Bergey's Manual of Systematic Bacteriology*, has called this a time of taxonomic transition. It references groups mainly by numerical taxonomy (Section 19.1) rather than by phylogeny. Our classification system does reflect evidence of evolutionary relationships for at least some bacterial groups.

The first life forms were prokaryotic. Similarities between Bacteria and Archaea have more ancient origins relative to the traits of eukaryotes.

Unlike the prokaryotes, all eukaryotic cells start out life with a DNA-enclosing nucleus and other membrane-bound organelles. Their chromosomes have many histones and other proteins attached. They include spectacularly diverse single-celled and multicelled species, which can reproduce by way of meiosis, mitosis, or both.

DOMAIN OF BACTERIA

KINGDOM BACTERIA The largest, and most diverse group of prokaryotic cells. Includes photosynthetic autotrophs, chemosynthetic autotrophs, and heterotrophs. All prokaryotic pathogens of vertebrates are bacteria.

PHYLUM AQIFACAE Most ancient branch of the bacterial tree. Gram-negative, mostly aerobic chemoautotrophs, mainly of volcanic hot springs. *Aquifex.*

PHYLUM DEINOCOCCUS-THERMUS Gram-positive, heat-loving chemoautotrophs. *Deinococcus* is the most radiation resistant organism known. *Thermus* occurs in hot springs and near hydrothermal vents.

PHYLUM CHLOROFLEXI Green nonsulfur bacteria. Gram-negative bacteria of hot springs, freshwater lakes, and marine habitats. Act as nonoxygen-producing photoautotrophs or aerobic chemoheterotrophs. *Chloroflexus.*

PHYLUM ACTINOBACTERIA Gram-positive, mostly aerobic heterotrophs in soil, freshwater and marine habitats, and on mammalian skin. *Propionibacterium, Actinomyces, Streptomyces.*

PHYLUM CYANOBACTERIA Gram-negative, oxygen-releasing photoautotrophs mainly in aquatic habitats. They have chlorophyll *a* and photosystem I. Includes many nitrogen-fixing genera. *Anabaena, Nostoc, Oscillatoria.*

PHYLUM CHLOROBIUM Green sulfur bacteria. Gram-negative nonoxygen-producing photosynthesizers, mainly in freshwater sediments. *Chlorobium.*

PHYLUM FIRMICUTES Gram-positive walled cells and the cell wall-less mycoplasmas. All are heterotrophs. Some survive in soil, hot springs, lakes, or oceans. Others live on or in animals. *Bacillus, Clostridium, Heliobacterium, Lactobacillus, Listeria, Mycobacterium, Mycoplasma, Streptococcus.*

PHYLUM CHLAMYDIAE Gram-negative intracellular parasites of birds and mammals. *Chlamydia.*

PHYLUM SPIROCHETES Free-living, parasitic, and mutualistic gram-negative spring-shaped bacteria. *Borelia, Pillotina, Spirillum, Treponema.*

PHYLUM PROTEOBACTERIA The largest bacterial group. Includes photoautotrophs, chemoautotrophs, and heterotrophs; free-living, parasitic, and colonial groups. All are gram-negative.

Class Alphaproteobacteria. *Agrobacterium, Azospirillum, Nitrobacter, Rickettsia, Rhizobium.*

Class Betaproteobacteria. *Neisseria.*

Class Gammaproteobacteria. *Chromatium, Escherichia, Haemopilius, Pseudomonas, Salmonella, Shigella, Thiomargarita, Vibrio, Yersinia.*

Class Deltaproteobacteria. *Azotobacter, Myxococcus.*

Class Epsilonproteobacteria. *Campylobacter, Helicobacter.*

DOMAIN OF ARCHAEA

KINGDOM ARCHAEA Prokaryotes that are evolutionarily between eukaryotic cells and the bacteria. Most are anaerobes. None are photosynthetic. Originally discovered in extreme habitats, they are now known to be widely dispersed. Compared with bacteria, the archaea have a distinctive cell wall structure and unique membrane lipids, ribosomes, and RNA sequences. Some are symbiotic with animals, but none are known to be animal pathogens.

PHYLUM EURYARCHAEOTA Largest archean group. Includes extreme thermophiles, halophiles, and methanogens. Others are abundant in the upper waters of the ocean and other more moderate habitats. *Methanocaldococcus, Nanoarchaeum.*

PHYLUM CRENARCHAEOTA Includes extreme theromophiles, as well as species that survive in Antarctic waters, and in more moderate habitats. *Sulfolobus, Ignicoccus.*

PHYLUM KORARCHAEOTA Known only from DNA isolated from hydrothermal pools. As of this writing, none have been cultured and no species have been named.

DOMAIN OF EUKARYOTES

KINGDOM "PROTISTA" A collection of single-celled and multicelled lineages, which does not constitute a monophyletic group. Some biologists consider the groups listed below to be kingdoms in their own right.

PARABASALIA Parabasalids. Flagellated, single-celled anaerobic heterotrophs with a cytoskeletal "backbone" that runs the length of the cell. There are no mitochondria, but a hydrogenosome serves a similar function. *Trichomonas, Trichonympha.*

DIPLOMONADIDA Diplomonads. Flagellated, anaerobic single-celled heterotrophs that do not have mitochondria or Golgi bodies and do not form a bipolar spindle at mitosis. May be one of the most ancient lineages. *Giardia.*

EUGLENOZOA Euglenoids and kinetoplastids. Free-living and parasitic flagellates. All with one or more mitochondria. Some photosynthetic euglenoids with chloroplasts, others heterotrophic. *Euglena, Trypanosoma, Leishmania.*

RHIZARIA Formaminiferans and radiolarians. Free-living, heterotrophic amoeboid cells that are enclosed in shells. Most live in ocean waters or sediments. *Pterocorys, Stylosphaera.*

ALVEOLATA Single cells having a unique array of membrane-bound sacs (alveoli) just beneath the plasma membrane.

Ciliata. Ciliated protozoans. Heterotrophic protists with many cilia. *Paramecium, Didinium.*

Dinoflagellates. Diverse heterotrophic and photosynthetic flagellated cells that deposit cellulose in their alveoli. *Gonyaulax, Gymnodinium, Karenia, Noctiluca.*

Apicomplexans. Single-celled parasites of animals. A unique microtubular device is used to attach to and penetrate a host cell. *Plasmodium.*

STRAMENOPHILA Stramenophiles. Single-celled and multicelled forms; flagella with tinsel-like filaments.

Oomycotes. Water molds. Heterotrophs. Decomposers, some parasites. *Saprolegnia, Phytophthora, Plasmopara.*

Chrysophytes. Golden algae, yellow-green algae, diatoms, coccolithophores. Photosynthetic. *Emiliania, Mischococcus.*

Phaeophytes. Brown algae. Photosynthetic; nearly all live in temperate marine waters. All are multicellular. *Macrocystis, Laminaria, Sargassum, Postelsia.*

RHODOPHYTA Red algae. Mostly photosynthetic, some parasitic. Nearly all marine, some in freshwater habitats. Most multicellular. *Porphyra, Antithamion.*

CHLOROPHYTA Green algae. Mostly photosynthetic, some parasitic. Most freshwater, some marine or terrestrial. Single-celled, colonial, and multicellular forms. Some biologists place the chlorophytes and charophytes with the land plants in a kingdom called the Viridiplantae. *Acetabularia, Chlamydomonas, Chlorella, Codium, Udotea, Ulva, Volvox.*

CHAROPHYTA Photosynthetic. Closest living relatives of plants. Include both single-celled and multicelled forms. Desmids, stoneworts. *Micrasterias, Chara, Spirogyra.*

AMOEBOZOA True amoebas and slime molds. Heterotrophs that spend all or part of the life cycle as a single cell that uses pseudopods to capture food. *Amoeba, Entoamoeba* (amoebas), *Dictyostelium* (cellular slime mold), *Physarum* (plasmodial slime mold).

KINGDOM FUNGI

Nearly all multicelled eukaryotic species with chitin-containing cell walls. Heterotrophs, mostly saprobic decomposers, some parasites. Nutrition based upon extracellular digestion of organic matter and absorption of nutrients by individual cells. Multicelled species form absorptive mycelia and reproductive structures that produce asexual spores (and sometimes sexual spores).

PHYLUM CHYTRIDIOMYCOTA Chytrids. Primarily aquatic; saprobic decomposers or parasites that produce flagellated spores. *Chytridium.*

PHYLUM ZYGOMYCOTA Zygomycetes. Producers of zygospores (zygotes inside thick wall) by way of sexual reproduction. Bread molds, related forms. *Rhizopus, Philobolus.*

PHYLUM ASCOMYCOTA Ascomycetes. Sac fungi. Sac-shaped cells form sexual spores (ascospores). Most yeasts and molds, morels, truffles. *Saccharomycetes, Morchella, Neurospora, Claviceps, Candida, Aspergillus, Penicillium.*

PHYLUM BASIDIOMYCOTA Basidiomycetes. Club fungi. Most diverse group. Produce basidiospores inside club-shaped structures. Mushrooms, shelf fungi, stinkhorns. *Agaricus, Amanita, Craterellus, Gymnophilus, Puccinia, Ustilago.*

"IMPERFECT FUNGI" Sexual spores absent or undetected. The group has no formal taxonomic status. If better understood, a given species might be grouped with sac fungi or club fungi. *Arthobotrys, Histoplasma, Microsporum, Verticillium.*

"LICHENS" Mutualistic interactions between fungal species and a cyanobacterium, green alga, or both. *Lobaria, Usnea.*

KINGDOM PLANTAE

Most photosynthetic with chlorophylls *a* and *b*. Some parasitic. Nearly all live on land. Sexual reproduction predominates.

BRYOPHYTES (NONVASCULAR PLANTS)

Small flattened haploid gametophyte dominates the life cycle; sporophyte remains attached to it. Sperm are flagellated; require water to swim to eggs for fertilization.

PHYLUM HEPATOPHYTA Liverworts. *Marchantia.*

PHYLUM ANTHOCEROPHYTA Hornworts.

PHYLUM BRYOPHYTA Mosses. *Polytrichum, Sphagnum.*

SEEDLESS VASCULAR PLANTS

Diploid sporophyte dominates, free-living gametophytes, flagellated sperm require water for fertilization.

PHYLUM LYCOPHYTA Lycophytes, club mosses. Small single-veined leaves, branching rhizomes. *Lycopodium, Selaginella.*

PHYLUM MONILOPHYTA

Subphylum Psilophyta. Whisk ferns. No obvious roots or leaves on sporophyte, very reduced. *Psilotum.*

Subphylum Sphenophyta. Horsetails. Reduced scalelike leaves. Some stems photosynthetic, others spore-producing. *Calamites* (extinct), *Equisetum.*

Subphylum Pterophyta. Ferns. Large leaves, usually with sori. Largest group of seedless vascular plants (12,000 species), mainly tropical, temperate habitats. *Pteris, Trichomanes, Cyathea* (tree ferns), *Polystichum.*

SEED-BEARING VASCULAR PLANTS

PHYLUM CYCADOPHYTA Cycads. Group of gymnosperms (vascular, bear "naked" seeds). Tropical, subtropical. Compound leaves, simple cones on male and female plants. Plants usually palm-like. Motile sperm. *Zamia, Cycas.*

PHYLUM GINKGOPHYTA Ginkgo (maidenhair tree). Type of gymnosperm. Motile sperm. Seeds with fleshy layer. *Ginkgo.*

PHYLUM GNETOPHYTA Gnetophytes. Only gymnosperms with vessels in xylem and double fertilization (but endosperm does not form). *Ephedra, Welwitchia, Gnetum.*

PHYLUM CONIFEROPHYTA Conifers. Most common and familiar gymnosperms. Generally cone-bearing species with needle-like or scale-like leaves. Includes pines (*Pinus*), redwoods (*Sequoia*), yews (*Taxus*).

PHYLUM ANTHOPHYTA Angiosperms (the flowering plants). Largest, most diverse group of vascular seed-bearing plants. Only organisms that produce flowers, fruits. Some families from several representative orders are listed:

BASAL FAMILIES

Family Amborellaceae. *Amborella.*
Family Nymphaeaceae. Water lilies.
Family Illiciaceae. Star anise.

MAGNOLIIDS

Family Magnoliaceae. Magnolias.
Family Lauraceae. Cinnamon, sassafras, avocados.
Family Piperaceae. Black pepper, white pepper.

EUDICOTS

Family Papaveraceae. Poppies.
Family Cactaceae. Cacti.
Family Euphorbiaceae. Spurges, poinsettia.
Family Salicaceae. Willows, poplars.
Family Fabaceae. Peas, beans, lupines, mesquite.
Family Rosaceae. Roses, apples, almonds, strawberries.
Family Moraceae. Figs, mulberries.
Family Cucurbitaceae. Squashes, melons, cucumbers.
Family Fagaceae. Oaks, chestnuts, beeches.
Family Brassicaceae. Mustards, cabbages, radishes.
Family Malvaceae. Mallows, okra, cotton, hibiscus, cocoa.
Family Sapindaceae. Soapberry, litchi, maples.
Family Ericaceae. Heaths, blueberries, azaleas.
Family Rubiaceae. Coffee.
Family Lamiaceae. Mints.
Family Solanaceae. Potatoes, eggplant, petunias.
Family Apiaceae. Parsleys, carrots, poison hemlock.
Family Asteraceae. Composites. Chrysanthemums, sunflowers, lettuces, dandelions.

MONOCOTS

Family Araceae. Anthuriums, calla lily, philodendrons.
Family Liliaceae. Lilies, tulips.
Family Alliaceae. Onions, garlic.
Family Iridaceae. Irises, gladioli, crocuses.
Family Orchidaceae. Orchids.
Family Arecaceae. Date palms, coconut palms.
Family Bromeliaceae. Bromeliads, pineapples.
Family Cyperaceae. Sedges.
Family Poaceae. Grasses, bamboos, corn, wheat, sugarcane.
Family Zingiberaceae. Gingers.

KINGDOM ANIMALIA

Multicelled heterotrophs, nearly all with tissues and organs, and organ systems, that are motile during part of the life cycle. Sexual reproduction occurs in most, but some also reproduce asexually. Embryos develop through a series of stages.

PHYLUM PORIFERA Sponges. No symmetry, tissues.

PHYLUM PLACOZOA Marine. Simplest known animal. Two cell layers, no mouth, no organs. *Trichoplax.*

PHYLUM CNIDARIA Radial symmetry, tissues, nematocysts.

Class Hydrozoa. Hydrozoans. *Hydra, Obelia, Physalia, Prya.*
Class Scyphozoa. Jellyfishes. *Aurelia.*
Class Anthozoa. Sea anemones, corals. *Telesto.*

PHYLUM PLATYHELMINTHES Flatworms. Bilateral, cephalized; simplest animals with organ systems. Saclike gut.

Class Turbellaria. Triclads (planarians), polyclads. *Dugesia.*
Class Trematoda. Flukes. *Clonorchis, Schistosoma.*
Class Cestoda. Tapeworms. *Diphyllobothrium, Taenia.*

PHYLUM ROTIFERA Rotifers. *Asplancha, Philodina.*

PHYLUM MOLLUSCA Mollusks.

Class Polyplacophora. Chitons. *Cryptochiton, Tonicella.*

Class Gastropoda. Snails, sea slugs, land slugs. *Aplysia, Ariolimax, Cypraea, Haliotis, Helix, Liguus, Limax, Littorina.*

Class Bivalvia. Clams, mussels, scallops, cockles, oysters, shipworms. *Ensis, Chlamys, Mytelus, Patinopectin.*

Class Cephalopoda. Squids, octopuses, cuttlefish, nautiluses. *Dosidiscus, Loligo, Nautilus, Octopus, Sepia.*

PHYLUM ANNELIDA Segmented worms.

Class Polychaeta. Mostly marine worms. *Eunice, Neanthes.*

Class Oligochaeta. Mostly freshwater and terrestrial worms, many marine. *Lumbricus* (earthworms), *Tubifex.*

Class Hirudinea. Leeches. *Hirudo, Placobdella.*

PHYLUM NEMATODA Roundworms. *Ascaris, Caenorhabditis elegans, Necator* (hookworms), *Trichinella.*

PHYLUM ARTHROPODA

Subphylum Chelicerata. Chelicerates. Horseshoe crabs, spiders, scorpions, ticks, mites.

Subphylum Crustacea. Shrimps, crayfishes, lobsters, crabs, barnacles, copepods, isopods (sowbugs).

Subphylum Myriapoda. Centipedes, millipedes.

Subphylum Hexapoda. Insects and sprintails.

PHYLUM ECHINODERMATA Echinoderms.

Class Asteroidea. Sea stars. *Asterias.*
Class Ophiuroidea. Brittle stars.
Class Echinoidea. Sea urchins, heart urchins, sand dollars.
Class Holothuroidea. Sea cucumbers.
Class Crinoidea. Feather stars, sea lilies.
Class Concentricycloidea. Sea daisies.

PHYLUM CHORDATA Chordates.

Subphylum Urochordata. Tunicates, related forms.

Subphylum Cephalochordata. Lancelets.

CRANIATES

Class Myxini. Hagfishes.

VERTEBRATES (SUBGROUP OF CRANIATES)

Class Cephalaspidomorphi. Lampreys.

Class Chondrichthyes. Cartilaginous fishes (sharks, rays, skates, chimaeras).

Class "Osteichthyes." Bony fishes. Not monophyletic (sturgeons, paddlefish, herrings, carps, cods, trout, seahorses, tunas, lungfishes, and coelocanths).

TETRAPODS (SUBGROUP OF VERTEBRATES)

Class Amphibia. Amphibians. Require water to reproduce.
Order Caudata. Salamanders and newts.
Order Anura. Frogs, toads.
Order Apoda. Apodans (caecilians).

AMNIOTES (SUBGROUP OF TETRAPODS)

Class "Reptilia." Skin with scales, embryo protected and nutritionally supported by extraembryonic membranes.

Subclass Anapsida. Turtles, tortoises.

Subclass Lepidosaura. *Sphenodon,* lizards, snakes.

Subclass Archosaura. Crocodiles, alligators.

Class Aves. Birds. In some classifications birds are grouped in the archosaurs.

Order Struthioniformes. Ostriches.
Order Sphenisciformes. Penguins.
Order Procellariiformes. Albatrosses, petrels.
Order Ciconiiformes. Herons, bitterns, storks, flamingoes.
Order Anseriformes. Swans, geese, ducks.
Order Falconiformes. Eagles, hawks, vultures, falcons.
Order Galliformes. Ptarmigan, turkeys, domestic fowl.
Order Columbiformes. Pigeons, doves.
Order Strigiformes. Owls.
Order Apodiformes. Swifts, hummingbirds.
Order Passeriformes. Sparrows, jays, finches, crows, robins, starlings, wrens.
Order Piciformes. Woodpeckers, toucans.
Order Psittaciformes. Parrots, cockatoos, macaws.

Class Mammalia. Skin with hair; young nourished by milk-secreting mammary glands of adult.

Subclass Prototheria. Egg-laying mammals (monotremes; duckbilled platypus, spiny anteaters).

Subclass Metatheria. Pouched mammals or marsupials (opossums, kangaroos, wombats, Tasmanian devils).

Subclass Eutheria. Placental mammals.

Order Edentata. Anteaters, tree sloths, armadillos.
Order Insectivora. Tree shrews, moles, hedgehogs.
Order Chiroptera. Bats.
Order Scandentia. Insectivorous tree shrews.
Order Primates.

Suborder Strepsirhini (prosimians). Lemurs, lorises.

Suborder Haplorhini (tarsioids and anthropoids).

Infraorder Tarsiiformes. Tarsiers.

Infraorder Platyrrhini (New World monkeys).

Family Cebidae. Spider monkeys, howler monkeys, capuchin.

Infraorder Catarrhini (Old World monkeys and hominoids).

Superfamily Cercopithecoidea. Baboons, macaques, langurs.

Superfamily Hominoidea. Apes and humans.

Family Hylobatidae. Gibbon.

Family "Pongidae." Chimpanzees, gorillas, orangutans.

Family Hominidae. Existing and extinct human species (*Homo*) and humanlike species, including the australopiths.

Order Lagomorpha. Rabbits, hares, pikas.

Order Rodentia. Most gnawing animals (squirrels, rats, mice, guinea pigs, porcupines, beavers, etc.).

Order Carnivora. Carnivores (wolves, cats, bears, etc.).

Order Pinnipedia. Seals, walruses, sea lions.

Order Proboscidea. Elephants, mammoths (extinct).

Order Sirenia. Sea cows (manatees, dugongs).

Order Perissodactyla. Odd-toed ungulates (horses, tapirs, rhinos).

Order Tubulidentata. African aardvarks.

Order Artiodactyla. Even-toed ungulates (camels, deer, bison, sheep, goats, antelopes, giraffes, etc.).

Order Cetacea. Whales, porpoises.

Appendix II. Annotations to A Journal Article

This journal article reports on the movements of a female wolf during the summer of 2002 in northwestern Canada. It also reports on a scientific process of inquiry, observation and interpretation to learn where, how and why the wolf traveled as she did. In some ways, this article reflects the story of "how to do science" told in section 1.5 of this textbook. These notes are intended to help you read and understand how scientists work and how they report on their work.

1 ARCTIC

2 VOL. 57, NO. 2 (JUNE 2004) P. 196–203

3 Long Foraging Movement of a Denning Tundra Wolf

4 Paul F. Frame,[1,2] David S. Hik,[1] H. Dean Cluff,[3] and Paul C. Paquet[4]

5 *(Received 3 September 2003; accepted in revised form 16 January 2004)*

6 **ABSTRACT** Wolves (*Canis lupus*) on the Canadian barrens are intimately linked to migrating herds of barren-ground caribou (*Rangifer tarandus*). We deployed a Global Positioning System (GPS) radio collar on an adult female wolf to record her movements in response to changing caribou densities near her den during summer. This wolf and two other females were observed nursing a group of 11 pups. She traveled a minimum of 341 km during a 14-day excursion. The straight-line distance from the den to the farthest location was 103 km, and the overall minimum rate of travel was 3.1 km/h. The distance between the wolf and the radio-collared caribou decreased from 242 km one week before the excursion to 8 km four days into the excursion. We discuss several possible explanations for the long foraging bout.

7 *Key words:* wolf, GPS tracking, movements, *Canis lupus*, foraging, caribou, Northwest Territories

8 **RÉSUMÉ** Les loups (*Canis lupus*) dans la toundra canadienne sont étroitement liés aux hardes de caribous des toundras (*Rangifer tarandus*). On a équipé une louve adulte d'un collier émetteur muni d'un système de positionnement mondial (GPS) afin d'enregistrer ses déplacements en réponse au changement de densité du caribou près de sa tanière durant l'été. On a observé cette louve ainsi que deux autres en train d'allaiter un groupe de 11 louveteaux. Elle a parcouru un minimum de 341 km durant une sortie de 14 jours. La distance en ligne droite de la tanière à l'endroit le plus éloigné était de 103 km, et la vitesse minimum durant tout le voyage était de 3,1 km/h. La distance entre la louve et le caribou muni du collier émetteur a diminué de 242 km une semaine avant la sortie à 8 km quatre jours après la sortie. On commente diverses explications possibles pour ce long épisode de recherche de nourriture.

Mots clés: loup, repérage GPS, déplacements, *Canis lupus*, recherche de nourriture, caribou, Territoires du Nord-Ouest

Traduit pour la revue *Arctic* par Nésida Loyer.

9 Introduction

Wolves (*Canis lupus*) that den on the central barrens of mainland Canada follow the seasonal movements of their main prey, migratory barren-ground caribou (*Rangifer tarandus*) (Kuyt, 1962; Kelsall, 1968; Walton et al., 2001). However, most wolves do not den near caribou calving grounds, but select sites farther south, closer to the tree line (Heard and Williams, 1992). Most caribou migrate beyond primary wolf denning areas by mid-June and do not return until mid-to-late July (Heard et al., 1996; Gunn et al., 2001). Consequently, caribou density near dens is low for part of the summer.

10 During this period of spatial separation from the main caribou herds, wolves must either search near the homesite for scarce caribou or alternative prey (or both), travel to where prey are abundant, or use a combination of these strategies.

11 Walton et al. (2001) postulated that the travel of tundra wolves outside their normal summer ranges is a response to low caribou availability rather than a pre-dispersal exploration like that observed in terri torial wolves (Fritts and Mech, 1981; Messier, 1985). The authors postulated this because most such travel was directed toward caribou calving grounds. We report details of such a long-distance excursion by a breeding female tundra wolf wearing a GPS radio collar. We discuss the relationship of the excursion to movements of satellite-collared caribou (Gunn et al., 2001), supporting the hypothesis that tundra wolves make directional, rapid, long-distance movements in response to seasonal prey availability.

[1] Department of Biological Sciences, University of Alberta, Edmonton, Alberta T6G 2E9, Canada
[2] Corresponding author: pframe@ualberta.ca
[3] Department of Resources, Wildlife, and Economic Development, North Slave Region, Government of the Northwest Territories, P.O. Box 2668, 3803 Bretzlaff Dr., Yellowknife, Northwest Territories X1A 2P9, Canada; Dean_Cluff@gov.nt.ca
[4] Faculty of Environmental Design, University of Calgary, Calgary, Alberta T2N 1N4, Canada; current address: P.O. Box 150, Meacham, Saskatchewan S0K 2V0, Canada

196

1 Title of the journal, which reports on science taking place in Arctic regions.

2 Volume number, issue number and date of the journal, and page numbers of the article.

3 Title of the article: a concise but specific description of the subject of study—one episode of long-range travel by a wolf hunting for food on the Arctic tundra.

4 Authors of the article: scientists working at the institutions listed in the footnotes below. Note #2 indicates that P. F. Frame is the *corresponding author*—the person to contact with questions or comments. His email address is provided.

5 Date on which a draft of the article was received by the journal editor, followed by date one which a revised draft was accepted for publication. Between these dates, the article was reviewed and critiqued by other scientists, a process called peer review. The authors revised the article to make it clearer, according to those reviews.

6 ABSTRACT: A brief description of the study containing all basic elements of this report. First sentence summarizes the *background* material. Second sentence encapsulates the *methods* used. The rest of the paragraph sums up the *results*. Authors introduce the main *subject* of the study—a female wolf (#388) with pups in a den—and refer to later *discussion* of possible explanations for her behavior.

7 Key words are listed to help researchers using computer databases. Searching the databases using these key words will yield a list of studies related to this one.

8 RÉSUMÉ: The French translation of the abstract and key words. Many researchers in this field are French Canadian. Some journals provide such translations in French or in other languages.

9 INTRODUCTION: Gives the background for this wolf study. This paragraph tells of known or suspected wolf behavior that is important for this study. Note that (a) major species mentioned are always accompanied by scientific names, and (b) statements of fact or *postulations* (claims or assumptions about what is likely to be true) are followed by references to studies that established those facts or supported the postulations.

10 This paragraph focuses directly on the wolf behaviors that were studied here.

11 This paragraph starts with a statement of the *hypothesis* being tested, one that originated in other studies and is supported by this one. The hypothesis is restated more succinctly in the last sentence of this paragraph. This is the *inquiry* part of the scientific process—asking questions and suggesting possible answers.

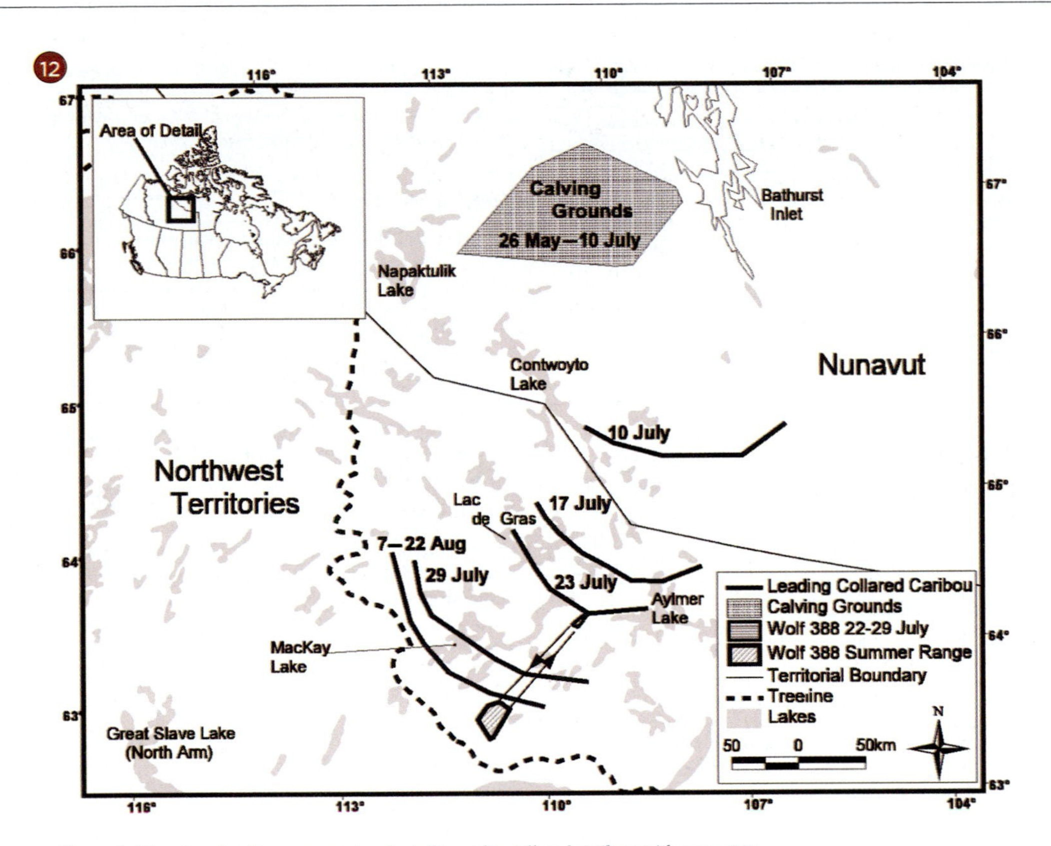

Figure 1. Map showing the movements of satellite radio-collared caribou with respect to female wolf 388's summer range and long foraging movement, in summer 2002.

12 This map shows the study area and depicts wolf and caribou locations and movements during one summer. Some of this information is explained below.

13 STUDY AREA: This section sets the stage for the study, locating it precisely with latitude and longitude coordinates and describing the area (illustrated by the map in Figure 1).

14 Here begins the story of how prey (caribou) and predators (wolves) interact on the tundra. Authors describe movements of these nomadic animals throughout the year.

15 We focus on the denning season (summer) and learn how wolves locate their dens and travel according to the movements of caribou herds.

13 Study Area

Our study took place in the northern boreal forest–low Arctic tundra transition zone (63° 30′ N, 110° 00′ W; Figure 1; Timoney et al., 1992). Permafrost in the area changes from discontinuous to continuous (Harris, 1986). Patches of spruce (*Picea mariana*, *P. glauca*) occur in the southern portion and give way to open tundra to the northeast. Eskers, kames, and other glacial deposits are scattered throughout the study area. Standing water and exposed bedrock are characteristic of the area.

14 *Details of the Caribou-Wolf System*

The Bathurst caribou herd uses this study area. Most caribou cows have begun migrating by late April, reaching calving grounds by June (Gunn et al., 2001; Figure 1). Calving peaks by 15 June (Gunn et al., 2001), and calves begin to travel with the herd by one week of age (Kelsall, 1968). The movement patterns of bulls are less known, but bulls frequent areas near calving grounds by mid-June (Heard et al., 1996; Gunn et al., 2001). In summer, Bathurst caribou cows generally travel south from their calving grounds and then, parallel to the tree line, to the northwest. The rut usually takes place at the tree line in October (Gunn et al., 2001). The winter range of the Bathurst herd varies among years, ranging through the taiga and along the tree line from south of Great Bear Lake to southeast of Great Slave Lake. Some caribou spend the winter on the tundra (Gunn et al., 2001; Thorpe et al., 2001).

15 In winter, wolves that prey on Bathurst caribou do not behave territorially. Instead, they follow the herd throughout its winter range (Walton et al., 2001; Musiani, 2003). However, during denning (May–

16 Other variables are considered—prey other than caribou and their relative abundance in 2002.

17 METHODS: There is no one scientific method. Procedures for each and every study must be explained carefully.

18 Authors explain when and how they tracked caribou and wolves, including tools used and the exact procedures followed.

19 This important subsection explains what data were calculated (average distance ...) and how, including the software used and where it came from. (The calculations are listed in Table 1.) Note that the behavior measured (traveling) is carefully defined.

20 RESULTS: The heart of the report and the *observation* part of the scientific process. This section is organized parallel to the Methods section.

21 This subsection is broken down by periods of observation. Pre-excursion period covers the time between 388's capture and the start of her long-distance travel. The investigators used visual observations as well as telemetry (measurements taken using the global positioning system (GPS)) to gather data. They looked at how 388 cared for her pups, interacted with other adults, and moved about the den area.

Table 1. Daily distances from wolf 388 and the den to the nearest radio-collared caribou during a long excursion in summer 2002.

Date (2002)	Mean distance from caribou to wolf (km)	Daily distance from closest caribou to den
12 July	242	241
13 July	210	209
14 July	200	199
15 July	186	180
16 July	163	162
17 July	151	148
18 July	144	137
19 July[1]	126	124
20 July	103	130
21 July	73	130
22 July	40	110
23 July[2]	9	104
29 July[3]	16	43
30 July	32	43
31 July	28	44
1 August	29	46
2 August[4]	54	52
3 August	53	53
4 August	74	74
5 August	75	75
6 August	74	75
7 August	72	75
8 August	76	75
9 August	79	79

[1] Excursion starts.
[2] Wolf closest to collared caribou.
[3] Previous five days' caribou locations not available.
[4] Excursion ends.

August, parturition late May to mid-June), wolf movements are limited by the need to return food to the den. To maximize access to migrating caribou, many wolves select den sites closer to the tree line than to caribou calving grounds (Heard and Williams, 1992). Because of caribou movement patterns, tundra denning wolves are separated from the main caribou herds by several hundred kilometers at some time during summer (Williams, 1990:19; Figure 1; Table 1).

16 Muskoxen do not occur in the study area (Fournier and Gunn, 1998), and there are few moose there (H.D. Cluff, pers. obs.). Therefore, alternative prey for wolves includes waterfowl, other ground-nesting birds, their eggs, rodents, and hares (Kuyt, 1972; Williams, 1990:16; H.D. Cluff and P.F. Frame, unpubl. data). During 56 hours of den observations, we saw no ground squirrels or hares, only birds. It appears that the abundance of alternative prey was relatively low in 2002.

17 Methods

Wolf Monitoring

18 We captured female wolf 388 near her den on 22 June 2002, using a helicopter net-gun (Walton et al., 2001). She was fitted with a releasable GPS radio collar (Merrill et al., 1998) programmed to acquire locations at 30-minute intervals. The collar was electronically released (e.g., Mech and Gese, 1992) on 20 August 2002. From 27 June to 3 July 2002, we observed 388's den with a 78 mm spotting scope at a distance of 390 m.

Caribou Monitoring

In spring of 2002, ten female caribou were captured by helicopter net-gun and fitted with satellite radio collars, bringing the total number of collared Bathurst cows to 19. Eight of these spent the summer of 2002 south of Queen Maud Gulf, well east of normal Bathurst caribou range. Therefore, we used 11 caribou for this analysis. The collars provided one location per day during our study, except for five days from 24 to 28 July. Locations of satellite collars were obtained from Service Argos, Inc. (Landover, Maryland).

Data Analysis

19 Location data were analyzed by ArcView GIS software (Environmental Systems Research Institute Inc., Redlands, California). We calculated the average distance from the nearest collared caribou to the wolf and the den for each day of the study.

Wolf foraging bouts were calculated from the time 388 exited a buffer zone (500 m radius around the den) until she re-entered it. We considered her to be traveling when two consecutive locations were spatially separated by more than 100 m. Minimum distance traveled was the sum of distances between each location and the next during the excursion.

We compared pre- and post-excursion data using Analysis of Variance (ANOVA; Zar, 1999). We first tested for homogeneity of variances with Levene's test (Brown and Forsythe, 1974). No transformations of these data were required.

Results 20

Wolf Monitoring

21 **Pre-Excursion Period:** Wolf 388 was lactating when captured on 22 June. We observed her and two other females nursing a group of 11 pups between 27 June and 3 July. During our observations, the pack consisted of at least four adults (3 females and 1 male) and 11 pups. On 30 June, three pups were moved to a location 310 m from the other eight and cared for by an uncollared female. The male was not seen at the den after the evening of 30 June.

Before the excursion, telemetry indicated 18 foraging bouts. The mean distance traveled during these bouts was 25.29 km (± 4.5 SE, range 3.1–82.5 km). Mean greatest distance from the den on foraging

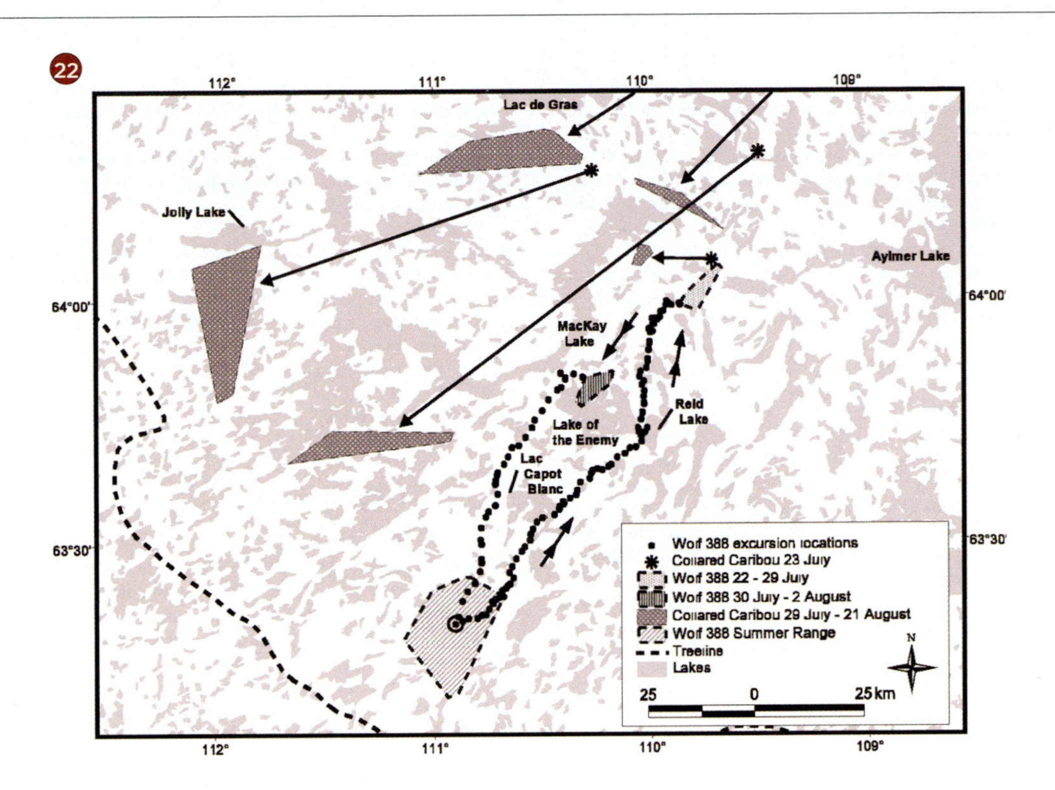

Figure 2. Details of a long foraging movement by female wolf 388 between 19 July and 2 August 2002. Also shown are locations and movements of three satellite radio-collared caribou from 23 July to 21 August 2002. On 23 July, the wolf was 8 km from a collared caribou. The farthest point from the den (103 km distant) was recorded on 27 July. Arrows indicate direction of travel.

22 The key in the lower right-hand corner of the map shows areas (shaded) within which the wolves and caribou moved, and the dotted trail of 388 during her excursion. From the results depicted on this map, the investigators tried to determine when and where 388 might have encountered caribou and how their locations affected her traveling behavior.

23 The wolf's excursion (her long trip away from the den area) is the focus of this study. These paragraphs present detailed measurements of daily movements during her two-week trip—how far she traveled, how far she was from collared caribou, her time spent traveling and resting, and her rate of speed. Authors use the phrase "minimum distance traveled" to acknowledge they couldn't track every step but were measuring samples of her movements. They knew that she went at least as far as they measured. This shows how scientists try to be exact when reporting results. Results of this study are depicted graphically in the map in Figure 2.

bouts was 7.1 km (± 0.9 SE, range 1.7–17.0 km). The average duration of foraging bouts for the period was 20.9 h (± 4.5 SE, range 1–71 h).

The average daily distance between the wolf and the nearest collared caribou decreased from 242 km on 12 July, one week before the excursion period, to 126 km on 19 July, the day the excursion began (Table 1).

23 **Excursion Period:** On 19 July at 2203, after spending 14 h at the den, 388 began moving to the northeast and did not return for 336 h (14 d; Figure 2). Whether she traveled alone or with other wolves is unknown. During the excursion, 476 (71%) of 672 possible locations were recorded. The wolf crossed the southeast end of Lac Capot Blanc on a small land bridge, where she paused for 4.5 h after traveling for 19.5 h (37.5 km). Following this rest, she traveled for 9 h (26.3 km) onto a peninsula in Reid Lake, where she spent 2 h before backtracking and stopping for 8 h just off the peninsula. Her next period of travel lasted 16.5 h (32.7 km), terminating in a pause of 9.5 h just 3.8 km from a concentration of locations at the far end of her excursion, where we presume she encountered caribou. The mean duration of these three movement periods was 15.7 h (± 2.5 SE), and that of the pauses, 7.3 h (± 1.5). The wolf required 72.5 h (3.0 d) to travel a minimum of 95 km from her den to this area near caribou (Figure 2). She remained there (35.5 km2) for 151.5 h (6.3 d) and then moved south to Lake of the Enemy, where she stayed (31.9 km^2) for 74 h (3.1 d) before returning to her den. Her greatest distance from the den, 103 km, was recorded 174.5 h (7.3 d) after the excursion

24 Post-excursion measurements of 388's movements were made to compare with those of the pre-excursion period. In order to compare, scientists often use *means*, or averages, of a series of measurements—mean distances, mean duration, etc.

25 In the comparison, authors used statistical calculations (F and df) to determine that the differences between pre- and post-excursion measurements were *statistically insignificant*, or close enough to be considered essentially the same or similar.

26 As with wolf 388, the investigators measured the movements of caribou during the study period. The areas within which the caribou moved are shown in Figure 2 by shaded polygons mentioned in the second paragraph of this subsection.

27 This subsection summarizes how distances separating predators and prey varied during the study period.

28 DISCUSSION: This section is the *interpretation* part of the scientific process.

29 This subsection reviews observations from other studies and suggests that this study fits with patterns of those observations.

30 Authors discuss a prevailing *theory* (CBFT) which might explain why a wolf would travel far to meet her own energy needs while taking food caught closer to the den back to her pups. The results of this study seem to fit that pattern.

began, at 0433 on 27 July. She was 8 km from a collared caribou on 23 July, four days after the excursion began (Table 1).

The return trip began at 0403 on 2 August, 318 h (13.2 d) after leaving the den. She followed a relatively direct path for 18 h back to the den, a distance of 75 km.

The minimum distance traveled during the excursionwas 339 km. The estimated overall minimum travel rate was 3.1 km/h, 2.6 km/h away from the den and 4.2 km/h on the return trip.

(24) **Post-Excursion Period:** We saw three pups when recovering the collar on 20 August, but others may have been hiding in vegetation.

Telemetry recorded 13 foraging bouts in the post-excursion period. The mean distance traveled during these bouts was 18.3 km (+ 2.7 SE, range 1.2–47.7 km), and mean greatest distance from the den was 7.1 km (+ 0.7 SE, range 1.1–11.0 km). The mean duration of these post-excursion foraging bouts was 10.9 h (+ 2.4 SE, range 1–33 h).

When 388 reached her den on 2 August, the distance to the nearest collared caribou was 54 km. On 9 August, one week after she returned, the distance was 79 km (Table 1).

Pre- and Post-Excursion Comparison

(25) We found no differences in the mean distance of foraging bouts before and after the excursion period ($F = 1.5$, $df = 1, 29$, $p = 0.24$). Likewise, the mean greatest distance from the den was similar pre- and post-excursion ($F = 0.004$, $df = 1, 29$, $p = 0.95$). However, the mean duration of 388's foraging bouts decreased by 10.0 h after her long excursion ($F = 3.1$, $df = 1, 29$, $p = 0.09$).

(26) *Caribou Monitoring*

Summer Movements: On 10 July, 5 of 11 collared caribou were dispersed over a distance of 10 km, 140 km south of their calving grounds (Figure 1). On the same day, three caribou were still on the calving grounds, two were between the calving grounds and the leaders, and one was missing. One week later (17 July), the leading radio-collared cows were 100 km farther south (Figure 1). Two were within 5 km of each other in front of the rest, who were more dispersed. All radio-collared cows had left the calving grounds by this time. On 23 July, the leading radio-collared caribou had moved 35 km farther south, and all of them were more widely dispersed. The two cows closest to the leader were 26 km and 33 km away, with 37 km between them. On the next location (29 July), the most southerly caribou were 60 km farther south. All of the caribou were now in the areas where they remained for the duration of the study (Figure 2).

A Minimum Convex Polygon (Mohr and Stumpf, 1966) around all caribou locations acquired during the study encompassed 85 119 km^2.

Relative to the Wolf Den: The distance from the nearest collared caribou to the den decreased from 241 km one week before the excursion to 124 km the day it began. The nearest a collared caribou came to the den was 43 km away, on 29 and 30 July. During the study, four collared caribou were located within 100 km of the den. Each of these four was closest to the wolf on at least one day during the period reported. (27)

(28) Discussion

Prey Abundance

Caribou are the single most important prey of tundra wolves (Clark, 1971; Kuyt, 1972; Stephenson and James, 1982; Williams, 1990). Caribou range over vast areas, and for part of the summer, they are scarce or absent in wolf home ranges (Heard et al., 1996). Both the long distance between radio-collared caribou and the den the week before the excursion and the increased time spent foraging by wolf 388 indicate that caribou availability near the den was low. Observations of the pups' being left alone for up to 18 h, presumably while adults were searching for food, provide additional support for low caribou availability locally. Mean foraging bout duration decreased by 10.0 h after the excursion, when collared caribou were closer to the den, suggesting an increase in caribou availability nearby. (29)

Foraging Excursion

One aspect of central place foraging theory (CPFT) deals with the optimality of returning different-sized food loads from varying distances to dependents at a central place (i.e., the den) (Orians and Pearson, 1979). Carlson (1985) tested CPFT and found that the predator usually consumed prey captured far from the central place, while feeding prey captured nearby to dependants. Wolf 388 spent 7.2 days in one area near caribou before moving to a location 23 km back towards the den, where she spent an additional 3.1 days, likely hunting caribou. She began her return trip from this closer location, traveling directly to the den. While away, she may have made one or more successful kills and spent time meeting her own energetic needs before returning to the den. Alternatively, it may have taken several attempts to make a kill, (30) which she then fed on before beginning her return trip. We do not know if she returned food to the pups, but such behavior would be supported by CPFT.

Other workers have reported wolves' making long round trips and referred to them as "extraterritorial" or "pre-dispersal" forays (Fritts and Mech, 1981; Messier, 1985; Ballard et al., 1997; Merrill and Mech, 2000). These movements are most often made by young wolves (1–3 years old), in areas where annual territories are maintained and prey are relatively sedentary (Fritts and Mech, 1981; Messier, 1985). The long excursion of 388 differs in that tundra wolves do not maintain annual territories (Walton et al., 2001), and the main prey migrate over vast areas (Gunn et al., 2001).

Another difference between 388's excursion and those reported earlier is that she is a mature, breeding female. No study of territorial wolves has reported reproductive adults making extraterritorial movements in summer (Fritts and Mech, 1981; Messier, 1985; Ballard et al., 1997; Merrill and Mech, 2001). However, Walton et al. (2001) also report that breeding female tundra wolves made excursions.

Direction of Movement

Possible explanations for the relatively direct route 388 took to the caribou include landscape influence and experience. Considering the timing of 388's trip and the locations of caribou, had the wolf moved northwest, she might have missed the caribou entirely, or the encounter might have been delayed.

A reasonable possibility is that the land directed 388's route. The barrens are crisscrossed with trails worn into the tundra over centuries by hundreds of thousands of caribou and other animals (Kelsall, 1968; Thorpe et al., 2001). At river crossings, lakes, or narrow peninsulas, trails converge and funnel towards and away from caribou calving grounds and summer range. Wolves use trails for travel (Paquet et al., 1996; Mech and Boitani, 2003; P. Frame, pers. observation). Thus, the landscape may direct an animal's movements and lead it to where cues, such as the odor of caribou on the wind or scent marks of other wolves, may lead it to caribou.

Another possibility is that 388 knew where to find caribou in summer. Sexually immature tundra wolves sometimes follow caribou to calving grounds (D. Heard, unpubl. data). Possibly, 388 had made such journeys in previous years and killed caribou. If this were the case, then in times of local prey scarcity she might travel to areas where she had hunted successfully before. Continued monitoring of tundra wolves may answer questions about how their food needs are met in times of low caribou abundance near dens.

Caribou often form large groups while moving south to the tree line (Kelsall, 1968). After a large aggregation of caribou moves through an area, its scent can linger for weeks (Thorpe et al., 2001:104). It is conceivable that 388 detected caribou scent on the wind, which was blowing from the northeast on 19–21 July (Environment Canada, 2003), at the same time her excursion began. Many factors, such as odor strength and wind direction and strength, make systematic study of scent detection in wolves difficult under field conditions (Harrington and Asa, 2003). However, humans are able to smell odors such as forest fires or oil refineries more than 100 km away. The olfactory capabilities of dogs, which are similar to wolves, are thought to be 100 to 1 million times that of humans (Harrington and Asa, 2003). Therefore, it is reasonable to think that under the right wind conditions, the scent of many caribou traveling together could be detected by wolves from great distances, thus triggering a long foraging bout.

Rate of Travel

Mech (1994) reported the rate of travel of Arctic wolves on barren ground was 8.7 km/h during regular travel and 10.0 km/h when returning to the den, a difference of 1.3 km/h. These rates are based on direct observation and exclude periods when wolves moved slowly or not at all. Our calculated travel rates are assumed to include periods of slow movement or no movement. However, the pattern we report is similar to that reported by Mech (1994), in that homeward travel was faster than regular travel by 1.6 km/h. The faster rate on return may be explained by the need to return food to the den. Pup survival can increase with the number of adults in a pack available to deliver food to pups (Harrington et al., 1983). Therefore, an increased rate of travel on homeward trips could improve a wolf's reproductive fitness by getting food to pups more quickly.

Fate of 388's Pups

Wolf 388 was caring for pups during den observations. The pups were estimated to be six weeks old, and were seen ranging as far as 800 m from the den. They received some regurgitated food from two of the females, but were unattended for long periods. The excursion started 16 days after our observations, and it is improbable that the pups could have traveled the distance that 388 moved. If the pups died, this would have removed parental responsibility, allowing the long movement.

Our observations and the locations of radio-collared caribou indicate that prey became scarce in

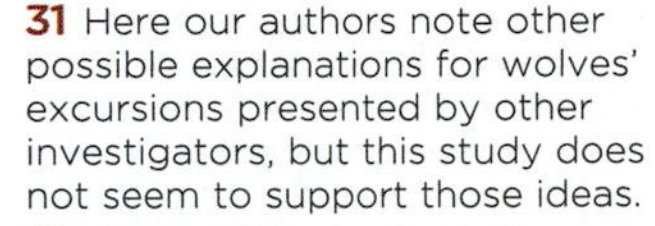

31 Here our authors note other possible explanations for wolves' excursions presented by other investigators, but this study does not seem to support those ideas.

32 Authors discuss possible reasons for why 388 traveled directly to where caribou were located. They take what they learned from earlier studies and apply it to this case, suggesting that the lay of the land played a role. Note that their description paints a clear picture of the landscape.

33 Authors suggest that 388 may have learned in traveling during previous summers where the caribou were. The last two sentences suggest ideas for future studies.

34 Or maybe 388 followed the scent of the caribou. Authors acknowledge difficulties of proving this, but they suggest another area where future studies might be done.

35 Authors suggest that results of this study support previous studies about how fast wolves travel to and from the den. In the last sentence, they speculate on how these observed patterns would fit into the theory of evolution.

36 Authors also speculate on the fate of 388's pups while she was traveling. This leads to . . .

37 Discussion of cooperative rearing of pups and, in turn, to speculation on how this study and what is known about cooperative rearing might fit into the animal's strategies for survival of the species. Again, the authors approach the broader theory of evolution and how it might explain some of their results.

38 And again, they suggest that this study points to several areas where further study will shed some light.

39 In conclusion, the authors suggest that their study supports the hypothesis being tested here. And they touch on the implications of increased human activity on the tundra predicted by their results.

40 ACKNOWLEDGEMENTS: Authors note the support of institutions, companies and individuals. They thank their reviewers ad list permits under which their research was carried on.

41 REFERENCES: List of all studies cited in the report. This may seem tedious, but is a vitally important part of scientific reporting. It is a record of the sources of information on which this study is based. It provides readers with a wealth of resources for further reading on this topic. Much of it will form the foundation of future scientific studies like this one.

the area of the den as summer progressed. Wolf 388 may have abandoned her pups to seek food for herself. However, she returned to the den after the excursion, where she was seen near pups. In fact, she foraged in a similar pattern before and after the excursion, suggesting that she again was providing for pups after her return to the den.

(37) A more likely possibility is that one or both of the other lactating females cared for the pups during 388's absence. The three females at this den were not seen with the pups at the same time. However, two weeks earlier, at a different den, we observed three females cooperatively caring for a group of six pups. At that den, the three lactating females were observed providing food for each other and trading places while nursing pups. Such a situation at the den of 388 could have created conditions that allowed one or more of the lactating females to range far from the den for a period, returning to her parental duties afterwards. However, the pups would have been weaned by eight weeks of age (Packard et al., 1992), so nonlactating adults could also have cared for them, as often happens in wolf packs (Packard et al., 1992; Mech et al., 1999).

Cooperative rearing of multiple litters by a pack could create opportunities for long-distance foraging movements by some reproductive wolves during summer periods of local food scarcity. We have recorded multiple lactating females at one or more tundra wolf dens per year since 1997. This reproductive strategy may be an adaptation to temporally and (38) spatially unpredictable food resources. All of these possibilities require further study, but emphasize both the adaptability of wolves living on the barrens and their dependence on caribou.

Long-range wolf movement in response to caribou (39) availability has been suggested by other researchers (Kuyt, 1972; Walton et al., 2001) and traditional ecological knowledge (Thorpe et al., 2001). Our report demonstrates the rapid and extreme response of wolves to caribou distribution and movements in summer. Increased human activity on the tundra (mining, road building, pipelines, ecotourism) may influence caribou movement patterns and change the interactions between wolves and caribou in the region. Continued monitoring of both species will help us to assess whether the association is being affected adversely by anthropogenic change.

(40) Acknowledgements

This research was supported by the Department of Resources, Wildlife, and Economic Development, Government of the Northwest Territories; the Department of Biological Sciences at the University of Alberta; the Natural Sciences and Engineering Research Council of Canada; the Department of Indian and Northern Affairs Canada; the Canadian Circumpolar Institute; and DeBeers Canada, Ltd. Lorna Ruechel assisted with den observations. A. Gunn provided caribou location data. We thank Dave Mech for the use of GPS collars. M. Nelson, A. Gunn, and three anonymous reviewers made helpful comments on earlier drafts of the manuscript. This work was done under Wildlife Research Permit – WL002948 issued by the Government of the Northwest Territories, Department of Resources, Wildlife, and Economic Development.

(41) References

BALLARD, W.B., AYRES, L.A., KRAUSMAN, P.R., REED, D.J., and FANCY, S.G. 1997. Ecology of wolves in relation to a migratory caribou herd in northwest Alaska. Wildlife Monographs 135. 47 p.

BROWN, M.B., and FORSYTHE, A.B. 1974. Robust tests for the equality of variances. Journal of the American Statistical Association 69:364–367.

CARLSON, A. 1985. Central place foraging in the red-backed shrike (*Lanius collurio* L.): Allocation of prey between forager and sedentary consumer. Animal Behaviour 33:664–666.

CLARK, K.R.F. 1971. Food habits and behavior of the tundra wolf on central Baffin Island. Ph.D. Thesis, University of Toronto, Ontario, Canada.

ENVIRONMENT CANADA. 2003. National climate data information archive. Available online: http://www.climate.weatheroffice.ec.gc.ca/Welcome_e.html

FOURNIER, B., and GUNN, A. 1998. Musk ox numbers and distribution in the NWT, 1997. File Report No. 121. Yellowknife: Department of Resources, Wildlife, and Economic Development, Government of the Northwest Territories. 55 p.

FRITTS, S.H., and MECH, L.D. 1981. Dynamics, movements, and feeding ecology of a newly protected wolf population in northwestern Minnesota. Wildlife Monographs 80. 79 p.

GUNN, A., DRAGON, J., and BOULANGER, J. 2001. Seasonal movements of satellite-collared caribou from the Bathurst herd. Final Report to the West Kitikmeot Slave Study Society, Yellowknife, NWT. 80 p. Available online: http://www.wkss.nt.ca/HTML/08_ProjectsReports/PDF/Seasonal MovementsFinal.pdf

HARRINGTON, F.H., and ASA, C.S. 2003. Wolf communication. In: Mech, L.D., and Boitani, L., eds. Wolves: Behavior, ecology, and conservation. Chicago: University of Chicago Press. 66–103.

HARRINGTON, F.H., MECH, L.D., and FRITTS, S.H. 1983. Pack size and wolf pup survival: Their relationship under varying ecological conditions. Behavioral Ecology and Sociobiology 13:19–26.

HARRIS, S.A. 1986. Permafrost distribution, zonation and stability along the eastern ranges of the cordillera of North America. Arctic 39(1):29–38.

HEARD, D.C., and WILLIAMS, T.M. 1992. Distribution of wolf dens on migratory caribou ranges in the Northwest

Territories, Canada. Canadian Journal of Zoology 70:1504–1510.
HEARD, D.C., WILLIAMS, T.M., and MELTON, D.A. 1996. The relationship between food intake and predation risk in migratory caribou and implication to caribou and wolf population dynamics. Rangifer Special Issue No. 2:37–44.
KELSALL, J.P. 1968. The migratory barren-ground caribou of Canada. Canadian Wildlife Service Monograph Series 3. Ottawa: Queen's Printer. 340 p.
KUYT, E. 1962. Movements of young wolves in the Northwest Territories of Canada. Journal of Mammalogy 43:270–271.
———. 1972. Food habits and ecology of wolves on barren-ground caribou range in the Northwest Territories. Canadian Wildlife Service Report Series 21. Ottawa: Information Canada. 36 p.
MECH, L.D. 1994. Regular and homeward travel speeds of Arctic wolves. Journal of Mammalogy 75:741–742.
MECH, L.D., and BOITANI, L. 2003. Wolf social ecology. In: Mech, L.D., and Boitani, L., eds. Wolves: Behavior, ecology, and conservation. Chicago: University of Chicago Press. 1–34.
MECH, L.D., and GESE, E.M. 1992. Field testing the Wildlink capture collar on wolves. Wildlife Society Bulletin 20:249–256.
MECH, L.D., WOLFE, P., and PACKARD, J.M. 1999. Regurgitative food transfer among wild wolves. Canadian Journal of Zoology 77:1192–1195.
MERRILL, S.B., and MECH, L.D. 2000. Details of extensive movements by Minnesota wolves (*Canis lupus*). American Midland Naturalist 144:428–433.
MERRILL, S.B., ADAMS, L.G., NELSON, M.E., and MECH, L.D. 1998. Testing releasable GPS radiocollars on wolves and white-tailed deer. Wildlife Society Bulletin 26:830–835.
MESSIER, F. 1985. Solitary living and extraterritorial movements of wolves in relation to social status and prey abundance. Canadian Journal of Zoology 63:239–245.
MOHR, C.O., and STUMPF, W.A. 1966. Comparison of methods for calculating areas of animal activity. Journal of Wildlife Management 30:293–304.
MUSIANI, M. 2003. Conservation biology and management of wolves and wolf-human conflicts in western North America. Ph.D. Thesis, University of Calgary, Calgary, Alberta, Canada.
ORIANS, G.H., and PEARSON, N.E. 1979. On the theory of central place foraging. In: Mitchell, R.D., and Stairs, G.F., eds. Analysis of ecological systems. Columbus: Ohio State University Press. 154–177.
PACKARD, J.M., MECH, L.D., and REAM, R.R. 1992. Weaning in an arctic wolf pack: Behavioral mechanisms. Canadian Journal of Zoology 70:1269–1275.
PAQUET, P.C., WIERZCHOWSKI, J., and CALLAGHAN, C. 1996. Summary report on the effects of human activity on gray wolves in the Bow River Valley, Banff National Park, Alberta. In: Green, J., Pacas, C., Bayley, S., and Cornwell, L., eds. A cumulative effects assessment and futures outlook for the Banff Bow Valley. Prepared for the Banff Bow Valley Study. Ottawa: Department of Canadian Heritage.
STEPHENSON, R.O., and JAMES, D. 1982. Wolf movements and food habits in northwest Alaska. In: Harrington, F.H., and Paquet, P.C., eds. Wolves of the world. New Jersey: Noyes Publications. 223–237.
THORPE, N., EYEGETOK, S., HAKONGAK, N., and QITIRMIUT ELDERS. 2001. The Tuktu and Nogak Project: A caribou chronicle. Final Report to the West Kitikmeot/Slave Study Society, Ikaluktuuttiak, NWT. 160 p.
TIMONEY, K.P., LA ROI, G.H., ZOLTAI, S.C., and ROBINSON, A.L. 1992. The high subarctic forest-tundra of northwestern Canada: Position, width, and vegetation gradients in relation to climate. Arctic 45(1):1–9.
WALTON, L.R., CLUFF, H.D., PAQUET, P.C., and RAMSAY, M.A. 2001. Movement patterns of barren-ground wolves in the central Canadian Arctic. Journal of Mammalogy 82:867–876.
WILLIAMS, T.M. 1990. Summer diet and behavior of wolves denning on barren-ground caribou range in the Northwest Territories, Canada. M.Sc. Thesis, University of Alberta, Edmonton, Alberta, Canada.
ZAR, J.H. 1999. Biostatistical analysis. 4th ed. New Jersey: Prentice Hall. 663 p.

Appendix III. Answers to Self-Quizzes and Genetics Problems

Italicized numbers refer to relevant section numbers

CHAPTER 44

1. d *44.1*
2. b *44.1*
3. c *44.3*
4. a *44.4*
5. a *44.4*
6. b *44.5*
7. d *44.6*
8. c *44.6*
9. e *44.7*
10. c *44.7*
11. True *44.7*
12. c *44.2*
    d *44.7*
    b *44.1, 44.2*
    a *44.2*
    e *44.4*

CHAPTER 45

1. a *45.1*
2. f *45.1*
3. c *45.2*
4. b *45.3*
5. a *45.3*
6. d *45.4*
7. d *45.5*
8. b *45.7*
9. d *45.9*
10. a *45.10*
11. c *45.4*
    d *45.3*
    a *45.3*
    e *45.4*
    b *45.4*

CHAPTER 46

1. d *46.1*
2. d *46.1*
3. d *46.1*
4. e *46.3*
5. b *46.3*
6. b *46.4*
7. a *46.4*
    b *46.2*
    c *46.1*
    d *46.6*
    e *46.3*
8. b *46.8*
9. False *46.6*
10. c *46.9*
    d *46.11*
    a *46.8*
    e *46.8*
    b *46.9*
    g *46.9*
    f *46.3*

CHAPTER 47

1. a *47.1*
2. d *47.1*
3. d *47.1*
4. d *47.1*
5. d *47.3*
6. d *47.4*
7. c *47.6*
8. b *47.7*
9. d *47.7*
10. d *47.8*
11. d *47.10*
12. c *47.10*
13. d *47.9, 47.10*
14. a *47.9*
15. d *47.1*
    e *47.1*
    c *47.1*
    b *47.1*
    a *47.1*
    f *47.4*

CHAPTER 48

1. d *48.1*
2. c *48.2*
3. d *48.3*
4. a *48.3*
5. a *48.1*
6. e *48.4*
7. d *48.4*
8. c *48.7*
9. a *48.11*
10. d *48.12, 48.16*
11. d *48.16*
12. d *48.11*
    h *48.7*
    e *48.6*
    c *48.7*
    b *48.14*
    g *48.10*
    a *48.8*
    f *48.16*

CHAPTER 49

1. True *49.1*
2. b *49.1*
3. b *49.1*
4. a *49.2*
5. b *49.2*
6. c *49.3*
7. c *49.4*
8. a *49.4*
9. True *49.5*
10. d *49.5*
11. a *49.7*
12. f *49.8*
    d *49.4*
    a *49.1*
    g *49.4*
    b *49.6*
    c *49.8*
    e *49.8*
    h *49.4*

Appendix IX. Units of Measure

Length

1 kilometer (km) = 0.62 miles (mi)
1 meter (m) = 39.37 inches (in)
1 centimeter (cm) = 0.39 inches

To convert	*multiply by*	*to obtain*
inches	2.25	centimeters
feet	30.48	centimeters
centimeters	0.39	inches
millimeters	0.039	inches

Area

1 square kilometer = 0.386 square miles
1 square meter = 1.196 square yards
1 square centimeter = 0.155 square inches

Volume

1 cubic meter = 35.31 cubic feet
1 liter = 1.06 quarts
1 milliliter = 0.034 fluid ounces = 1/5 teaspoon

To convert	*multiply by*	*to obtain*
quarts	0.95	liters
fluid ounces	28.41	milliliters
liters	1.06	quarts
milliliters	0.03	fluid ounces

Weight

1 metric ton (mt) = 2,205 pounds (lb) = 1.1 tons (t)
1 kilogram (kg) = 2.205 pounds (lb)
1 gram (g) = 0.035 ounces (oz)

To convert	*multiply by*	*to obtain*
pounds	0.454	kilograms
pounds	454	grams
ounces	28.35	grams
kilograms	2.205	pounds
grams	0.035	ounces

Temperature

Celcius (°C) to Fahrenheit (°F) :

$$°F = 1.8\,(°C) + 32$$

Fahrenheit (°F) to Celsius:

$$°C = \frac{(°F - 32)}{1.8}$$

	°C	°F
Water boils	100	212
Human body temperature	37	98.6
Water freezes	0	32

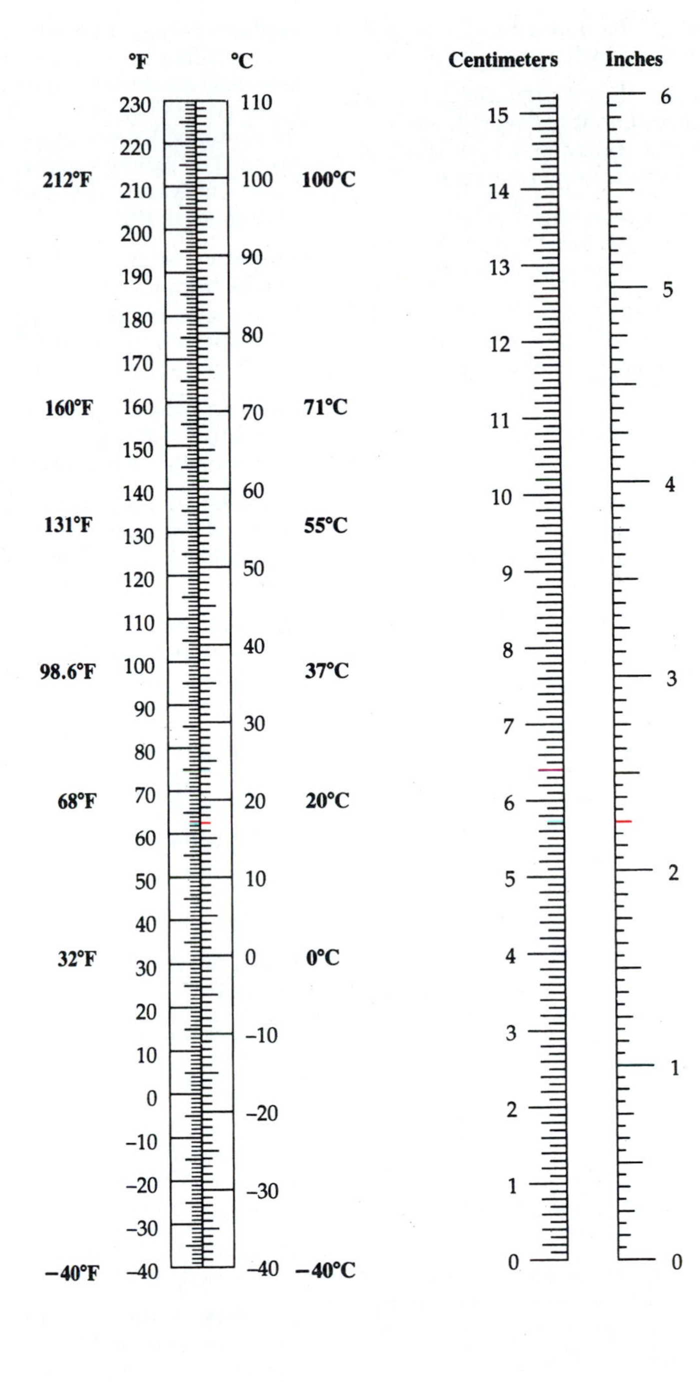

Glossary of Biological Terms

acid rain Rain or snow made acidic by airborne oxides of sulfur or nitrogen. **865**

age structure Of a population, the number of individuals in each age category. **798**

alpine tundra Biome prevailing at high altitudes throughout the world; even in the summer snow persists in shaded areas, but there is no permafrost. **877**

altruistic behavior Social behavior that can lower an individual's reproductive success but improve that of others. **792**

ammonification Process by which bacteria and fungi break down nitrogen-containing organic material and release ammonia and ammonium ions. **855**

aquifer Permeable rock layers that hold water. **848**

arctic tundra Biome prevailing at high latitudes, where short, cool summers alternate with long, cold winters; it forms between the polar ice cap and the belts of boreal forests in the Northern Hemisphere. **877**

area effect Biogeographical pattern; larger islands support more species than smaller ones at equivalent distances from sources of colonizing species. **835**

benthic province Oceanic zone comprised of the ocean bottom—its rocks, and sediments. **884**

biodiversity Variety of forms of life, in terms of genetic diversity, species diversity, and ecosystem diversity. **896**

biogeochemical cycle Slow movement of an element from environmental reservoirs, through food webs, then back. **847**

biogeographic realm One of many vast expanses of land defined by the presence of certain types of plants and animals. **868**

biogeography Study of patterns in the geographic distribution of species and communities. **834**

biological magnification A pesticide or other chemical that becomes increasingly concentrated in the tissues of organisms at higher trophic levels. **846**

biomass pyramid Chart in which tiers of a pyramid depict biomass (dry weight) in each of an ecosystem's trophic levels. **844**

biome A subdivision of a biogeographic realm; usually described in terms of the dominant plants; e.g., tropical broadleaf forest, grassland, tundra. **868**

biotic potential The maximum rate of increase per individual for a population growing under ideal conditions. **801**

camouflage Body coloration, patterning, form, or behavior that helps predators or prey blend with the surroundings and possibly escape detection. **824**

capture–recapture method Individuals of a mobile species are captured (or selected) at random, marked, then released so they can mix with unmarked individuals. One or more samples are taken. The ratio of marked to unmarked individuals is used to estimate total population size. **799**

carbon cycle Atmospheric cycle. Carbon moves from its environmental reservoirs (sediments, rocks, the ocean), through the atmosphere (mostly as CO_2), food webs, and back to the reservoirs. **850**

carrying capacity Maximum number of individuals of a species that a particular environment can sustain. **802**

character displacement Modifications of a trait of one species in a way that lowers intensity of competition with another species; occurs over generations. **821**

classical conditioning An animal's involuntary response to a stimulus becomes associated with another stimulus that is presented at the same time. **785**

climate Prevailing weather conditions of a region; e.g., temperature, cloud cover, wind speed, rainfall, and humidity. **862**

coevolution The joint evolution of two closely interacting species; each species is a selective agent that shifts the range of variation in the other. **818**

cohort A group of individuals of the same age. **804**

commensalism An interspecific interaction in which one species benefits and the other is neither helped nor harmed. **818**

communication signal A social cue that is encoded in stimuli, such as the body's surface coloration or patterning, odors, sounds, and postures. **786**

community All populations of all species in a habitat. **818**

competition, interspecific Interaction in which the individuals of different species compete for a limited resource; suppresses population size of both species. **818**

competitive exclusion When two species require the same limited resource to survive or reproduce, the better competitor will drive the less competitive one to extinction in the shared habitat. **820**

coniferous forest Biome dominated by conifers, which tolerate cold and drought, and poor soils. **876**

consumer Heterotroph that gets energy and carbon by feeding on tissues, wastes, or remains of other organisms. **840**

coral reef A formation consisting mainly of calcium carbonate secreted by reef-building corals. **882**

decomposer One of the prokaryotic or fungal heterotrophs that obtains carbon and energy by breaking down wastes or remains of organisms. **840**

demographics Statistics that describe a population; e.g., size, age structure. **798**

demographic transition model Model that correlates changes in population growth with stages of economic development. **812**

denitrification Conversion of nitrate or nitrite to gaseous nitrogen (N_2) or nitrogen oxide (NO_2) by soil bacteria. **855**

density-dependent factor A factor that slows population growth, and either appears or worsens with crowding; e.g., disease, competition for food. **803**

density-independent factor A factor that slows population growth; its likelihood of occurring and magnitude of effect does not vary with population density. **803**

desalinization The removal of salt from saltwater. **849**

desert Biome of areas where evaporation greatly exceeds rainfall, where soil is thin and vegetation sparse. **871**

desertification Conversion of grassland or irrigated or rain-fed cropland to desertlike conditions. **900**

detrital food chain Food chain in which energy flows from producers to detritivores and decomposers (rather than herbivores). **843**

detritivore Any animal that feeds on small particles of organic matter; e.g., a crab or earthworm. **840**

distance effect A biogeographic pattern. Islands distant from a mainland have fewer species than those closer to the potential source of colonists. **835**

doubling time The time it takes for a population to double in size. **801**

dry shrubland Biome of areas that get less than 25 to 60 centimeters of rain; short, multibranched woody shrubs dominate. **873**

dry woodland Biome of areas that get about 40 to 100 centimeters of rain; may have many tall trees but no dense canopy. **873**

ecoregion Broad land or ocean province influenced by abiotic and biotic factors. **897**

ecosystem Community interacting with its environment through a one-way flow of energy and cycling of materials. **840**

El Niño Eastward displacement of warm surface waters of the western equatorial Pacific. Recurs, alters global climates. **885**

emigration Permanent move of one or more individuals out of a population. **800**

endangered species A species endemic (native) to a habitat, found nowhere else, and highly vulnerable to extinction. **893**

endemic species A species that is confined to the limited area in which it evolved. It is more likely to go extinct than a species with a more widespread distribution. **894**

energy pyramid Diagram that depicts the energy stored in the tissues of organisms at each trophic level in an ecosystem. Lowest tier of the pyramid, representing primary producers, is always the largest. **845**

equilibrium model of island biogeography A model describing the number of species expected to inhabit a habitat island of a particular size and distance from mainland as source of colonists. **835**

estuary Partly enclosed coastal region where seawater mixes with fresh water and runoff from land, as in rivers. **880**

eutrophication Nutrient enrichment of a body of water; promotes population growth of phytoplankton. **857**

evergreen broadleaf forest Biome between latitudes 10° north and south of the equator with rainfall averages of 130 to 200 centimeters each year; tropical rain forest. **874**

exotic species Species that has become established in a new community after dispersing from its home range. **831**

exponential growth Population increases in size by the same proportion of its total in each successive interval. **800**

fall overturn During the fall, waters of a temperate zone mix. Upper, oxygenated water cools, gets dense, and sinks; nutrient-rich water from the bottom moves up. **879**

fixed action pattern A series of instinctive movements, triggered by a simple stimulus, that continues no matter what else is going on in the environment. **784**

food chain Linear sequence of steps by which energy stored in autotroph tissues enters higher trophic levels. **841**

food web Cross-connecting food chains consisting of producers, consumers, and decomposers, detritivores, or both. **842**

global warming Long-term increase in temperature of Earth's lower atmosphere; rising levels of greenhouse gases contribute to the increase. **853**

grassland Biome dominated by grasses and other nonwoody plants; common in interiors of continents with warm summers, cool winters, recurring fires, and 25–100 centimeters of rain. **872**

grazing food chain Food chain in which energy flows from producers to herbivores. **842**

greenhouse effect Some atmospheric gases absorb infrared wavelengths (heat) from the sun-warmed surface, and then radiate some back toward Earth, warming it. **852**

groundwater Water contained in soil and in aquifers. **848**

habitat Place where an organism or species lives; described by physical and chemical features and array of species. **818**

habituation An animal learns through experience not to respond to a stimulus that has neither positive or negative effects. **785**

hydrologic cycle *See* water cycle.

hydrothermal vent Underwater fissure where superheated, mineral-rich water is forced out under pressure. **884**

immigration One or more individuals move and take up residence in another population of its species. **800**

imprinting A form of learning triggered by exposure to sign stimuli; time-dependent, usually occurs during a sensitive period while an animal is young. **784**

indicator species Any species which, by its abundance or scarcity, is a measure of the health of its habitat. **896**

instinctive behavior Behavior performed without having first been learned. **784**

intermediate disturbance hypothesis An explanation of community structure; holds that species richness is greatest in habitats where disturbances are moderate in intensity, frequency, or both. **829**

interspecific competition *See* competition, interspecific.

keystone species A species that has a disproportionately a large effect on community structure, relative to its own abundance. **830**

***K*-selection** Selection for traits that make offspring better competitors; occurs in a population near carrying capacity. **805**

La Niña Climatic event in which Pacific waters become cooler than average. **886**

lake A body of standing fresh water. **878**

learned behavior Enduring modification of a behavior as an outcome of experience in the environment. **784**

life history pattern Of a species, pattern of when and how many offspring are produced during a typical lifetime. **804**

limiting factor Any essential resource that limits population growth when scarce. **802**

logistic growth Population growth pattern. A population grows exponentially when small, then levels off in size once carrying capacity has been reached. **802**

migration Of many animals, a recurring pattern of movement between two or more regions in response to seasonal change or other environmental rhythms. **800**

mimicry Evolutionary convergence of body form; a close resemblance between species. A defenseless species may look like a well-defended one, or several well-defended species may all look alike. **824**

monsoon Wind and weather pattern that changes seasonally and is caused by differential heating of a continental interior and the nearby ocean. **867**

mutualism An interspecific interaction that benefits both participants. **818**

niche A species' unique ecological role; it is described in terms of the conditions, resources, and interactions necessary for survival and reproduction. **818**

nitrification One stage of the nitrogen cycle. Soil bacteria break down ammonia or ammonium to nitrite; then other bacteria break down nitrite to nitrate, which plants can absorb. **855**

nitrogen cycle An atmospheric cycle. Nitrogen moves from its largest reservoir (the atmosphere), then through the ocean, ocean sediments, soils, and food webs, then back to the atmosphere. **854**

nitrogen fixation Conversion of gaseous nitrogen to ammonia. **854**

observational learning One animal acquires a new behavior by observing and imitating behavior of another. **785**

operant conditioning A type of learning in which an animal's voluntary behavior is modified by the consequences of that behavior. **785**

ozone layer An atmospheric layer of high ozone concentration. **864**

parasite An organism that obtains some or all the nutrients it needs from a living host, which it usually does not kill outright. **826**

parasitism Interaction in which a parasitic species benefits as it exploits and harms (but usually does not kill) the host. **818**

parasitoid A type of insect that, in a larval stage, grows inside a host (usually another insect), feeds on its tissues, and kills it. **826**

pelagic province The open waters of the ocean. **884**

per capita growth rate The rate obtained by subtracting a population's per capita death rate from per capita birth rate. **800**

permafrost A perpetually frozen layer that underlies arctic tundra. **877**

pheromone Signaling molecule secreted by one individual that affects another of the same species; has roles in social behavior. **786**

phosphorus cycle A sedimentary cycle. Phosphorus (mainly phosphate) moves from land, through food webs, to ocean sediments, then back to land. **856**

pioneer species An opportunistic colonizer of barren or disturbed habitats. Adapted for rapid growth and dispersal. **828**

pollutant A natural or synthetic substance released into soil, air, or water in greater than normal amounts; it disrupts natural processes because organisms evolved in its absence, or are adapted to lower levels. **864**

population density Number of individuals of a population in a specified volume or area of a habitat. **798**

population distribution The pattern in which individuals of a population are dispersed through their habitat. **798**

population size The number of individuals that actually or potentially contribute to the gene pool of a population. **798**

predation Ecological interaction in which a predator kills and eats prey. **818**

predator A heterotroph that eats other living organisms (its prey). **822**

prey An organism that a predator kills and eats. **822**

primary producer An autotroph at the first trophic level of an ecosystem. **840**

primary productivity The rate at which an ecosystem's primary producers secure and store energy in tissues. **844**

primary succession A community arises and species arrive and replace one another over time in an environment that was without soil, such as a newly formed island. **828**

quadrat One of a number of sampling areas of the same size and shape used to estimate population size. **799**

rain shadow Reduction in rainfall on the side of a high mountain range facing away from prevailing wind; results in arid or semiarid conditions. **867**

reproductive base The number of actually and potentially reproducing individuals of a population. **798**

resource partitioning Use of different parts of a resource; permits two or more similar species to coexist in a habitat. **821**

riparian zone The narrow corridor of vegetation along a stream or river. **903**

***r*-selection** Selection that favors traits that maximize number of offspring; operates when the population is well below its carrying capacity. **805**

runoff The water that flows into streams when the ground is saturated. **848**

salinization Salt buildup in soil. **849**

seamount Extinct seafloor volcano. **884**

secondary succession A community arises and changes over time in a habitat where another community existed previously. **828**

selfish herd Animal group that forms when individuals each attempt to hide in the midst of others. **790**

semi-evergreen forest Biome in the humid tropics that includes a mix of broadleaf trees that retain leaves year round, and deciduous broadleaf trees that shed once a year in the cold or dry season. **874**

sexual selection Mode of natural selection in which some individuals outreproduce others of a population because they are better at securing mates. **788**

smog Atmospheric condition in which winds cannot disperse airborne pollutants trapped by a thermal inversion. **864**

social parasite An animal that takes advantage of its host's behavior, thus harming it; e.g., cuckoo. **826**

soil profile Distinct soil layers that form over time in a biome. **870**

spring overturn In temperate zone lakes, a downward movement of oxygenated surface water and an upward movement of nutrient-rich water in spring. **879**

stimulus A specific form of energy that activates a sensory receptor able to detect it; e.g., pressure. **782**

survivorship curve Plot of age-specific survival of a cohort, from the time of birth until the last individual dies. **804**

symbiosis Ecological interaction in which members of two species live together or otherwise interact closely; e.g., mutualism, parasitism, commensalism. **818**

temperate deciduous forest Biome with 50 to 150 centimeters of precipitation throughout the year, warm summers and cold winters. Dominant vegetation is trees that shed leaves in fall. **874**

theory of inclusive fitness The idea that genes associated with altruism are adaptive if they cause behavior that promotes the reproductive success of an altruist's closest relatives. **792**

thermocline Thermal stratification in a large body of water; a cool midlayer stops vertical mixing between warm surface water above it and cold water below it. **879**

threatened species Species likely to be endangered in the near future. **893**

total fertility rate (TFR) For humans, the average number of children born to a female during her lifetime. **810**

trophic level All organisms that are the same number of transfer steps away from the energy input into an ecosystem. **840**

tropical deciduous forest Equatorial biome where less than 2.5 centimeters of rain falls in the dry season. Most trees shed leaves at the start of the dry season. **874**

upwelling Upward movement of cool water from the depths, as when winds blow surface water away from a coast. **885**

vector An insect or some other animal that carries a pathogen between hosts; e.g., a mosquito that transmits malaria. **826**

warning coloration In many toxic species and their mimics, bright colors, patterns, and other signals that predators learn to recognize and avoid. **824**

water cycle The process by which water moves among ocean, the atmosphere, and freshwater reservoirs. **848**

watershed A region of any specified size in which all precipitation drains into one stream or river. **848**

zero population growth No net increase or decrease in population size during a specified interval. **800**

Art Credits and Acknowledgments

This page constitutes an extension of the book copyright page. We have made every effort to trace the ownership of all copyrighted material and secure permission from copyright holders. In the event of any question arising as to the use of any material, we will be pleased to make the necessary corrections in future printings. Thanks are due to the following authors, publishers, and agents for permission to use the material indicated.

TABLE OF CONTENTS **Page iv** from left, © Monty Sloan, www.wolfphotography.com; © Charles Lewallen. **Page v** from left, © Doug Peebles/ Corbis; Graphic created by FoodWeb3D program written by Rich Williams courtesy of the Webs on the Web project (www.foodwebs.org); NOAA. **Page vi** from left, © James Randklev/ Corbis; Image provided by GeoEye and NASA SeaWIFS Project; © Dr. John Hilty; © Billy Grimes.

Page 779 Unit VII, © Minden Pictures.

CHAPTER 44 **44.1** © Scott Camazine. **Page 781** From top, Reprinted from *Trends in Neuroscience*, Vol. 21, Issue 2, 1998, L.J.Young, W. Zuoxin, T.R. Insel, "Neuroendocrine bases of monogamy," Pages 71–75, ©1998, with permission from Elsevier Science; © Kevin Schafer/ Corbis; © B. Borrell Casals/ Frank Lane Picture Agency/ Corbis; © Australian Picture Library/ Corbis. **44.2** (a) Eugene Kozloff; (b) Stevan Arnold. **44.4** Left, © Robert M. Timm & Barbara L. Clauson, University of Kansas; (a,b) Reprinted from *Trends in Neuroscience*, Vol. 21, Issue 2, 1998, L.J.Young, W. Zuoxin, T.R. Insel, "Neuroendocrine bases of monogamy," Pages 71–75, ©1998, with permission from Elsevier Science. **44.5** (a) Eric Hosking; (b) © Stephen Dalton/ Photo Researchers, Inc. **44.6** © Nina Leen/ TimePix/ Getty Images; inset, © Robert Semeniuk/ Corbis. **44.7** © Professor Jelle Atema, Boston University. **44.8** © Bernhard Voelkl. **44.9** Left, Robert Maier/ Animals Animals. **44.10** (a) © Tom and Pat Leeson, leesonphoto.com; (b) © Kevin Schafer/ Corbis; (c) © Monty Sloan, www.wolfphotography.com. **44.11** © Stephen Dalton/Photo Researchers, Inc. **44.12** (a) John Alcock, Arizona State University; (b,c) © Pam Gardner/ Frank Lane Picture Agency/ Corbis; (d) © D. Robert Franz/ Corbis. **44.13** Michael Francis/ The Wildlife Collection. **44.14** (a) © B. Borrell Casals/ Frank Lane Picture Agency/ Corbis; (b) © Steve Kaufman/ Corbis; (c) © John Conrad/ Corbis. **44.15** (a) © Tom and Pat Leeson, leesonphoto.com; (b) John Alcock, Arizona State University; (c) © Paul Nicklen/ National Geographic/ Getty Images. **44.16** © Jeff Vanuga/ Corbis. **44.17** © Steve Bloom/ stevebloom.com. **44.18** © Eric and David Hosking/ Corbis. **44.19** (a) © Australian Picture Library/ Corbis; (b) © Alexander Wild; (c) © Professor Louis De Vos. **44.20** (a) Kenneth Lorenzen; (b) © Peter Johnson/ Corbis; (c) © Nicola Kountoupes/ Cornell University. **Page 794** © Lynda Richardso/ Corbis.

CHAPTER 45 **45.1** © David Nunuk/ Photo Researchers, Inc. **Page 797** From top, © Amos Nachoum/ Corbis; © David Scharf, 1999. All rights reserved; © G. K. Peck; © Joe McDonald/ Corbis; © Don Mason/ Corbis. **45.2** (a) © Amos Nachoum/ Corbis; (b) © Corbis; (c) A. E. Zuckerman/ Tom Stack & Associates. **45.3** E. R. Degginger; inset, Jeff Fott Productions/ Bruce Coleman, Ltd. **45.4** (a) © Cynthia Bateman, Bateman Photography; (b) © Tom Davis. **45.5** © Jeff Lepore/ Photo Researchers, Inc. **45.6** © David Scharf, 1999. All rights reserved. **45.7** (a) © G. K. Peck; (b) © Rick Leche, www.flickr.com/photos/ rick_leche. **45.9** © Charles Lewallen. **45.10** (a) © Joe McDonald/ Corbis; (b) © Wayne Bennett/ Corbis; (c) Estuary to Abyss 2004. NOAA Office of Ocean Exploration. **45.11** (a,b) Upper David Reznick/ University of California—Riverside; computer enhanced by Lisa Starr; (a,b) lower Hippocampus Bildarchiv; (c) Helen Rodd. **Page 807** © Bruce Bornstein, www .captbluefin.com. **45.13** Left, © Mark Harmel/ Photo Researchers, Inc.; right, AP Images. **45.14** NASA. **45.17** Left © Adrian Arbib/ Corbis; right, © Don Mason/ Corbis. **45.19** © Polka Dot Images/ SuperStock. **Page 814** U.S. Department of the Interior, U.S. Geological Survey. **45.20** © Reinhard Dirscherl/ www.bciusa.com.

CHAPTER 46 **46.1** (a) Photography by B. M. Drees, Texas A&M University. Http://fireant.tamu; (b) Scott Bauer/ USDA; (c) USDA. **Page 817** From top, © Martin Harvey, Gallo Images/ Corbis; © Nigel Jones; © Pat O'Hara/ Corbis; © Pierre Vauthey/ Corbis Sygma. **46.2** Left, Donna Hutchins; (a) © B. G. Thomson/ Photo Researchers, Inc.; (b) © Len Robinson, Frank Lane Picture Agency/ Corbis; (c) Martin Harvey, Gallo Images/ Corbis. **46.3** Upper, Harlo H. Hadow; lower, Bob and Miriam Francis/ Tom Stack & Associates. **46.4** © Thomas W. Doeppner. **46.5** © Pekka Komi. **46.6a** (a) Left, © Michael Abbey/ Photo Researchers, Inc.; right, © Eric V. Grave/ Photo Researchers, Inc. **46.7** Stephen G. Tilley. **46.8** Upper, © Joe McDonald/ Corbis; lower, left, © Hal Horwitz/ Corbis; right, © Tony Wharton, Frank Lane Picture Agency/ Corbis. **46.9** © W. Perry Conway/ Corbis. **46.10** Left, © Ed Cesar/ Photo Researchers, Inc.; right, © Robert McCaw, www.robertmccaw. com. **46.11** (a) © JH Pete Carmichael; (b) Edward S. Ross; (c) W. M. Laetsch. **46.12** (a–c) Edward S. Ross; (d) © Nigel Jones. **46.13** (a,b) Thomas Eisner, Cornell University; (c) David Burdick/ NOAA; (d) © Bob Jensen Photography. **46.14** (a) MSU News Service, photo by Montata Water Center; (b) © Karl Andree. **46.15** Left, © The Samuel Roberts Noble Foundation, Inc.; right, Courtesy of Colin Purrington, Swarthmore College. **46.16** © C. James Webb/ Phototake USA. **46.17** © Peter J. Bryant/ Biological Photo Service. **46.18** (a) © Richard Price/Getty Images; (b) © E.R. Degginger/ Photo Researchers, Inc. **46.19** (a) © Doug Peebles/ Corbis; (b) © Pat O'Hara/ Corbis; (c,d) © Tom Bean/ Corbis; (e) © Duncan Murrell/ Taxi/ Getty Images. **46.20** (a) R. Barrick/ USGS; (b) USGS; (c) P. Frenzen, USDA Forest Service. **46.21** (a,c) Jane Burton/ Bruce Coleman, Ltd.; (b) Heather Angel. **46.22** (a) © Pr. Alexande Meinesz, University of Nice-Sophia Antipolis; (b) © Angelina Lax/ Photo Researchers, Inc.; right, © The University of Alabama Center for Public TV. **46.23** © John Carnemolla/ Australian Picture Library. **46.24** After W. Dansgaard et al., *Nature*, 364:218–220, July 15, 1993; D. Raymond et al., *Science*, 259:926–933, February 1993; W. Post, *American Scientist*, 78:310–326, July–August 1990. **46.25** © Pierre Vauthey/ Corbis Sygma. **Page 836** John Kabashima.

CHAPTER 47 **47.1** Left, © C. C. Lockwood/ Cactus Clyde Productions; right, Diane Borden-Bilot, U.S. Fish and Wildlife Service. **Page 839** From top, © D. A. Rintoul; Graphic created by FoodWeb3D program written by Rich Williams courtesy of the Webs on the Web project (www.foodwebs.org); NASA's Earth Observatory; USDA Forest Service, Northeastern Research Station. **47.3** Right top, © Lloyd Spitalnik/ lloydspitalnikphotos.com; right bottom, © Van Vives; all others, © D. A. Rintoul. **47.4** from left, top row, © Bryan & Cherry Alexander/ Photo Researchers, Inc.; © Dave Mech; © Tom & Pat Leeson, Ardea London Ltd.; 2nd row, © Tom Wakefield/ Bruce Coleman, Inc.; © Paul J. Fusco/ Photo Researchers, Inc.; © E. R. Degginger/ Photo Researchers, Inc.; 3rd row, © Tom J. Ulrich/ Visuals Unlimited; © Dave Mech; © Tom McHugh/ Photo Researchers, Inc.; mosquito, Photo by James Gathany, Centers for Disease Contro; flea, © Edward S. Ross; 4th row, © Jim Steinborn; © Jim Riley; © Matt Skalitzky; earthworm, © Peter Firus, flagstaffotos. com.au. **47.5** Left, Courtesy of Dr. Chris Floyd; right, Graphic created by FoodWeb3D program written by Rich Williams courtesy of the Webs on the Web project (www.foodwebs.org). **47.6** NASA. **47.9** Upper, U.S. Department of the Interior, National Park Service; lower, Gary Head. **47.10** Jack Scherting, USC&GS, NOAA. **47.12** USDA Forest Service, Northeastern Research Station. **47.14** Lisa Starr after Paul Hertz; photograph © Photodisc/ Getty Images. **47.15, 47.16** Lisa Starr and Gary Head, based on NASA photographs from JSC Digital Image Collection. **47.17** © Yann Arthus-Bertrand/ Corbis. **47.19** © Jeff Vanuga/ Corbis. **47.20** © Frederica Georgia/ Photo Researchers, Inc. **47.21** Art, Gary Head and Lisa Starr; photograph, © Photodisc/ Getty Images. **47.22** Fisheries & Oceans Canada, Experimental Lakes Area. **Page 858** U.S. Department of Transportation, Federal Highway Administration. **47.24** Courtesy of NASA's Terra satellite, supplied by Ted Scambos, National Snow and Ice Data Center, University of Colorado, Boulder.

CHAPTER 48 **48.1** © Hank Fotos Photography; inset, © Wolfgang Kaehler/ Corbis. **Page 861** From second from top, NASA; © James Randklev/ Corbis; Jack Carey; Raghu Rai/ Magnum Photos. **48.5** NASA. **48.7** (a) Adapted from *Living in the Environment* by G. Tyler Miller, Jr., p. 428. © 2002 by Brooks/Cole, a division of Thomson Learning; (b) © Ted Spiegel/ Corbis. **48.8** After M. H. Dickerson, "ARAC: Modeling an Ill Wind," in *Energy and Technology Review*, August 1987. Used by permission of University of California Lawrence Livermore National Laboratory and U.S. Dept. of Energy. **48.9** NASA. **48.10** Left, © Sally A. Morgan, Ecoscene/ Corbis; right, © Bob Rowan, Progressive Image/ Corbis. **48.12** NASA. **48.13** Above, © Yves Bilat, Ardea London Ltd.; below, © Eagy Landau/ Photo Researchers, Inc. **48.15** (a,b) Courtesy of Jim Deacon, The University of Edinburgh; (c) Jeff Servos, US Fish & Wildlife Service; (d) Bill Radke, US Fish & Wildlife Service. **48.16** (a) Ray Wagner/ Save the Tall Grass Prairie, Inc.; (b) © Tom Bean Photography; (c) Jonathan Scott/ Planet Earth Pictures. **48.17** (a) © John C. Cunningham/ Visuals Unlimited; (b) Jack Wilburn/ Animals Animals; (c) AP Images; (d) © Richard W. Halsey, California Chaparral Institute. **48.18** left, © James Randklev/ Corbis; all others, © Randy Wells/ Corbis. **48.19** Upper, © Franz Lanting/ Minden Pictures; inset, Edward S. Ross; lower, Hans Renner; inset, © Adolf Schmidecker/ FPG/ Getty Images. **48.20** (a) © Thomas Wiewandt/ ChromoSohm Media Inc./ Photo Researchers, Inc.; (b) © Raymond Gehman/ Corbis; inset, Donna Dewhurst, US Fish & Wildlife Service. **48.21** © Darrell Gulin/ Corbis; inset, Thomas D. Mangelsen/ Images of Nature. **48.22** © Pat O'Hara/ Corbis. **48.24** Jack Carey. **48.26** Ocean Arks International. **48.27** (a) © Annie Griffiths Belt/ Corbis; (b) © Douglas Peebles/ Corbis. **48.28** (a) © Nancy Sefton; (b) Courtesy of J. L. Sumich, *Biology of Marine Life*, 7th ed., W. C. Brown, 1999; (c) © Paul A. Souders/ Corbis. **Page 882** © Sea Studios/ Peter Arnold, Inc. **48.29** (a) C. B. & D. W. Frith/ Bruce Coleman, Ltd.; (b) Douglas Faulkner/ Sally Faulkner Collection; (c) © Douglas Faulkner/ Photo Researchers, Inc.; (e) lionfish, Douglas Faulkner/ Sally Faulkner Collection; all others, © John Easley, www.johneasley. com. **48.30** © Dr. Ray Berkelmans, Australian Institute of Marine Science. **48.31** (b) NOAA. **48.32** (a) Courtesy of © Montery Bay Aquarium Research Institute; (b) © Peter Herring/ imagequestmarine.com; (c) Image courtesy of NOAA and MBARI; (d–f) © Peter Batson/ imagequestmarine.com. **48.35** NASA Goddard Space Flight Center Scientific Visualization Studio. **48.36** (a) CHAART, at NASA Ames Research Center; (b) © Eye of Science/ Photo Researchers, Inc.; (c) Courtesy of Dr. Anwar Huq and Dr. Rita Colwell, University of Maryland; (d) Raghu Rai/ Magnum Photos. **Page 888** © Jose Luis Pelaez, Inc./ Corbis.

CHAPTER 49 **49.1** U.S. Navy photo by Chief Yeoman Alphanso Braggs. **Page 891 Top,** © George M. Sutton/ Cornell Lab of Ornithology; From third from top, NOAA; Bureau of Land Management. **49.3** Mansell Collection/ Time, Inc./ Getty Images. **49.4** © George M. Sutton/ Cornell Lab of Ornithology. **49.5** Jeffrey Sylvester/ FPG/ Getty Images. **49.6** (a) © Dr. John Hilty; (b) Joe Fries, U.S. Fish & Wildlife Service. **49.8** © Billy Grimes. **49.10** NOAA. **49.11** Image provided by GeoEye and NASA SeaWIFS Project. **49.12** Claire Fackler/ NOAA. **49.13** © PhotoDisc/ Getty Images. **49.14** Bureau of Land Management. **Page 904** © Dan Guravich/ Corbis.

Index

Page numbers followed by an *f* or *t* indicate figures and tables. ■ indicate applications. Bold terms indicate major topics.

A

A horizon, 870, 870*f*
■ Abortion
  induced, 811
Absorption, of nutrients and water
  mycorrhizae and, 819
Acid rain, 855, 865, 865*f*, 894
Acoustical communication
  signals, 786
Acquired Immune Deficiency Syndrome. *See* AIDS
Adaptation, evolutionary
  in birds
    bill, 821
Adaptive behavior
  defined, 786, 786*f*
Adaptive radiation
  after mass extinctions, 892
ADH. *See* Antidiuretic hormone
Aerobic respiration
  in carbon cycle, 850*f*–851*f*, 851
Africa
■ AIDS in, 810, 812
■ desertification and dust storms, 900, 900*f*
■ economic development in, 812
  savanna, 872*f*, 873
■ total fertility rate, 810
Africanized honeybees, 780, 780*f*, 794, 795, 795*f*
Agav7e, 871
Age structure, defined, 798
Age structure diagrams, 810, 811*f*
■ Aging
  of population in developed nations, 813
■ Agriculture. *See also* Fertilizers; Pesticides
  and carrying capacity, 808–809, 813
  chemicals, environmental effects, 846, 846*f*
  crop rotation, 855
  and desertification, 900, 900*f*
  on Great Plains, 872
  and nitrogen cycle, 855
  phosphorus cycle and, 856
  and plant diversity, 894
  water use, 848–849
AIDS (Acquired Immune Deficiency Syndrome)
  in Africa, 810, 812
Air circulation patterns, 862–863, 863*f*
■ Air pollution
  and acid rain, 855, 865, 865*f*, 894
  global circulation of, 865, 865*f*
  health effects, 865
  mercury, 846
  nitrogen oxides, 855, 855*f*, 864–865
  and ozone hole, 864, 864*f*
  particulates, 865
Alarm calls, 790, 790*f*
Alarm pheromone, 780, 780*f*, 786
Albatross, 901*f*
Alder trees, 828*f*
Algae. *See also* Brown algae; Green algae; Red algae
  algal blooms, 855, 857, 886, 886*f*
  in coastal zones, 830–831, 830*f*, 880, 881, 881*f*
■ as exotic species, 832, 832*f*
  as pioneer species, 829
Alpine tundra, 877, 877*f*
Altruistic behavior 792–793, 792*f*
Ammonia (NH_3)
  in nitrogen cycle, 854, 854*f*
Ammonification, 854*f*, 855
Ammonium (NH_4^+), in nitrogen cycle, 854–855, 854*f*
Amphibian(s)
  species
    diversity, over time, 892*f*
  threatened and endangered species, 896*t*
Anemone fish, 819, 819*f*
Angiosperm(s)
  as pioneer species, 828, 828*f*, 835
  species
    diversity of, 892*f*
  threatened or endangered species, 894, 896*t*
Angler fish, 884, 885*f*
Animal(s)
  communication, 786–787, 786*f*, 787*f*, 790, 790*f*
Ant(s), 792, 792*f*
Antarctica
■ ice shelf shrinkage, 859, 859f
■ ozone hole over, 864, 864f
Antidiuretic hormone (ADH; vasopressin)
  action, 783
Aphids, 827, 827*f*
Aquatic ecosystems
■ eutrophication, 857, 857*f*
  food chains in, 843, 845
■ nitrogen enrichment, 855
■ phosphorus levels in, 856–857
■ water pollution and, 849
Aquifers
  defined, 848
■ threats to, 849, 849*f*, 894
Arctic
  food web, 842–843, 842*f*
■ ice, shrinkage of, 859, 859*f*, 890, 904
■ pollution in, 890, 905, 905*f*
Arctic fox, 842*f*
Arctic tundra, 877, 877*f*
Arctic willow, 842*f*
Arctic wolf, 842*f*
Area effect, 835, 835*f*
Arnold, Steven, 782
Asia
■ total fertility rate, 810
Asian long-horned beetle, 831, 836
■ Asthma, 865
Atacama Desert, 871
Atlantic codfish (*Gadus morhua*), 807, 807*f*
Atmosphere. *See also* Air pollution
  as biosphere component, 860
  in carbon cycle, 850–851, 850*f*–851*f*
  carbon dioxide in, 851, 859, 859*f*
  greenhouse effect, 852, 852*f*
  greenhouse gases and, 851, 852–853, 852*f*, 853*f*, 894
  ozone layer, 864
    hole in, 864, 864*f*
  rain forest loss and, 875
  as water reservoir, 848*t*
Atolls, 882*f*
Australia
■ age structure diagram, 811*f*
■ exotic species in, 833, 833*f*
  Great Barrier Reef, 882, 882*f*
Australian realm, 868*f*–869*f*
■ Autism, 793
Autumn equinox, 862*f*
■ Avian malaria, 895

B

B horizon, 870*f*
Baboon(s) , 786, 786*f*
■ Baby boomers, 810, 811*f*, 813
Bacterium (bacteria). *See also* Cyanobacteria; Prokaryotes
  biotic potential of, 801
  as decomposer, 840, 843, 855
  denitrifying, 855
  nitrogen-fixing, 819, 819*f*, 828, 828*f*, 847, 854, 854*f*, 855, 857
  photosynthetic. *See* Cyanobacteria
  as pioneer species, 835
  population growth
    exponential, 801, 801*f*
    limiting factors, 802
Bamboo, 805
Banana slugs, 782, 782*f*
Banded coral shrimp, 883*f*
Bangladesh, 810, 887
Bark beetle, 831*t*
Barrier reefs, 882*f*
Bat(s)
  lesser long-nosed bats, 871*f*
  as pollinator, 871*f*
Bathyal zone, 884*f*
Bay of Bengal, 887, 887*f*
Bears, 789*f*, 877, 877*f*
Beaver, 831
Bee(s). *See* Honeybees
Beetle (coleopteran)
  defenses, 825, 825*f*
  exotic species, 831, 831*t*, 836
Behavior
  adaptive
    defined, 786, 786*f*
    vs. morality, 793
  conditioned responses, 785, 785*f*
  genetic basis of, 782–783, 782*f*, 783*f*
  habituation, 785
  human
    evolutionary basis, 793
    factors affecting, 793
  instinctive, 784, 784*f*
  learned, 784–785, 784*f*, 785*f*
  observational learning, 785, 785*f*
Benthic province, 884, 884*f*
Bergerub, Toha, 780
■ Bicarbonate (HCO_3^-)
  in carbon cycle, 850–851, 850*f*–851*f*
Big bluestem grass, 841, 841*f*
Bill, bird
  adaptations in, 821
■ Biodiesel, 875
Biodiversity
■ agriculture and, 894
  of algae, factors of, 830–831, 830*f*
  assessment of, 896–897
■ as biological wealth, 902
  in coral reefs, 882*f*–883*f*
  and extinctions, 892 (*See also* Extinction(s))
  within habitat, as self-reinforcing, 834
  on islands, 834–835, 834*f*, 835*f*
■ and land development, 898
  levels of, 896
  maintaining, 902–903, 902*f*, 903*f*
  over time, 892*f*
■ and resource consumption, 898
Biogeochemical cycles
  carbon cycle, 850–851, 850*f*–851*f*
  defined, 847, 847*f*
  nitrogen cycle, 854–855, 854*f*
  phosphorus cycle, 856–857, 856*f*
  water cycle, 847–848, 847*f*, 848*f*, 856
Biogeographic realms, 868, 868*f*–869*f*
Biogeography, 834
■ Biological controls, 816, 827, 827*f*, 837, 837*f*
■ Biological magnification, 846, 846*f*
■ Biological wealth, 902
Biomass, 850
Biomass pyramids, 844, 844*f*
Biomes, 868–869, 868*f*–869*f*. *See also specific biomes*
  preservation efforts, 897
  soils, 870, 870*f*
Biosphere
  defined, 860
■ human impact on, 890 (*See also* Pollution)
    assessment of, 896–897
    current threats, 894–895
    deforestation, 848, 849*f*, 855
    desertification, 900, 900*f*
    effects of development and consumption, 898–899, 898*f*
    endangered ecoregions, 897, 897*f*, 897*t*
    endangered or threatened species, 893, 894–895, 894*f*, 896, 896*t*, 897, 898
    extinctions, 893, 893*f*
    severity of, 890
    trash, 901, 901*f*
    ubiquity of, 890, 890*f*
Biotic potential, 801
Bird(s)
  behavior
    cannibalism, 791
    instinctive, 784, 784*f*
    learned, 784–785, 784*f*
    mating, 786*f*, 788–789, 788
    migration , 880
    nesting , 791, 791*f*, 798, 798*f*
    parental care, 789, 789*f*
  bill, 821
  defenses, 824, 824*f*
  diversity, 892*f*
  of Galápagos Islands, 821
  of Hawaiian Islands, 895
  human impacts on , 846, 846*t*, 890, 896*t*
  as seed dispersers, 831
  survivorship curve, 805
Birth rate, human, 800–801, 801*f*, 804, 804*t*, 810
Bison (American buffalo), 789*f*, 827, 827*f*, 872*f*
Bitterns, 824, 824*f*
Black walnut tree, 820
Blue jays, 820
Bluegill sunfish, 790
Bony Fish (Osteichthyes)
  species diversity, over time, 892*f*
Boreal (northern) coniferous forests (taigas), 868*f*–869*f*, 876, 876*f*
Bradshaw, Ken, 860
Brazil, demographic data, 810*f*
Bristly foxtail, 821*f*
Brittle star, 884, 885*f*
Broadleaf forest, 868*f*–869*f*, 874, 879*f*
Brown algae
  threatened species, 896*t*
Brown bear, 877, 877*f*
Brown tree snake, 895
■ Bubonic plague, 803, 809
Budworm, 846
Bunkley-Williams, Lucy, 882
Butterfly
  endangered species, 898

C

C horizon, 870*f*
C4 plants, 871
Cactus
  as CAM plant, 871
Caffeine
  as plant defense, 824
Calcium
  in soil, 855
Calcium carbonate, in coral reefs, 882
Calicivirus, 833
CAM plants, 871
Cambrian period
  major events, 892*f*
Camouflage, 824, 824*f*, 825, 825*f*
Canada, age structure diagram, 811*f*
Canadian lynx, 822–823, 823*f*
■ Cancer
  treatment, 875
Capture–recapture methods, 799
Carbon cycle, 850–851, 850*f*–851*f*
  atmospheric carbon dioxide and, 851
■ human disruption of, 851
  photosynthesis and, 850*f*–851*f*, 851
Carbon dioxide
  atmospheric, 851, 859, 859*f*
  in carbon cycle, 850–851, 850*f*–851*f*
■ as greenhouse gas, 852, 853*f*, 875
Carboniferous period
  major events, 892*f*
Caribou/reindeer (*Rangifer tarandus*), 803*f*, 822, 822*f*
Carnivores
  as consumers, 840, 844, 844*f*, 845*f*
Carrot (*Daucus carota*), 786, 786*f*
Carrying capacity, 802–803, 802*f*, 803*f*
■ and economic development, 812–813, 813*f*
■ and human population growth, 808–809, 808*f*, 809*f*, 812–813, 813*f*
Carson, Rachel, 846
Castor-oil plant (*Ricinus communis*), 824
Caterpillar, defenses, 824, 824*f*
Caulerpa taxifolia (invasive alga), 832, 832*f*
Cellulose
  digestion of, 845
Cenozoic era, major events, 892*f*
Cestode. *See* Tapeworm
CFCs. *See* Chlorofluorocarbons
Chaparral, 873, 873*f*, 897*f*
Character(s)
  character displacement, 821
Cheetah, 825
Chemoautotrophs, 884
■ Chemotherapy, drugs, sources of, 875
■ Chernobyl nuclear plant disaster, 865*f*
■ Chestnut blight fungus, 831*t*
Chile, Atacama Desert, 871
Chimpanzee(s)
  cultural traits, transmission of, 790–791, 791*f*
China
■ age structure diagram, 811*f*
■ carbon emissions, 858
■ desertification in, 900
■ and economic development, 813
■ population, 810
■ population control programs, 810–811
■ total fertility rate, 811
■ Chlorofluorocarbons (CFCs), 852, 853*f*, 864
■ Cholera, 809, 886–887, 887*f*
Cichlids, 806*f*, 807, 807*f*
Clark, Larry, 786
Classical conditioning, 785
Clay, as soil, characteristics of, 870
Climate. *See also* Global warming
  and biomes, 868–869
  defined, 862
  desert, 871
  dry shrubland, 873
  evergreen broadleaf forest, 874
  grasslands, 872
  greenhouse gases and, 852–853, 852*f*, 853*f*
  ocean currents and, 866, 866*f*
  regional, origins of, 862
  savannas, 873
  tropical deciduous forests, 874
Clouds, origin of, 848
Clumped population distribution, 798, 798*f*
Cnidarians (Cnidaria)
  in ocean food chain, 884

Coal
and air pollution, 846
Coast redwoods, 876
Coastal breezes, 867, 867f
Coastal zones, 880–881, 881f
Coastlines, rocky and sandy, 881, 881f
Codfish, Atlantic (*Gadus morhua*), 807, 807f, 894
Coevolution
defined, 818
Coho salmon, 897
Cohort, defined, 804
Coleopteran. *See* Beetle
Colwell, Rita, 887, 887f
Commensalism, 818, 818t
Communication. *See also* Language
animal
alarm calls, 790, 790f
signal types, 786–787, 786f, 787f
Community
defined, 818
ecological succession in, 828–829, 828f, 829f
species interactions in. *See* Species, interaction
structure, biogeography patterns in, 834–835, 834f, 835f
structure of, factors in, 818
Competition, 818, 818t, 820–821, 820f, 821f. *See also* Natural selection
competitive exclusion, 820
effects of, 820–821
exploitative, 820
interference, 820, 820f
Conditioned responses, 785, 785f
Coniferous forest, 876, 876f
boreal (taigas), 868f–869f, 876, 876f
endangered ecoregions, 897f
montane, 876, 876f
northern, 868f–869f
soil, 870, 870f
temperate, 868f–869f
tropical, 868f–869f
Conservation biology, 896–897
Consumers
in food web, 842f
role in ecosystem, 840, 840f
Consumption of resources, biosphere impact, 898–899, 899f
Contraception
and population control, 810
Copepods, 878, 887
Coral, 882
reefs, 882–883, 882f–883f
threatened species, 896t
Coral banks, 882f
Coral bleaching, 882, 882f
Cormorants, 791, 791f
Corn (*Zea mays*)
wild maize (*Zea diploperennis*), 902
Cortisol
regulation of, 793
Costa Rican owl butterfly (*Caligo*), 799f
Cowbird (*Molothrus ater*), 827, 827f
Crater Lake, 878f
Creosote bush (*Larrea*), 799f, 871, 871f
Cretaceous period
major events, 892f
Crocodilians
parental care in, 789
Crustaceans
threatened species, 896t
Cuckoo bird, 784, 784f, 826
Cultural traits, transmission of, in animals, 790–791, 791f
Currents, ocean, 866, 866f
Curtis, Tom, 896
Cyanobacteria
in lakes, 878
nitrogen fixation, 854

D

Dale, B. W., 822f
Dam(s), ecological impact, 899
Damaraland mole-rats, 795
Darwin, Charles
on natural selection, 820
DDT (diphenyl-trichloroethane)
biological magnification of, 846, 846f
Death rate
human, 810
life tables and, 804, 804t
and population growth, 800–801, 801f
Deciduous forest
soil, 870, 870f
temperate, 868f–869f, 874, 874f
tropical, 874
Deciduous trees, in semi-evergreen forest, 874
Decomposers
bacteria as, 840, 843, 855
fungi as, 840, 843, 855
in lakes, 879
role of, 840, 842f, 843, 844f, 845f, 855
Deer
and exploitative competition, 820
Florida Key deer, 799f
overpopulation of, 796, 814
white-tailed (*Odocoileus virginianus*), 796, 814
Defense(s), 824–825, 824f, 825f
camouflage and subterfuge, 824, 824f
poisons and venom, 780, 816, 824, 832
secretions and ejaculations, 824, 825, 825f
spines and stingers, 819, 819f, 824
Deforestation, 848, 849f, 855
Demographic transition model, 812, 812f
Demographics
age structure diagrams, 810, 811f
defined, 798
fertility rates, 810, 810f
life history patterns, 804
population growth and decline
limiting factors, 802–803, 802f, 803f, 808–809
types and terminology, 800–801, 800f, 801f
survivorship curves, 804–805, 805f
terminology, 798
Denitrification, 854–855, 854f
Density-dependent limiting factors, 803, 809
Density-independent limiting factors, 803
Desalinization, of water, 849
Desert, 871, 871f
global distribution, 862, 868f–869f
soil, 870, 870f
Desert kangaroo rat, 871
Desertification, 900, 900f
Detritivores, 840, 842f, 843, 844f, 845f
Detritus, carbon held in, 850
Developed nations
population, aging of, 813
resource use by, 812–813
Developing nations
and carrying capacity, 813
environmental impact, 899
Devonian period
major events, 892f
Diatom(s)
in lakes, 878
Dinoflagellates, 882
Disease
as density-dependent limiting factor, 803, 808–809
parasites and, 826
social behaviors and, 791
Distance effect, 835, 835f
Diversity. *See* Biodiversity
Dodders (*Cuscuta*), 826, 826f
Dodo (*Raphus cucullatus*), 893
Dog(s)
communication in, 787
classical conditioning of, 785
Doldrums, 863f
Dominance hierarchies, 791, 791f
Dormancy
in plants, 874
Drosophila melanogaster
courtship behaviors, 783
feeding behavior, 782–783, 783f
foraging (*for*) gene, 782–783, 782f
fruitless (*fru*) gene, 783
Drug(s), prescription
drug research, 875, 902
natural sources of, 875, 902
Dry shrubland, 868f–869f, 873, 873f
Dry woodland, 868f–869f, 873
Duncan, Ruth, 832f
Dust Bowl, 872, 900, 900f
Dutch elm disease, 831t, 846

E

Earth. *See also* Atmosphere
air circulation patterns, 862–863, 863f
crust
carbon in, 850
and phosphorus cycle, 856, 856f
resources, limits on, 890
rotation, and seasons, 862, 862f
Easter Island, 796, 796f
Ecological pyramids, 844–845, 844f, 845f
Ecological succession, 828–829, 828f, 829f
Economic development, and population growth, 812, 812f
Ecoregions
critical or endangered, 897, 897f, 897t
marine, 868f–869f
Ecosystem(s)
biomass pyramids, 844, 844f
defined, 840
element cycle in. *See* Biogeochemical cycles
energy flow in, 840, 840f, 844–845, 844f, 845f
energy pyramids, 845, 845f
freshwater, 878–879, 878f, 879f
nutrient cycling in, 840, 840f
oceans as, 884–885, 884f
saltwater, 881–885
trophic levels, 840–843, 841f, 842f
Ecotourism, 902–903
Ectotherms, 845
Edwards Aquifer, 894
El Niño, 815, 860, 860f, 885–887, 886f
Elephant(s)
survivorship curve, 804
Emigration
and population size, 800
Endangered ecoregions, 897, 897f, 897t
Endangered species, 894, 894f, 896, 896t, 897
current threats to, 894–895
defined, 893
Endemic species, 894
Endler, John, 806
Endosymbiosis
defined, 818
Energy
flow, through ecosystem, 840, 840f, 844–845, 844f, 845f
for human use (*See also* Fossil fuels)
ecological impact, 898–899, 899f
sunlight as source of, 840, 845f (*See also* Photosynthesis)
Energy pyramids, 845, 845f
Enteromorpha (green alga), 830f, 831
Environment
carrying capacity of, 802–803, 802f, 803f
and economic development, 812–813, 813f
and human population growth, 808–809, 808f, 809f, 812–813, 813f
and population growth, limiting factors, 802–803, 802f, 803f, 808–809
Equilibrium model of island biogeography, 835
Ermine, 842f
Erosion
and desertification, 900
logging and, 903
Estuary, 880, 881f
Ethical issues
Arctic resources, exploitation of, 891, 904
deer hunting as population control, 797, 814
El Niño research funding, 861, 888
exotic species, inspection of freight for, 817, 836
fuel economy standards, 839, 858
killer bee research, 781, 794
tropical rain forest, loss of, 875
Ethiopian realm, 868f–869f
Eucalyptus tree, 820
Eusocial insects, 792, 792f
Eutrophic lakes, 878, 878f
Eutrophication, 857, 857f, 878
Evaporation
in water cycle, 848, 848f
Evergreen broadleaf forest, 874
Evolution. *See also* Natural selection
altruistic behavior, 792–793
coevolution
defined, 818
humans, 793
mitochondria, 819
Exotic pathogens, 895
Exotic species
impact of, 831–833, 831t, 832f, 833f, 895
increasing numbers of, 836
Experiment(s). *See* Scientific research
Exploitative competition, 820
Exponential growth, of population, 800–801, 800f, 801f
Extinction(s)
current rate of, 890
mass, 892, 892f
current, 893
on geological time scale, 892f
verification of, 893

F

Fall overturn, 879, 879f
Feeding behavior
genetic basis of, 782–783, 782f
group hunting, 790, 791f
Female(s)
sexual selection and, 788
Ferns
threatened species, 896t
Fertility rates, 810, 810f
Fertilizers
ammonia, 855
as pollutant, 856, 856f, 865, 894
Fiddler crab, 785, 788, 788f
Fiji, reefs, 883f
Fir trees, 876
Fire ants (*Solenopsis invicta*), 816, 816f, 831, 837, 837f
Firefly, North American (*Photinus pyralis*), 787
Fish
acid rain and, 865
bony (osteichthyes)
species diversity, over time, 892f
habitat loss, 880
mercury and, 846
parasites, 826, 826f
survivorship curve, 805
threatened species, 896t
Fixed action patterns, 784
Florida Everglades, 881f
Florida Key deer, 799f
Florida sandhill crane, 898
Flukes (trematodes), 826
Flytrap anemone, 885f
Food chains, 841, 841f
aquatic, 843
and biological magnification, 846, 846f
detrital, 843, 880, 884
grazing, 842–843
Food webs, 842–843, 842f, 843f
Foraging (*for*) gene, 782–783, 782f
Foraminiferans (forams), 850
Forams. *See* Foraminiferans
Forest(s)
broadleaf
endangered ecoregions, 897f
evergreen, 874
tropical, 868f–869f
coniferous, 876, 876f
boreal (taigas), 868f–869f, 876, 876f
endangered ecoregions, 897f
montane, 876, 876f
soil, 870, 870f
temperate, 868f–869f
tropical, 868f–869f
deciduous
soil, 870, 870f
temperate, 868f–869f, 874, 874f
tropical, 874
deforestation, 848, 849f, 855
rain forest
endangered ecoregions, 897t
tropical, 870, 870f, 874, 875, 875f
semi-evergreen, 874
temperate, 875
coniferous, 868f–869f
deciduous, 868f–869f, 874, 874f
endangered ecoregions, 897t
tropical dry, 868, 868f–869f
tropical forest
broadleaf, 868f–869f
coniferous, 868f–869f
deciduous, 874
dry, 868, 868f–869f
rain forest, 870, 870f, 874, 875, 875f
Fossil fuels. *See also* Coal
and carbon cycle, 850f–851f, 851
ecological impact, 898–899, 899f
and global warming, 838
and greenhouse gases, 853, 855
and Industrial Revolution, 809
known reserves, 850
Fox, 842f
Freshwater ecosystems, 878–879, 878f, 879f
Fringing reefs, 882f
Frogs
communication in, 787
endangered species, 898
Fru gene. *See Fruitless* (*fru*) gene
Fruit fly. *See Drosophila melanogaster*
Fruitless (*fru*) gene, 783
Fungus (fungi)
as decomposer, 840, 843, 855
endangered species, 896
mycorrhizae, 819
as pioneer species, 835
survivorship curve, 805
threatened species, 896t

G

Galápagos Islands
finches on, 821
iguana population, 815, 815f
reefs, 883
sea lions, 860, 860f
Garter snakes, 782, 782f, 784
Gause, G., 820
Gene(s)
and behavior, 782–783, 782f, 783f
Gene flow
habitat fragmentation and, 894
Genetic disorders
autism, 793
Geographic dispersal, 831
Geological time scale, 892f

Geology
  geological time scale, 892*f*
Giant panda, 894, 894*f*
■ Global warming, 853, 853*f*
  impact of, 838, 838*f*, 866, 877, 882, 904
  rain forest loss and, 875
Gobi Desert, 900
Goose (geese), 784, 784*f*
Grant, Peter, 821
Grant, Rosemary, 821
Grasshopper(s), 841, 841*f*
Grasshopper mouse, 825, 825*f*
Grassland, 872–873, 872*f*
  mountain, 868*f*–869*f*
  shortgrass prairie, 872, 872*f*
  soil, 870, 870*f*
  tallgrass prairie, 872, 872*f*
  temperate, 868, 868*f*–869*f*, 869*f*
  warm, 868*f*–869*f*
Gray squirrel (*Sciurus carolinensis*), 833
Grazers, 843
Great Plains, American, Dust Bowl crisis, 872, 900, 900*f*
Great Smoky Mountain National Park, 855, 855*f*
Green algae
  in lakes, 878
  threatened species, 896*t*
Greenhouse effect, 852, 852*f*
Greenhouse gases
  carbon cycle and, 851
■ and climate change, 852–853, 852*f*, 853*f*
■ and habitat degradation, 894
Gross primary production, 844
Groundwater
■ contamination of, 849, 849*f*
  defined, 848
  as water reservoir, 848*t*
Growth
  plant
    phosphorus and, 856–857
Growth rate per capita, of population, 800
■ Guam, introduced species on, 895
Gulf Stream, 866, 866*f*
Guppies (*Poecilia reticulata*), natural selection in, 806–807, 806*f*, 807*f*
Gymnosperms
  species
    diversity, over time, 892*f*
  threatened species, 896*t*
Gyrfalcon, 842*f*

H

Habitat
  carrying capacity of, 802–803, 802*f*, 803*f*
    and economic development, 812–813, 813*f*
    and human population growth, 808–809, 808*f*, 809*f*, 812–813, 813*f*
  characteristics of, and community structure, 818
  characterization of, 798
■ damage, indicator species, 896
  definition of, 818
  ecological succession in, 828–829, 828*f*, 829*f*
■ fragmentation, impact of, 894, 894*f*
■ loss, impact of, 894, 894*f*
  and population growth, limiting factors, 802–803, 802*f*, 803*f*, 808–809
  species interactions in. *See* Species, interaction
Habitat islands, 835
Habituation, 785
Hairston, Nelson, 820–821
Hamilton, William, 792–793
Hangingflies (*Harpobittacus*), 788, 788*f*
Hare, in arctic food web, 842*f*, 843
Hartshorn, Gary, 903
Hawaiian honeycreepers, 895
Hawaiian Islands
  biological controls on, 827
  birds of, 895
  exotic pathogens, 895
  insects, 827
  reefs, 882*f*, 883
  waves, 860, 860*f*
Hawk(s), 841, 841*f*
■ Health care, and aging of populations, 813
Herbivores, role in ecosystems, 840, 844*f*, 845*f*
Herd, selfish, 790
Herring gulls, 791
Honeybees
  Africanized, 780, 780*f*, 794, 795, 795*f*
  altruistic behavior in, 792, 792*f*
  communication in, 780, 786, 786*f*, 787, 787*f*
  European, 795
  queen, 792, 793*f*
Horizons, soil, 870, 870*f*
Hormone(s), **human**
  and behavior, 793
Hot spots
■ of habitat destruction, 896
Houston toad, 898
Hrdy, Sarah Blaffer, 793
Hubbard Brook watershed, 848, 849*f*
Huber, Ludwig, 785
Human(s)
  in arctic food web, 842*f*
  behavior
    evolutionary basis, 793
    factors affecting, 793
  and carbon cycle, 851
  evolution of, 793
  exotic species and, 831–833, 831*t*, 832*f*, 833*f*
  impact on biosphere. *See* Biosphere, human impact on
  language
    advantages of, 808
  nitrogen cycle and, 855, 855*f*
  population and demographics
    agriculture and, 808–809, 813
    biotic potential, 801
    carrying capacity, 812–813, 813*f*
    census, 799
    collapses of, 796, 796*f*, 813, 813*f*
    economic development and population growth, 812, 812*f*
    economic effects of population growth, 812–813, 812*f*, 813*f*
    growth, 808–809, 808*f*, 809*f*, 810
    life table, 804*t*
    limiting factors, 808–809
    population control programs, 810–811
    survivorship curve, 804
    total fertility rate, 810
Humboldt Current, 885, 885*f*
Huq, Anwarul, 887
■ Hurricane Katrina, 838
Hydrogen
  solar–hydrogen energy, 863
Hydrosphere, defined, 860
Hydrothermal vents, 884, 885*f*

I

Ice
■ Arctic, shrinkage of, 859, 859*f*, 890, 904
Ice cover, perpetual, 868*f*–869*f*
Iguanas, 815, 815*f*
Immigration
  and population size, 800
Imprinting, 784, 784*f*
Inbreeding
  in eusocial species, 793
Inclusive fitness, theory of, 792
India
  age structure diagram, 811*f*
  and economic development, 813
  population, 810
Indian mallow, 821*f*
Indicator species, 896
Indonesia, population, 810
Industrial Revolution, 809
Industrial smog, 864
Industrial stage, of demographic transition model, 812, 812*f*
■ Industrialization, and resource use, 812–813
Infant mortality rate, global decline in, 810
Infanticide
  in animals, 793
  in China, 811
Infrared radiation
  and greenhouse effect, 852, 852*f*
Insect(s). *See also* Mosquito
  altruistic behavior in, 792, 792*f*
  as detritivore, 843
  as disease vector, 846
  eusocial, 792, 792*f*
  parasitoids, 816, 826–827, 827*f*
  plant defenses against, 824
  species diversity, over time, 892*f*
  survivorship curve, 805
  threatened species, 896*t*
Insecticides. *See* Pesticides
Instinctive behavior, 784, 784*f*
Interference competition, 820, 820*f*
Intermediate disturbance hypothesis, 829
■ Introduced species, 895. *See also* Exotic species
Invertebrate(s)
  marine, survivorship curve, 805
  threatened species, 896*t*
■ Irrigation, 848–849
Isaacson, Peter, 846*f*
■ Island(s), biodiversity patterns on, 834–835, 834*f*, 835*f*
IUNC Red List of Threatened Species, 896, 896*t*
Ivory-billed woodpecker (*Campephilus principalis*), 893, 893*f*

J

Japan, 813
Japanese beetle, 831*t*
Japanese honeysuckle, 895
Jordan's salamanders (*Plethodon jordani*), 820–821, 821*f*
J-shaped curve, 800*f*, 801
Jump dispersal, 831
Jurassic period
  major events, 892*f*

K

Kangaroo rat, 871
Key Largo, reefs, 883
Keystone species, 830–831, 831*f*
Killer bees. *See* Africanized honeybees
Killer whale (orca), 899
Killifish, 806*f*, 807, 807*f*
Klamath–Siskiyou forest, 897, 897*t*
Klein, David, 803*f*
Knockout experiments, 783
Koa bugs, 827
Koala (*Phascolarctos cinereus*), 825
Krebs, Charles, 823
K-selection, 805
Kudzu (*Pueraria montana*), 832–833, 832*f*, 895

L

La Niña, 886, 886*f*, 889, 889*f*
Lakes
■ acid rain and, 865, 865*f*
  as ecosystem, 878–879, 878*f*
  as water reservoir, 848*t*
Lampreys
  characteristics of, 831*t*
Land development, biosphere impact, 898
Land ecosystems, food webs, 843
Land plants
  threatened species, 896*t*
  water-conserving adaptations, 871, 876
Language
  advantages of, 808
Latitude, community structure, 834, 834*f*
Laysan albatross, 901*f*
Leaching, of soil, 855
Learned behavior, 784–785, 784*f*, 785*f*
Lek, 788
Lemmings, in arctic food web, 842*f*, 843
Leopard frog (*Rana pipiens*), 898
Lesser long-nosed bats, 871*f*
Lichens
  as indicator species, 896
  as mutualistic relationship, 819
  as pioneer species, 828, 828*f*, 835
  threatened species, 896*t*
Life history patterns, 804
  natural selection and, 806–807, 806*f*, 807*f*
  uses of, 807
Life tables, 804, 804*t*
Light. *See* Sunlight
Lignin
  in plant structures, 845
Limiting factors, on population, 802–803, 802*f*, 803*f*, 808–809
Limnetic zone, of lake, 878, 878*f*
Limpets, 885*f*
Lion, 779*f*, 790, 793
Lionfish, 883
Lithops, 824, 824*f*
Lithosphere, defined, 860
Littoral zones
  of coastlines, 881
  of lake, 878, 878*f*
Lizards
  survivorship curve, 805
Loblolly pines, 876
Lobster (*Homarus americanus*), 785, 785*f*
■ Logging, sustainable, 902*f*, 903
Logistic growth, 802–803, 802*f*
Longnose hawkfish, 883*f*
Lorenz, Konrad, 784*f*
Louisiana, marshes, 838, 838*f*
Louisiana Shrimp and Petroleum Festival, 838
Lower littoral zone, 881
Lubchenco, Jane, 830

M

MacArthur, Robert H., 835
Maize. *See* Corn
Malaria
  avian, 895
■ control of, 846
Male(s)
  sexual selection and, 788
Mammal(s)
  monogamy in, 789
  parental care in, 789, 789*f*
  species diversity, over time, 892*f*
  survivorship curve, 805
  threatened species, 896*t*
Mangrove swamp/wetlands, 868*f*–869*f*, 880, 881*f*
Marbled murlet, 897
Marine ecoregions, 868*f*–869*f*
Marmosets, 785, 785*f*
Marsh grass (*Spartina*), 881*f*
Martinez, Neo, 843
Mason, Russell, 786
Mating behavior
  guppies, 806
  hormones and, 783, 783*f*
  parental care and, 789, 789*f*
  and sexual selection, 788–789, 788*f*, 789*f*
  visual signals, 786*f*, 787
Meadow vole (*Microtus pennsylvanicus*), 783
Menstrual cycle
  human, synchronization of, 793
Mercury
  in fish, 846
  in polar bears, 890
Mesozoic era
  major events, 892*f*
Mesquite, 871
Methane
■ as greenhouse gas, 852, 853*f*
Mexico, age structure diagram, 811*f*
Mexico City, 852*f*
Microfilaments, 878
Midlittoral zone, 881
Midwife toad, 789, 789*f*
Migration
  and population size, 800
Mimicry, as defense, 824–825, 825*f*
■ Mining, surface, impact of, 899
Mission Blue butterfly, 898
Mississippi River Basin, 848
Mistletoe, 826
Mite, 786, 786*f*
Mitochondrion (**mitochondria**)
  origin, 819
Mole-rats
  Damaraland, 795
  naked (*Heterocephalus glaber*), 792, 793*f*
Mollusks (**Mollusca**)
  threatened species, 896*t*
Monogamy, in mammals, 789
Monsoons, 867
Montane coniferous forests, 876, 876*f*
Monteverde Cloud Forest Reserve, 902–903
Moose, 876, 876*f*
Moray eel, 883*f*
Mosquito
  as disease vector, 846
Mosses
  as pioneer species, 828, 828*f*, 835
  threatened species, 896*t*
Moths
  as pollinator, 819, 819*f*
■ Motor vehicles. *See also* Fossil fuels
  and air pollution, 864–865
  and carbon emissions, 858
■ Mount Saint Helens, and ecological succession, 829, 829*f*
Mountain(s)
  as biome, 868*f*–869*f*
  rain shadows, 866–867, 867*f*
  seamounts, 884, 884*f*
Mountain avens (*Dryas*), 828, 828*f*
Mountain grassland, 868*f*–869*f*
Mountain vole (*Microtus montanus*), 783, 783*f*
Mouse (**mice**)
  behavior, genetic basis of, 783
  population growth, exponential, 800–801, 800*f*
  as predator, 825, 825*f*
Mushrooms
  threatened species, 896*t*
Musk oxen (*Ovibos moschatus*), 790–791, 791*f*
Mussels, 830
Mutualism, 818, 818*t*, 819, 819*f*
  mycorrhizae, 819
Mycorrhiza(e), 819
Myxobolus cerebralis, 826, 826*f*
Myxoma virus, 833
Myxomatosis, 833

N

Naked mole-rat (*Heterocephalus glaber*), 792, 793*f*
National Institute of Biodiversity (Costa Rica), 902
National Oceanographic and Atmospheric Administration (NOAA), 889

Natural resources. *See also* Carrying capacity
consumption, biosphere impact, 898–899, 899*f*
industrialization and, 812–813, 813*f*
Natural selection
character displacement, 821
Darwin on, 820
and life histories, 806–807, 806*f*, 807*f*
in parasites, 826
predation and, 806–807, 806*f*, 807*f*, 823–826, 824*f*, 825*f*
selection pressures
human activity as, 807
Nearctic realm, 868*f*–869*f*
Nematocysts, 819, 824
Nematodes. *See* Roundworms
Neotropical realm, 868*f*–869*f*
Neritic zone, 884, 884*f*
Net primary production, 844, 844*f*
Niche(s)
characteristics of, 818
defined, 818
Nicotine
as plant defense, 824
Nigeria
demographic data, 810*f*
population, 810
Niña, La. *See* La Niña
Niño, El. *See* El Niño
Nitrate (NO_3–), in nitrogen cycle, 854–855, 854*f*
Nitric oxide (NO), 864
Nitrification, 854*f*, 855
Nitrites, 854–855, 854*f*
Nitrogen cycle, 854–855, 854*f*
Nitrogen dioxide, 864–865
Nitrogen fixation
nitrogen-fixing bacteria, 819, 819*f*, 828, 828*f*, 847, 854, 854*f*, 855, 857
Nitrogen oxides, and global warming, 855, 855*f*
Nitrous oxide, as greenhouse gas, 852, 853*f*
NOAA. *See* National Oceanographic and Atmospheric Administration
North Atlantic codfish (*Gadus morhua*), 807, 807*f*
Northern (boreal) coniferous forests (taigas), 868*f*–869*f*, 876, 876*f*
Northern fur seals, 860
Northern spotted owl, 804, 897
Nudibranch, 883*f*
Nutria, 831*t*
Nutrient(s)
absorption
mycorrhizae and, 819
cycling of, in ecosystems, 840, 840*f* (*See also* Biogeochemical cycles)
plant
mycorrhizae and, 819
in soil, deforestation and, 848, 849*f*

O

O horizon, 870, 870*f*
Observational learning, 785, 785*f*
Ocean(s)
in carbon cycle, 850–851, 850*f*–851*f*
currents, 866, 866*f*
ecosystems, 884–885, 884*f*
primary productivity, 844
upwelling in, 885, 885*f*
as water reservoir, 848*t*
zones, 884, 884*f*
Oceanic zone, 884, 884*f*
Ogallala aquifer, 849
Oligotrophic lakes, 878, 878*f*
Omnivores, 840, 841
Operant conditioning, 785
Opossum, 825
Orca (killer whale), 899
Orchid
endangered, 894, 895*f*
Ordovician Period
major events, 892*f*
Oriental realm, 868*f*–869*f*
Osprey (*Pandion haliaetus*), 846*f*
OT. *See* Oxytocin
Overharvesting, 894
Owls
northern spotted, 804, 897
snowy, 842*f*
Oxytocin (OT)
action, 783, 783*f*, 793
Ozone layer, 864
hole in, 864, 864*f*

P

Pacific Northwest, climate, 866
Pacific salmon, 805
Paine, Robert, 830
Pakistan, population, 810
Palearctic realm, 868*f*–869*f*
Paleozoic era, 892*f*
Papua New Guinea
infanticide in, 793
pigeon species in, 818*f*
Paramecium
competition among, 820, 821*f*
Parasites and parasitism, 818, 818*t*, 826–827, 826*f*, 827*f*
as consumers, 840
as density-dependent limiting factor, 803
natural selection in, 826
parasitoids, 816, 826–827, 827*f*
social group behaviors and, 791
social parasitism, 826
Parasitoids, 816, 826–827, 827*f*
Parental care
and courtship behavior, 789, 789*f*
and survivorship curve, 805
Passenger pigeons, 894
Pathogen(s)
exotic, 895
Pavlov, Ivan, 785
PCBs (polychlorinated biphenyls), 905, 905*f*
Peat bog, 851
Pelagic province, 884, 884*f*
Penguins, 786*f*, 798*f*
Per capita, defined, 800
Periwinkle (*Littorina littorea*), 830–831, 830*f*
Permafrost, 850, 877
Permian period
major events, 892*f*
Pesticides
and biological magnification, 846, 846*f*
as pollutant, 894
residue
in animal tissue, 890
Pet(s)
ecological impacts of trade in, 883
pH
of acid rain, 865, 865*f*
Pheromones, 786, 786*f*
alarm, 780, 780*f*, 786
and human behavior, 793
priming, 786
sex, 788, 792
Phlox
life table, 804*t*
survivorship curve, 805
Phorid flies, 816, 816*f*, 826
Phosphate
in phosphorus cycle, 856
uses by organisms, 856
Phosphorus cycle, 856–857, 856*f*
Photochemical smog, 864
Photosynthesis
carbon cycle and, 850*f*–851*f*, 851
and greenhouse gases, 852, 853*f*
Photosynthetic bacteria. *See* Cyanobacteria
Photosynthetic dinoflagellates, 882
Phytophthora, 897
Phytoplankton, 880
Pied imperial pigeon, 818*f*
Pig(s)
parasites, 826*f*
Pigeons
in Papua New Guinea, 818*f*
passenger pigeons, 894
Pine (*Pinus*)
in coniferous forest, 876
Pink anemone fish (*amphiprion perideraion*), 819
Pioneer species, 828–829, 828*f*
Pitch pines, 876
PKG, 782–783, 782*f*
Plankton. *See also* Diatom(s); Foraminiferans
in lakes, 878
phytoplankton, 880
Plant(s). *See also specific types*
annual, survivorship curve, 805
defenses, 824, 824*f*
desert, 871, 871*f*
extinctions, 893
growth
phosphorus and, 856–857
interference competition in, 820, 820*f*
nitrogen cycle and, 854–855
nutrients
mycorrhizae and, 819
phosphorus cycle and, 856
pioneer species, 828, 828*f*, 835
resource partitioning in, 821, 821*f*
water and
water-conserving adaptations, 871, 876
Plastic, ecological impact, 901, 901*f*
Plate tectonics theory
and geographic dispersal, 831
Play bow, 786*f*, 787
Poaching, 894–895
Poison and venom
as defense, 780, 816, 824, 832
Polar bear (*Ursus maritimus*), 825, 890, 890*f*, 905
Polar region
as desert, 862
ice
shrinkage of, 859, 859*f*, 890, 904
as water reservoir, 848*t*
Pollination
as mutualistic relationship, 819, 819*f*
Pollinator(s)
bats as, 871*f*
Pollution. *See also* Air pollution; Water pollution
agricultural chemicals, 846, 846*f*
in polar regions, 890, 905, 905*f*
pollutant, defined, 864
Polychaetes, 885*f*
Polychlorinated biphenyls (PCBs), 905, 905*f*
Pompei-worm, 885*f*
Population(s)
aging of, 813
biotic potential of, 801
collapse of, 796, 796*f*, 802, 803, 803*f*, 807, 813, 813*f*
defined, 798
ecological impact of, 796, 796*f*
human. *See* Human(s)
Population density
defined, 798
and reproductive strategy, 805
Population distribution, types of, 798, 798*f*
Population ecology
carrying capacity, 802–803, 802*f*, 803*f*
and economic development, 812–813, 813*f*
and human population growth, 808–809, 808*f*, 809*f*, 812–813, 813*f*
focus of, 796
population growth, economic effects, 812–813, 812*f*, 813*f*
terminology, 798
Population growth and decline
exponential, 800–801, 800*f*, 801*f*
human, 808–809, 808*f*, 809*f*, 810
economic development and, 812, 812*f*
economic effects of, 812–813, 812*f*, 813*f*
limiting factors, 802–803, 802*f*, 803*f*, 808–809
types and terminology, 800–801, 800*f*, 801*f*
zero, defined, 800
Population size
defined, 798
doubling time, 801
estimation of, 799, 799*f*
gains and losses in, 800–801, 800*f*, 801*f*
Postindustrial stage, of demographic transition model, 812, 812*f*
Post-reproductive individuals, 798
Prairie
endangered ecoregions, 897*t*
shortgrass, 872, 872*f*
tallgrass, 872, 872*f*
Prairie dogs, 790, 790*f*
Prairie fringed orchids (*Platanthera*), 894, 895*f*
Prairie vole (*Microtus ochrogaster*), 783, 783*f*
Praya dubia (siphonophore), 885*f*
Praying mantis, 825, 825*f*
Precambrian period, 892*f*
Precipitation
acid rain, 855, 865, 865*f*, 894
origin of, 862
in water cycle, 848, 848*f*
Predators and predation, 818, 818*t*
adaptive responses, 825, 825*f*
defined, 822
and natural selection, 806–807, 806*f*, 807*f*, 823–826, 824*f*, 825*f*
predator-prey interaction models, 822–823, 822*f*
social groups as defense against, 790, 790*f*
Pregnancy. *See also* Contraception
DDT exposure and, 846
mercury exposure and, 846
Preindustrial stage, of demographic transition model, 812, 812*f*
Pre-reproductive individuals, 798
Prey. *See also* Predators and predation
defenses, 824–825, 824*f*, 825*f*
defined, 822
Primary producers, 840, 842*f*, 847
Primary production, 844, 844*f*
in coastal zones, 880, 881, 881*f*
in lakes, 878, 879
in oceans, 884, 886, 886*f*
Primary succession, 828, 828*f*
Priming pheromones, 786
Producers
in food web, 842*f*
role in ecosystem, 840, 840*f*, 844, 844*f*, 845*f*, 847
Profundal zone, of lake, 878, 878*f*
Prokaryotes
chemoautotrophic, 884
Protists
and carbon cycle, 850
endangered species, 896
threatened species, 896*t*
Protozoans
species diversity, over time, 892*f*
Purple saxifrage, 842*f*
Purple tube sponge, 883*f*

Q

Quadrats, 799
Quaternary period, major events, 892*f*
Queen(s), in eusocial species, 792, 793*f*
Quinine, 875

R

Rabbits
European (*Oryctolagus curiculus*), 833, 833*f*
Radioactive fallout, from Chernobyl nuclear plant disaster, 865*f*
Rain forests
endangered ecoregions, 897*t*
tropical
characteristics of, 874
loss of, 875, 875*f*
soil, 870, 870*f*
Rain shadows, 866–867, 867*f*
Ranching, environmental impact, reduction of, 903, 903*f*
Random population distribution, 798, 798*f*
Rats
as disease vector, 803
as exotic species, 895
kangaroo rat, 871
Recycling, 901
Red algae
in coral reefs, 882, 882*f*
threatened species, 896*t*
Red mangroves (*Rhizophora*), 881*f*
Red sea fan, 883*f*
Red squirrel (*Sciurus vulgaris*), 833
Reefs, coral, 882–883, 882*f*–883*f*
Reindeer, 803*f*
Reproduction. *See also* Sexual reproduction
strategies, environmental factors affecting, 805
Reproductive base, defined, 798
Reptile(s)
species
diversity over time, 892*f*
threatened species, 896*t*
Research. *See* Scientific research
Resource partitioning, 821
Reznick, David, 806, 806*f*
Rhinoceros, 895
Rhizobium, 854
Ricin, 824
Riparian zones, preservation of, 903, 903*f*
River(s)
as ecosystem, 879
as water reservoir, 848*t*
Rocky coastlines, 881, 881*f*
Root(s)
resource partitioning and, 821, 821*f*
Rosy periwinkle (*Catharanthus roseus*), 875
Rotifers (Rotifera)
in lakes, 878
Roundworms (nematodes), 826, 826*f*
as detritivore, 843
r-selection, 805
Run off, 848, 855, 856, 856*f*

S

Sage grouse (*Centrocercus urophasianus*), 788–789, 788*f*
Sagebrush, 820, 829
Sahara desert, 900, 900*f*
St. Matthew's Island, 803*f*
Salamander
competition among, 820–821, 821*f*
Salinization, of soil, 849
Salmon, 805, 826, 897, 899
Saltwater ecosystems, 881–885
Sampling error, 799
Sand, in soil, 870
Sandy coastlines, 881, 881*f*Savanna
characteristics, 872*f*, 873
global distribution, 868*f*–869*f*
Sawfly caterpillars, 790, 790*f*

■ Science, moral issues. *See* Ethical issues
■ Scientific research
■   accuracy of conclusions, 823*f*
  on biological controls, 837, 837*f*
■   on prescription drugs, 875, 902
Scorpionfish, 825, 825*f*
Scrub oaks, 876
Sea anemones, 819, 819*f*
Sea lions, 860, 860*f*
Sea stars
  as keystone species, 830
Sea turtles, 898
Sea urchins
  survivorship curve, 805, 805*f*
Seagulls, 834*f*, 835, 846*f*
Seals
  northern fur seals, 860
Seamounts, 884, 884*f*
Season(s)
  cause of, 862, 862*f*
  in lake ecosystems, 878–879, 879*f*
  variation in primary production, 844, 844*f*
Secondary succession, 828, 828*f*
Seed(s)
  dispersal of, 818*f*, 831
Selection. *See* Natural selection
Selection pressure(s)
  human activity as, 807
Selfish herd, 790
Self-sacrifice. *See* Altruistic behavior
Semi-evergreen forest, 874
■ Sewage, 855, 865
Sex pheromones, 788, 792
Sexual selection
  courtship behavior and, 788–789, 788*f*, 789*f*
Shortgrass prairie, 872, 872*f*
■ Shrimp farms, 880
Shrubland, dry, 868*f*–869*f*, 873, 873*f*
Signal pheromones, 786, 786*f*
Silent Spring (Carson), 846
Sillén-Tullberg, Birgitta, 790, 790*f*
Silurian period
  major events, 892*f*
Silver Springs, Florida, ecosystem, 844, 844*f*
Siphonophores, 885*f*
Sitka spruce, 876
Slimy salamanders (*Plethodon glutinosus*), 820–821, 821*f*
■ Smartweed, 821*f*
■ Smog, 852*f*, 864–865, 864*f*
Snails
  in coastal zones, 881
  periwinkle (*Littorina littorea*), 830–831, 830*f*
Snakes
  feeding behaviors, 782, 782*f*
  as introduced species, 895
  learned behavior in, 784
Snowshoe hare, as prey, 822–823, 823*f*
Snowy egret, 805*f*
Snowy owl, 842*f*
Social groups, costs and benefits of, 790–791, 790*f*, 791*f*
Social parasites, 826
■ Social Security programs, and aging of population, 813
Sodium cyanide, 883
Soil
  of biomes, major, 870, 870*f*
  coniferous forest, 870, 870*f*
  desert, 871
  ecological succession and, 828
  evergreen broadleaf forest, 874
  horizons, 870, 870*f*
■   leaching of nutrients from, 855
  in nitrogen cycle, 854–855, 854*f*
■   runoff from, 848, 855, 856, 856*f*
  salinization of, 849
  in tundra, 877
  as water reservoir, 848*t*
Soil profiles, 870, 870*f*
Sokolowski, Marla, 782
Solar energy, 863
■ Solar-aquatic wastewater treatment system, 880, 880*f*
■ Solar–hydrogen energy, 863
Sonoran Desert, Arizona, 871, 871*f*
Sonoran desert tortoise, 871*f*
Sparrow, 784, 841, 841*f*
Species
  biotic potential of, 801
  diversity. *See* Biodiversity
  endemic, 894
  indicator, 896
  interaction (*See also* Commensalism; Competition; Parasites and parasitism; Predators and predation; Symbiosis)
    and community stability, 830–831, 830*f*, 831*t*
    types, 818
  life history pattern of, 804
Spiders
  as predator, 822
Spines and stingers, as defense, 819, 819*f*, 824
Spring equinox, 862*f*
Spring overturn, 879, 879*f*
Spruce trees, 876
Squirrel(s)
  as exotic species, 833
Squirrelfish, 798*f*
S-shaped curve, in logistic growth, 802*f*, 803
Starfish. *See* Sea stars
Starling(s), 786, 786*f*
Steinbeck, John, 872
Sterility
  in eusocial animals, 792–793
  parasites and, 826
Stimulus
  defined, 782
  detection and response to, 782
Stinkbugs, 827
Stoma(ta)
  of CAM plants, 871
  of conifers, 876
Streams, as ecosystem, 879
■ Strip logging, 902*f*, 903
Succession
  in lakes, 878
  primary, 828, 828*f*
  secondary, 828, 828*f*
■ Sudden oak death, 831
Sulfur
  dioxides, 865
Summer solstice, 862*f*
Sunfish, bluegill, 790
Sunlight
  as energy source, 840, 845*f* (*See also* Photosynthesis)
  intensity, variation with latitude, 862, 862*f*
  and ocean currents, 866
  and wind, 862, 863*f*
Superb crowned fruit pigeon, 818*f*
Surf grass, 829
■ Surface mining, impact of, 899
Surtsey, 834–835, 834*f*
Survivorship curves, 804–805, 805*f*
Symbiosis
  defined, 818

T

Tactile displays, 787, 787*f*
Tahiti, reefs, 882*f*
Taigas (boreal coniferous forests), 868*f*–869*f*, 876, 876*f*
Tallgrass prairie, 841, 841*f*, 872, 872*f*
Tapeworms (cestode), 826
Temperate forest, 875
  coniferous, 868*f*–869*f*
  deciduous, 868*f*–869*f*, 874, 874*f*
  endangered ecoregions, 897*t*
Temperate grassland, 868, 868*f*–869*f*, 869*f*
Termites (*Nasutitermes*)
  altruistic behavior in, 792, 792*f*
  king, 792
Tertiary period
  major events, 892*f*
Texas blind salamander (*typhlomolge rathbuni*), 894, 895*f*
Texas horned lizard, 816
TFR. *See* Total fertility rate
Thelohania solenopsae, 837, 837*f*
Thermal inversions, 864, 864*f*
Thermocline, 879, 879*f*
Thomson's gazelle, 825
Threatened species, 893, 898
Tiger, 825
Toads
  Houston, 898
  midwife, 789, 789*f*
Tobacco plant (*Nicotiana tabacum*)
  defenses, 824
Todd, John, 880, 880*f*
Total fertility rate (TFR), 810
Toyon (*Heteromeles arbutifolia*), 873*f*
Transitional stage, of demographic transition model, 812, 812*f*
Transpiration, 848
■ Trash, ecological impact, 901, 901*f*
■ Trawling, ecological impact of, 884
Tree(s)
■   acid rain and, 865
Trematode. *See* Flukes
Trenches
  deep-sea, 884*f*
Triassic period
  major events, 892*f*
Trilobites, 892*f*
Trinidad, guppy natural selection in, 806–807, 806*f*, 807*f*
Trophic levels, 840–843, 841*f*, 842*f*
Tropical forest
  broadleaf, 868*f*–869*f*
  coniferous, 868*f*–869*f*
  deciduous, 874
  dry, 868, 868*f*–869*f*
  rain forest
    characteristics of, 874
■   loss of, 875, 875*f*
    soil, 870, 870*f*
Trout, 826, 826*f*, 895, 896
Tsunami, Indonesia (4), 803
Tube worms, 884, 885*f*
Tundra, 868*f*–869*f*, 877, 877*f*
Tungara frogs, 787
■ Typhus, 803, 846

U

Ultraviolet radiation. *See* UV (ultraviolet) radiation
Uniform population distribution, 798, 798*f*
United Nations
  Population Division, 810
United States
■   age structure diagram, 810, 811*f*
■   carbon emissions, 858
■   population and demographics
■   age of population, 813
■   current population, 810
■   demographic data, 810*f*
■   life table, 804*t*
■   resource consumption of natural resources, 898, 899*f*
■   use of natural resources, 812–813
■ United States Department of Agriculture (USDA)
■   biological controls research, 837, 837*f*
Upper littoral zone, 881
Upwelling, in oceans, 885, 885*f*
USDA. *See* United States Department of Agriculture
UV (ultraviolet) radiation
  ozone layer and, 864
  for water purification, 880

V

Vasopressin. *See* Antidiuretic hormone
Vector(s)
  disease, 846
Venom
  fire ants, 816
  honeybees, 780
Vertebrate(s)
  threatened species, 896*t*
Vibrio cholerae, 886–887, 887*f*
Victoria crowned pigeon, 818*f*
Vinblastine, 875
Vincristine, 875
Visual communication signals, 786–787, 786*f*
Voelkel, Bernhard, 785
Voles
  in arctic food web, 842*f*, 843
  hormonal basis of mating behavior in, 783, 783*f*

W

Warm grassland, 868*f*–869*f*
Warning coloration, 824
Wasps
  mimics of, 825, 825*f*
■ Wastewater treatment, 880, 880*f*
Water
  desalinization, 849
  drinking
■   pollution and, 849
■   purification of, 880, 880*f*
  environmental reservoirs of, 848, 848*t*
  evaporation
    in water cycle, 848, 848*f*
■   fresh, global shortage of, 848–849, 849*f*
  groundwater
■   contamination of, 849, 849*f*
    defined, 848
    as water reservoir, 848*t*
■   wastewater treatment, 880, 880*f*
  water cycle, 847–848, 847*f*, 848*f*, 856
Water hyacinth, 831*t*
■ Water pollution
  drinking water, 849
  fertilizer runoff, 856
  groundwater, 849, 849*f*
  impact on aquatic ecosystems, 849
  in estuaries, 880
  mercury, 846
  sources of, 880
  trash as, 901, 901*f*
Water-conserving adaptations
  animals, 871
  plants, 871, 876
Watersheds, 848
Weather
  cycles in, 888
  El Niño and, 886
  monsoons, 867
  rain shadows, 866–867, 867*f*
Whales
  killer whale (orca), 899
■ Whirling disease, 826, 826*f*, 831
White abalone, 894
White-tailed deer (*Odocoileus virginianus*), 796, 814
Wikelsky, Martin, 815
Wild maize (*Zea diploperennis*), 902
Wildebeest, 872*f*, 873
Williams, Ernest, 882
Wilson, Edward O., 835
Wind(s)
  coastal breezes, 867, 867*f*
  origin and global prevailing patterns, 862–863, 863*f*
  power from, 863
Winter solstice, 862*f*
Wolf (*Canis lupus*), 786*f*
  arctic, 842*f*
  communication in, 787
  pack behaviors, 790, 791, 791*f*
  as predator, 822, 822*f*
Wolf spider, 798, 798*f*
Wood ducks (*Aix sponsa*), 802*f*
Woodland, dry, 868*f*–869*f*, 873
Woodpecker, ivory-billed (*Campephilus principalis*), 893, 893*f*
Woodwell, George, 846*f*
World Conservation Union, 893, 896
World War I, 809
World War II, 803*f*, 810
World Wildlife Fund, 897
Worm, Boris, 894
Wurster, Charles, 846*f*

Y

Yellow jacket, 825*f*
Yucca (*Yucca*), 819, 819*f*

Z

Zebra finches, 784–785
Zero population growth, 800
Zooplankton, in lakes, 878